D0778644

DEC 18 2002

WITHDRAWN

FEB 2 1 2023

DAVID O. McKAY LIBRARY
BYU-IDAHO

PROPERTY OF:
DAVID O. McKAY LIBRARY
BYU-IDAHO
REXBURG ID 83460-0405

Climates: Past and Present

Geological Society Special Publications

Series Editors

A. J. HARTLEY

R. E. HOLDSWORTH

A. C. MORTON

M. S. STOKER

Special Publication reviewing procedures

The Society makes every effort to ensure that the scientific and production quality of its books matches that of its journals. Since 1997, all book proposals have been refereed by specialist reviewers as well as by the Society's Publications Committee. If the referees identify weaknesses in the proposal, these must be addressed before the proposal is accepted.

Once the book is accepted, the Society has a team of series editors (listed above) who ensure that the volume editors follow strict guidelines on refereeing and quality control. We insist that individual papers can only be accepted after satisfactory review by two independent referees. The questions on the review forms are similar to those for *Journal of the Geological Society*. The referees' forms and comments must be available to the Society's series editors on request.

Although many of the books result from meetings, the editors are expected to commission papers that were not presented at the meeting to ensure that the book provides a balanced coverage of the subject. Being accepted for presentation at the meeting does not guarantee inclusion in the book.

Geological Society Special Publications are included in the ISI Science Citation Index, but they do not have an impact factor, the latter being applicable only to journals.

More information about submitting a proposal and producing a Special Publication can be found on the Society's web site: www.geolsoc.org.uk.

GEOLOGICAL SOCIETY SPECIAL PUBLICATION NO. 181

Climates: Past and Present

EDITED BY

MALCOLM B. HART

Department of Geological Sciences, University of Plymouth, UK

2000
Published by
The Geological Society
London

THE GEOLOGICAL SOCIETY

The Geological Society of London was founded in 1807 and is the oldest geological society in the world. It received its Royal Charter in 1825 for the purpose of 'investigating the mineral structure of the Earth' and is now Britain's national society for geology.

Both a learned society and a professional body, the Geological Society is recognized by the Department of Trade and Industry (DTI) as the chartering authority for geoscience, able to award Chartered Geologist status upon appropriately qualified Fellows. The Society has a membership of 9099, of whom about 1500 live outside the UK.

Fellowship of the Society is open to persons holding a recognized honours degree in geology or a cognate subject and who have at least two years' relevant postgraduate experience, or not less than six years' relevant experience in geology or a cognate subject. A Fellow with a minimum of five years' relevant postgraduate experience in the practice of geology may apply for chartered status. Successful applicants are entitled to use the designatory postnominal CGeol (Chartered Geologist). Fellows of the Society may use the letters FGS. Other grades of membership are available to members not yet qualifying for Fellowship.

The Society has its own Publishing House based in Bath, UK. It produces the Society's international journals, books and maps, and is the European distributor for publications of the American Association of Petroleum Geologists (AAPG), the Society for Sedimentary Geology (SEPM) and the Geological Society of America (GSA). Members of the Society can buy books at considerable discounts. The Publishing House has an online bookshop (*http://bookshop.geolsoc. org.uk*).

Further information on Society membership may be obtained from the Membership Services Manager, The Geological Society, Burlington House, Piccadilly, London W1V 0JU (Email: *enquiries@geolsoc.org.uk;* tel: + 44 (0)207 434 9944).

The Society's Web Site can be found at *http://www.geolsoc.org.uk/*.The Society is a Registered Charity, number 210161.

Published by The Geological Society from:
The Geological Society Publishing House
Unit 7, Brassmill Enterprise Centre
Brassmill Lane
Bath BA1 3JN, UK

(*Orders*: Tel. + 44 (0)1225 445046
        Fax + 44 (0)1225 442836)
Online bookshop: *http://bookshop.geolsoc.org.uk*

The publishers make no representation, express or implied, with regard to the accuracy of the information contained in this book and cannot accept any legal responsibility for any errors or omissions that may be made.

© The Geological Society of London 2000. All rights reserved. No reproduction, copy or transmission of this publication may be made without written permission. No paragraph of this publication may be reproduced, copied or transmitted save with the provisions of the Copyright Licensing Agency, 90 Tottenham Court Road, London W1P 9HE. Users registered with the Copyright Clearance Center, 27 Congress Street, Salem, MA 01970, USA: the item-fee code for this publication is 0305-8719/00/$15.00.

British Library Cataloguing in Publication Data
A catalogue record for this book is available from the British Library.

ISBN 1-86239-075-4

Typeset by Bath Typesetting, UK
Printed by Alden Press, Oxford, UK

Distributors

USA
    AAPG Bookstore
    PO Box 979
    Tulsa
    OK 74101-0979
    USA
    Orders: Tel. + 1 918 584-2555
            Fax + 1 918 560-2652
            Email *bookstore@aapg.org*

Australia
    Australian Mineral Foundation Bookshop
    63 Conyngham Street
    Glenside
    South Australia 5065 Australia
    Orders: Tel. + 61 88 379-0444
            Fax + 61 88 379-4634
            Email *bookshop@amf.com.au*

India
    Affiliated East-West Press PVT Ltd
    G-1/16 Ansari Road, Daryaganj,
    New Delhi 110 002 India
    Orders: Tel. + 91 11 327-9113
            Fax + 91 11 326-0538

Japan
    Kanda Book Trading Co.
    Cityhouse Tama 204
    Tsurumaki 1-3-10
    Tama-shi
    Tokyo 206-0034 Japan
    Orders: Tel. + 81 (0)423 57-7650
            Fax + 81 (0)423 57-7651

Contents

It is recommended that reference to all or part of this book should be made in one of the following
ways:

Hart, M. B. (ed.) 2000. *Climates: Past and Present*. Geological Society, London, Special
Publications, **181**.
Beerling, D. J. 2000. Global terrestrial productivity in the Mesozoic era. *In*: Hart, M. B. (ed.)
Climates: Past and Present. Geological Society, London, Special Publications, **181**, 17–32.

Introduction

Plymouth Environmental Research Centre and Department of Geological Sciences, University of Plymouth, Plymouth PL4 8AA, U.K.

Abstract: In recent years media attention has focused on global climate change and the impacts on the human experience that this may cause. This interest has stimulated earth scientists into a re-appraisal of the rates of change seen in the geological record and which, for years, had been taken for granted. Climatic modelling and other palaeoclimate assessments have reversed the geological adage that the 'present is the key to the past' into 'the past may be the key to the future'. This renewed interest was the driving force behind the European Palaeontological Association's Congress in Vienna.

The climate of the early Earth was probably very warm; the direct result of the accumulation of gravitational energy, the loss of kinetic energy of incoming planetary fragments and the heat generated by short-lived radio-isotopes as the planet was forming. The intense volcanicity of the early Earth would also have created a 'greenhouse' planet that was only slowly modified by the arrival of the single-celled marine algae about 3500 million years ago. Despite the hypothetical increase in solar luminosity through geological time the temperature of the Earth has, in general, reduced since the Archean (Fig. 1), apart from a particularly warm interval during the Cretaceous. The main controls on global climate through the Phanerozoic appear to be:

- the bio-regulatory effect of photosynthetic marine plants and algae;
- the movements of the crustal plates and the rate of sea-floor spreading; and
- the increasing solar luminosity (?).

Punctuating the general evolution of global climate are the major glaciations, when large terrestrial and marine ice sheets covered substantial areas of the globe. Between these major glacial expansions (shown in Fig. 1) there may have been intervals with little (see Miller *et al.* 1999; Price 1999), or no, polar ice.

Global climate change

Climate is changing globally, with the world being about 0.6°C warmer than one hundred years ago (Hulme & Jenkins 1998). The warmest years ever recorded have been (in increasing order) 1990, 1995, 1997 and 1999; the last year being the warmest since about 1204 (Lamb 1995). In the UK five of the six warmest years in the 340-year Central England Temperature Series have occurred since 1988.

Estimates of the rate of global warming over the next century range from 0.16°C per decade to 0.35°C per decade. In the same time period global sea-level is predicted to rise from between 2.4 cm per decade to 10.0 cm per decade (see Hart 1997; Hart & Hart in press). Even to geologists these figures are alarming as they imply rates of change well beyond those normally associated with the geological record. We now appreciate that some changes in this record (e.g. temperature changes during the Younger Dryas Event; Alley *et al.* 1993) were much quicker than previously imagined and this is causing a major re-assessment of the rates of future change. Following the last glacial maximum (*c.* 10 000 years ago) sea-levels have risen significantly, but at a rate of approximately 2 mm per annum; not 10.0 cm per decade as some forecasters are predicting!

As a result of this interest in global climate change and the likely impact on sea-levels, the biosphere and human activity, many geologists and palaeontologists are re-assessing their data on both historical and geological changes as demonstrated by the fossil record. It was this interest that stimulated the Council of the European Palaeontological Association to organize the Vienna Congress in July 1997.

The 2nd European Palaeontological Congress.

The Congress was held at the Geozentrum of the University of Vienna, 10–12 July 1997. It was

From: HART, M. B. (ed.) *Climates: Past and Present.* Geological Society, London, Special Publications, **181**, 1 4, 1-86239-075-4/$15.00

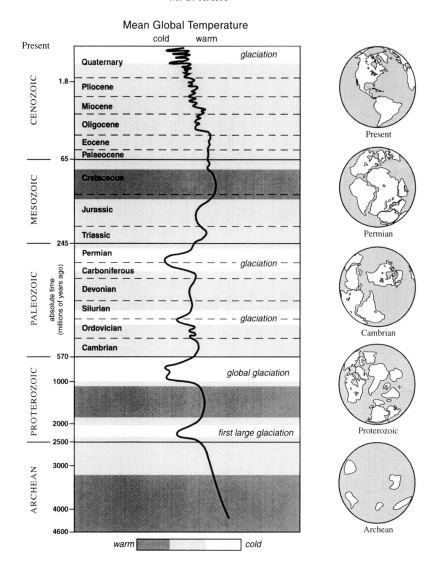

Fig. 1. Generalized temperature history of the planet (adapted from Merritts *et al.* 1998).

jointly organized by Professor David Ferguson (University of Vienna) and Dr Heinz Kollman (Natural History Museum, Vienna). While the papers and posters presented during the meeting covered the whole of the geological succession, the great majority of the contributors addressed the climatic changes of the Quaternary. Much climate change research involves proxies of one form or another (isotopes, microfossils, macro-fossils, leaf morphology, sedimentation patterns, etc.) and the total range of methods employed in such work covers the whole spectrum of geological research. These varied contributions

give examples of a range of approaches to climatic problems. They have been arranged in geological order, ranging from the Permo-Carboniferous to the post-glacial recovery of the last 18 000 years.

Clausing & Boy, using phytoplankton-generated laminations in lake sediments attempt to reconstruct Permo-Carboniferous and Palaeogene environments by using Recent environments as a benchmark. Such comparisons rely on a clear understanding of the taphonomy of the phytoplankton in modern environments in order to fully appreciate the composition of the fossil assemblages. These locally applied models are quite different to the global-scale modelling of **Beerling** who presents General Circulation Model (GCM) data for both the Cretaceous and Jurassic. The GCM data are compared to the results of vegetation models, leaf area measurements and predicted global CO_2 levels. This theme is extended to the marine environment by **Hart**, who uses the distribution of planktonic foraminifera in the Cretaceous oceans to identify palaeo-zoogeographical regions. The distribution of the planktonic foraminifera parallels the climatic provinces identified by the vegetation, although some of the newly forming mid-Cretaceous oceans (South Atlantic, Indian, etc.) appear to have been characterized by semi-isolated plankton assemblages. The vegetation and floral assemblages of NE Asia are described by **Golovneva** who indicates that the Maastrichtian was characterized by relatively high precipitation and a cold-month average temperature of 3–4°C. There is also a suggestion that, even in winter, the temperature was rarely below 0°C. This information is clearly important in the current debate on Maastrichtian climates (Miller *et al.* 1999; Price 1999).

After the Palaeogene temperature maximum in the Eocene, global temperatures (Fig. 1) began falling as ice began to accumulate, amalgamate into sheets and advance over Antarctica. The next series of papers describe the changes during this time of climatic deterioration. **Chira** *et al.*, in an investigation of the Miocene sediments of the Transylvanian Depression of Romania, uses a range of integrated palaeontological data to assess the climatic variability of the area. Warming and cooling episodes are identified with the last significant warm interval being in the Late Miocene. The Miocene is also described by **Jurkschat** *et al.*, who have investigated the pre-Messinian diatom assemblages of the Lorca Basin, SE Spain. The variations in the diatom floras (holoplanktonic vs meroplanktonic, etc.) are used to assess the progressive environmental stress suffered by the Lorca Basin as the climatic changes led towards evaporite formation during the Messinian 'crisis'. **Korpás-Hódi** *et al.* have also investigated the Late Miocene interval, using boreholes from the sedimentary basins of western Hungary. In this area palaeobotanical and vertebrate palaeontological data provide evidence of climatic cyclicity which, in turn, controls sedimentation in the area. Climatic oscillations and the climatic decline in Central Europe are also described by **Kvacek** who uses plant assemblages, leaf morphology and plant morphology to document these changes in the Late Oligocene and Miocene.

The climatic changes that were associated with the Pleistocene glacial maximum received a great deal of attention at the congress and the following authors, using a variety of palaeontological data, have attempted to describe the results of the climatic variability. **Amore** *et al.* combine palynological and micropalaeontological data to interpret the Late Pleistocene to Recent climatic changes recorded in the sediments of the Tyrrhenian Sea. The Pleistocene palaeontology of Sicily is used by **Di Geronimo** *et al.* to investigate the arrival of 'boreal guests'; cooler-water molluscan migrants that appear in Sicily during the cooler intervals of the Pleistocene. The climatic changes of the Late Glacial and post-Glacial are described by **Nagy-Bodor** *et al.* who use pollen analysis to document the vegetation changes in Hungary during that interval. They recognize a variety of influences on the flora (sub-Mediterranean from the south, Atlantic–Alpine from the west, etc.) of this Central European area, each of which represents a particular type of climatic change in the area. Palynology is also used by **Russo Ermoli** to document the mid-Pleistocene vegetation changes in southern Italy. These changes in the vegetation can be matched to glacial/interglacial cycles in other parts of Europe. In the marine environment **Rohling** *et al.* show that, by using GCMs and analytical data from sapropels, climatic changes can be recorded from the sediments of the Mediterranean Sea. The numerical models generated by this research are then used to test the viability of proxy-based conceptual models.

In terrestrial environments the remains of vertebrates and hominids can be used to determine whether evolution/migration events are a direct response to climate change. In the final section of the book a number of papers deal with these issues. **Azanza** *et al.*, working in the western Mediterranean area, have studied the diversity changes and speciation events for large mammals during the Neogene. The major events are coincident with the latest Cenozoic glaciations, with a major, total rejuvenation event approximately 1 million years ago. This is coincident with the arrival of *Homo* species in the western Mediterranean area. Quaternary vertebrates from Sicily are used by **Bonfiglio** *et al.* for the recognition of four major dispersal events during the Quaternary. These are associated with changes in endemism and diversity. **Fladerer**, using avian and mammalian fossils from caves in SE Austria of Middle to Late Pleistocene age, demonstrates how these communities for the late glacial period are similar to the boreal/tundra associations of today. Shifts in the species content of the cave faunas correlate with known climate changes (as detected by palynological data). **Chaline** et al. contend that climate, while not being responsible for the appearance of bipedalism and hominization in hominids, exerted a significant influence on the development of ecological niches which allowed a common ancestor to develop into a variety of sub-species.

References

ALLEY, R. B., MEESE, D. A., SHUMAN, C. A., GOW, A. J., TAYLOR, K. C. et al. 1993. Abrupt increase in Greenland snow accumulation at the end of the Younger Dryas event. Nature, **362**, 527–529.

HART, A. B. 1997. A risk assessment of the inland area around Charmouth, West Dorset. MSc thesis, University of Portsmouth.

HART, M. B. & HART, A. B. (in press). Global climate change: a geological perspective. Geoscience in south-west England, **10**.

HULME, M. & KENKINS, G. J. 1998. Climate Change Scenarios for the U.K.: Scientific Report. UKCIP Technical Report No.1, Climate Research Unit, Norwich.

LAMB, H. H. 1995. Climate History and the Modern World. Routledge, London.

MERRITTS, D., DE WET, A. & MENKING, K. 1998. Environmental Geology: an earth system Science approach. W. H. Freeman & Company, New York.

MILLER, K. G., BARRERA, E., OLSSON, R. K., SUGARMAN, P. J. & SAVIN, S. M. 1999. Does ice Drive early Maastrichtian eustacy. Geology, **27**, 783–786.

PRICE, G. D. 1999. The evidence and implications of polar ice during the Mesozoic. Earth Science Reviews, **48**, 183–210.

Lamination and primary production in fossil lakes: relationship to palaeoclimate in the Carboniferous–Permian transition

ANDREAS CLAUSING[1] & JÜRGEN A. BOY[2]

[1]*Institut für Geologische Wissenschaften und Geiseltalmuseum, Domstr. 5, D-06108 Halle (Saale), Germany (e-mail: clausing@geologie.uni-halle.de)*
[2]*Institut für Geowissenschaften—Paläontologie, Saarstr. 21, D-55099 Mainz, Germany*

Abstract: Comparative studies are made on the development of lamination in Recent, Paleogene and Permian lake sediments. The study concentrates on the relationship between primary production and lamina formation. The main phytoplankton development in lakes is correlated with environmental changes and it is often preserved in the lamination in chronological succession. Seasonal patterns may be especially represented. Morphological identification and classification of laminites is achieved by studies of Recent lake sediments. This leads to possible comparison, even in badly preserved ancient examples. Paleogene lamination can be clearly correlated with the Recent lamination by the relatively tenuous relationships of the phytoplankton. Permian lake sediments, however, are difficult to decipher, as the phytoplankton species are different or absent as a result of taphonomy. Nevertheless, morphological comparisons suggest periodic sedimentation and lamination development that is clearly similar to that in lakes today. As the same physical properties apply, this seasonal sedimentation can also be used as an additional tool for the reconstruction of environmental and climatic changes in the Carboniferous–Permian boundary strata.

Microscopic changes in the composition of fine-grained sediments can produce laminations in the marine and lacustrine environments. Laminated lake sediments are known from the beginning of the Palaeozoic era. As Zolitschka (1990) noted, lamination may be caused by: a changing clastic input during sedimentation; variation in the organic production of planktonic and benthic organisms; by physico-chemical processes (e.g. evaporation). Gradual mixing of these main processes also occurs and leads to more blurred models for laminae formation. The development and nature of laminated lake sediments have been mainly studied in Recent lakes of various genetic and physical types (see Zolitschka 1998).

Environmental factors are the main controls on rhythmical layering and sedimentary cycles, which result in lithological changes (Einsele *et al.* 1991). Precipitation of water and solar insolation are of particular interest. Rhythmical layering may be episodic or periodic. The pleistocene varves, with their annual significance, are well-known examples of laminites. These annually produced varves are so-called periodic laminites. They are mainly formed by an allochthonous inorganic input (glacial varves) or they result from autochthonous biological productivity and carbonate precipitation (Junge 1998). In tropical lakes annual varve couplets are also formed by changes between dry and wet climatic periods (Glenn & Kelts 1991). Lambert & Hsü (1979) demonstrated that laminations may develop independently of annual events. In that case, detailed observations were required to prove a climatic or non-climatic control.

Compared with the marine realm, lakes are relatively short lived. They are laterally limited in their extent, and their sediments respond quickly to environmental changes. As a result of rapid deposition, the sediments formed can preserve complete lake histories. Lake sediments are often characterized by laminated horizons that are composed of lighter and darker layers (Talbot & Allen 1996). As mentioned above, lake laminations may correspond to the original primary production and thus reflect the direct influence of changing environmental factors on organisms. As these factors are mostly coupled to the climate, the study of fossil lake sediments, especially laminites, contributes directly to climatic reconstructions of fossil environments (Kemp 1996). Sedimentological and mineralogical aspects of lamina formation are often studied but research on the effects of organic production is comparatively rare, mainly because of problems of diagenetic overprint and taphonomic modification in fossil lakes.

This study compares the development of

From: HART, M. B. (ed.) *Climates: Past and Present.* Geological Society, London, Special Publications, **181**, 5–16, 1-86239-075-4/$15.00

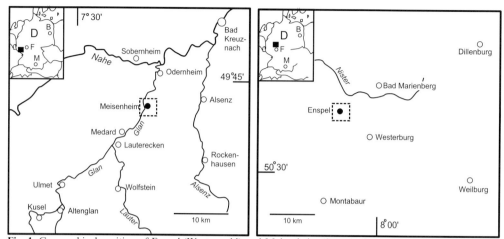

Fig. 1. Geographical position of Enspel (Westerwald) and Meisenheim (Saar–Nahe Basin) in Central Germany. B, Berlin; D, Germany; F, Frankfurt and M, Munich.

organically influenced lamination in Recent and fossil lakes. On the basis of information from modern lakes, physical and biological patterns of lake lamination are identified. The identified features are correlated with their origin, applied to the Paleogene example and then transferred to Permian lake facies (for location, see Fig. 1). Climatically relevant processes for the distribution of the organic components of the lamination are investigated for and preliminary models given.

Aspects of lamination in modern lakes

Schwoerbel (1993) described the development of a phytoplankton succession from the Bodensee (Überlinger See, Germany) in detail. This succession was repeated annually and may serve as an example of the seasonal development of phytoplankton associations. Principal components of the phytoplankton are green algae, diatoms, chrysophytes and dinoflagellates (Fig. 2). The species composition is relatively consistent and continuity is reported for a great number of years. Variation in the lifespan of species and bloom intensity are caused by ecological changes, such as a decrease in nutrient supply or variations in mean temperature (Schmidt 1996). Not all components of the phytoplankton can be equally fossilized; the mineralized frustules of diatoms and cysts of chrysophytes are preferentially preserved. Complete organic remains are rarely buried and may become synsedimentary or diagenetically degraded if they are not composed of sporopollenine.

The depth and morphology of the lake floor may control the preservation, or non-preservation, of laminated sediments, as observed in Lake Tanganyika (Huc et al. 1990). Laminae are also preserved when bioturbation is lacking or rare, a condition commonly encountered in a dysaerobic to anaerobic environment (Sageman & Biga 1997). When currents near the lake floor are absent, resuspension of allochthonous sediment is impossible and autochthonous laminae are produced (Anderson & Dean 1988). This generally occurs in areas below the wave base, except for shallow-water environments, where benthic microbial mats may stabilize the sediment. In meromictic or di- or monomictic lakes, undisturbed sedimentation occurs in basinal areas lacking circulation (Anderson & Dean 1988).

Small and very deep lakes, such as maar lakes (Negendank et al. 1990), seem to favour the preservation of organic structures. In the modern Eifel maar lakes of Germany, nearly completely preserved successions of laminated sediments have been obtained by drill cores. They contain a distinct lamination with a large number of lamina couplets which represent at least a 15 ka succession. A composite varve of a eutrophic maar lake shows the seasonal distribution of algae (Zolitschka 1989, 1990). The laminations consist of layers that can be attributed to the seasonal phytoplankton development (Fig. 3). A succession of mainly siliceous fossils is preserved. Chrysophytes prevail in early spring layers, small-sized centric diatoms occur in spring, and larger ones in summer. The autumn layer is mainly formed by organic matter and vivianite minerals, whereas winter sedimentation is dominated by clay particles.

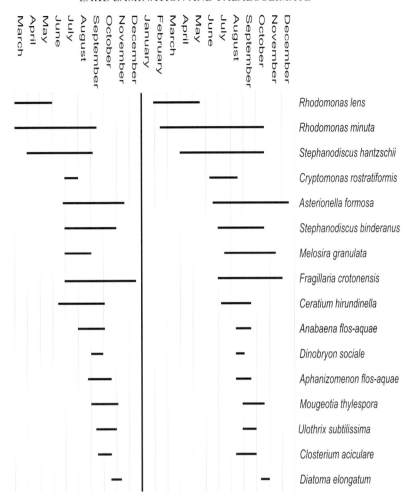

Fig. 2. Example of an annual phytoplankton succession in modern Lake Constance (modified from Schwoerbel 1993).

Palaeogene lake strata

During Palaeogene time many of the phytoplankton organisms were similar or identical to present-day taxa. Thus, a direct comparison between the occurrences in Recent, Neogene and Palaeogene sediments is possible.

In Germany there are a large number of ancient lake deposits that are famous for their fossils, such as the Eocene (Lutetian) Lake Messel near Darmstadt in the upper Rhine valley. The contents of the Messel oil shale have been extensively studied from various standpoints, including petrology and formation of the laminites. Goth (1990) showed that seasonal blooms of the green alga *Tetraedron* were the

principal component of the lamina couplets. This observation was also confirmed by geochemical means (Goth *et al.* 1988). The green alga *Botryococcus* was also found but, to date, evidence for chrysophytes is rare, for diatoms dubious, and for dinoflagellates completely lacking.

A recently discovered fossil lake site of Late Oligocene age is located at Enspel in the Westerwald area (Figs 1 and 4). It contains various vertebrate remains, macroplants and microscopic fossils (Storch *et al.* 1996; Clausing 1998). Like the Messel site, the sediments of Lake Enspel are rich in organic remains, including various groups of algae, which formed the autochthonous phytoplankton. Chryso-

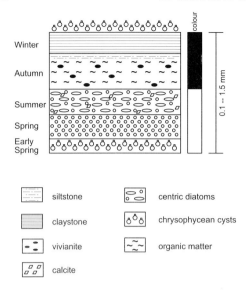

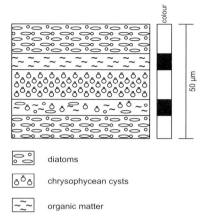

Fig. 5. Idealized type of lamination in Lake Enspel (Upper Oligocene sequence), a fossil maar lake in the Rhenish Slate Mountains, Germany.

Fig. 3. Composite varve from modern Lake Holzmaar, Eifel region, Germany (after Zolitschka 1989).

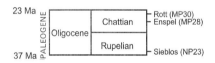

Fig. 4. Stratigraphic subdivision of the Oligocene epoch in Central Europe and relative position of Lake Enspel and other lake horizons.

phytes, various diatoms, a dinoflagellate species and the green algae *Botryococcus* and *Tetraedron* have been found. Benthic microbial mats in the Enspel sediments are preserved as alginite B (Wuttke & Radtke 1993), and may be derived from cyanobacteria (Tyson 1995).

Two types of lamination are recognized in the Lake Enspel sediments; a rhythmical change between layers formed by chrysophyte cysts, diatoms and organic matter, and laminae composed of dinoflagellates, organic and inorganic matter are recognized (Figs 5 & 8a). These laminated horizons are similar to successions recognized in modern eutrophic lakes, but the algal composition is not completely identical. The idealized lamination types represent the lake laminites preserved at the top of the sedimentary succession, exposed during fossil excavation.

It is assumed that this variable development is caused by rapidly changing environmental conditions around Lake Enspel in a volcanically active area. Laminites with organic remains were seemingly formed very quickly in a time of increased nutrient input correlated with volcanic events (Schulz *et al.* 1997). The laminites composed of siliceous algal remains represent the undisturbed 'normal' cyclical development. For Enspel it is not yet clear if the different lamination types and variability of phytoplankton organisms may also reflect annual or seasonal climatic changes (Clausing 1998; Gaupp & Wilke 1998), as was observed at other localities.

The Eocene Eckfeld maar lake deposits exhibit lamina couplets that are composed of diatoms and siltstone layers. These occur rhythmically and are interpreted as annual varves (Mingram 1994). The existence of solar periodicities in the laminated oil shales is given as further evidence of seasonal cyclicity. A brief study of samples from the Randecker Maar provided laminites that also contain rhythmically occurring phytoplankton organisms. The lakes of the Eocene Green River Formation (Western Interior, USA) are an example from outside Europe where laminites produced in different environments are thought to be annual varves (Crowley *et al.* 1986).

Lower Permian (Autunian) lakes

The German Saar–Nahe Basin (SNB) is one of the largest Rotliegend basins in the Variscan mountain chain of central Europe (Fig. 1). During Late Carboniferous and Early Permian times the basin developed along the Southern Hunsrück Fault (Henk 1993a,b) with sediment accumulations reaching more than 4000 m

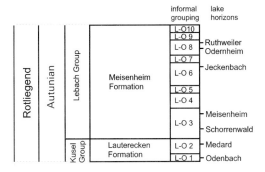

Fig. 6. Stratigraphic subdivision of a part of the Rotliegend in the Saar–Nahe Basin with the position of various lake horizons.

(Schäfer 1989). They are mainly of fluvio-lacustrine origin (Stollhofen & Stanistreet 1994) and also include numerous volcaniclastic layers (Stollhofen 1994). About 30 lacustrine units of Rotliegend age exist in the basin (Clausing 1996a). From these, Schäfer & Sneh (1983) described laminated sediments with winter and summer couplets. Non-annual lamina couplets were also reported (Schäfer & Stamm 1989). There are variations in thickness and completeness of the lamination patterns. In most of the Rotliegend lakes the laminite units are thin, and are commonly interlayered with non-laminated units. This may result from phases of extensive or frequent circulation in the lakes. The laminites can contain abundant organic matter of different composition (dark) and alternate with purely inorganic (light) layers. Our investigations suggest two principal types of laminae for the Autunian lakes; graded laminae and alginite laminae. Three different lake units from the Saar–Nahe Basin are presented (Fig. 6) as examples.

Graded laminae are preserved from Lake Ruthweiler. The profundal facies consists of fine-grained to coarser siltstones. The grain size of the dark laminae is coarser than that of the light laminae. Lamination is extremely fine, part graded and shows slightly erosional features. Microbioturbation is often visible and single laminae of 2–12 mm thickness are interbedded. They possess a concordant base and are commonly graded (coarse- to fine-grained silt) and bioturbated. A few layers exhibit inverse grading. Detrital carbonate and extremely thin laminae (thickness about 100 μm) are absent. Coarse plant debris is rare; seeds and relatively well-preserved plant remains are abundant, and document a mainly hygro- to mesophilous environment. Abundant autochthonous primary

production is documented by yellowish hetero-geneous amorphous matter as interpreted by palynofacies studies (Clausing 1996b). Al-lochthonous material, represented by structured matter, pollen and spores, is present in the lower and upper parts of the section.

Lake Ruthweiler was small but deep. Deltaic influx was stronger in the profundal facies as in larger lakes. This led to an increased, often pulsating, sedimentary input and more aerobic conditions on the lake floor. This type of lamination is similar to that of Late Carbonifer-ous delta-influenced lakes, such as that described by Haszeldine (1984), lamina formation by turbiditic underflows correlating with flooding events of a nearby river system. This kind of underflow caused by increased density of silt-laden river water is even possible in thermally stratified lakes (Lambert & Hsü 1979). By comparison, we believe that the Ruthweiler-type lamination can be interpreted as non-annual varves, *sensu* Lambert & Hsü (1979).

As opposed to the lamination described above, alginite-containing laminae are inter-preted as being annual varves. This type of lamination is present, for instance, in Lake Odernheim (Boy 1995). The lake floor was anoxic, as benthic organisms and disturbances of the laminations and bioturbation, even on a very small scale, are completely lacking. The laminations are composed of organic matter (alginite B) in the dark layers and mineral-rich light layers. This suggests rhythmical, probably annually controlled primary production in the eutrophic lake. For ecological reasons and for analysis of the primary production, identifica-tion of taxa would be of interest, but producers have not yet been identified, as a result of poor preservation. Phytoplankton organisms are rare as compared with those of the phytobenthos (Clausing 1996a). Cyanobacteria are the prob-able producers (Clausing 1992), but it could not be determined if the alginite B is a residue of planktonic or benthic microbes or algae. Single layers of ultralaminites (< 100 μm), which in-cluded carbonate grains, may be derived from microbial mats similar to those described by Gerdes et al. (1991). These mats can also exist in an anaerobic environment, which would favour benthic microbial communities in those lakes where carbonate grains occur.

An anoxic bottom zone, without strong currents and varve-like lamination, is typical of meromictic lakes. These are lakes with enduring stratification, consisting of a lower monimolim-nion with increased salinity, and an upper mixolimnion of normal salinity which is mixed from time to time. Anoxic conditions may also

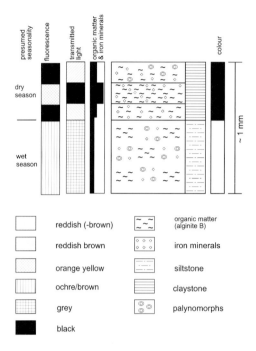

Fig. 7. Composite lamina couplet from Lake Meisenheim (Rotliegend, Lower Permian, Germany), modified from Clausing (1997).

occur in mono- or dimictic lakes and thermally stratified lakes with one or two circulations per year, if these are deep (Anderson & Dean 1988). One difference is that meromictic lakes are less productive in the mixolimnion, because the nutrients derived from microbial degradation remain in the monimolimnion (Anderson *et al.* 1985). In the mono- or dimictic lakes the support of nutrients from the hypolimnion is supported by the circulation events, which means that these lakes are more productive. Unfortunately, the kind of water column conditions that existed cannot be deduced directly from the sediment.

The laminations of Lake Meisenheim have been studied in great detail and a model for lamination development proposed (Clausing *et al.* 1992; Clausing 1996a). The lake sediments mainly are laminated sediments, which are characteristic of the deep-water facies (Allen & Collinson 1986) and which are distributed over an area > 40 km². The reconstruction of the three developmental stages in the evolution of the lake is based on the results of an excavation (Clausing *et al.* 1992). The so-called phase III marks the lake's greatest depth at the end of lake development. At that time a pro-delta facies can be noted prograding from the southwestern

area, while deep-water facies were still being deposited in the northeastern part. This profundal facies is characterized by a cyclicity of light grey and dark laminae (Figs 7 and 8b).

The light laminae represent phases of continuous sedimentation. Remains of macroplants, spores, plant debris and fish remains are oriented parallel to the bedding. Locally significant slightly coarser layers, without grading, with increased quartz, feldspar and plant debris are included. These layers are interpreted as remains of distal turbidites. The dark laminae consist of a central black layer between two reddish brown layers. Compared with the black layers, the reddish brown layers possess a nearly identical but somewhat lower content of organic matter. Remnants of calcareous algae can be identified as small calcitic bodies in the organic matter which mainly consists of alginite B. The content of iron minerals and organic matter decreases towards the overlying light laminae. From this evidence the black layers are suggested to be indicators of a biologically induced hypertrophy of the lake sediments (Schmidt 1978), produced by periodic algal blooms.

Phase III of the evolution of Lake Meisenheim represents a deep lake with a normally stratified water body (Clausing *et al.* 1992). According to the fossil content, warmer water, sufficient oxygen and continuous water circulation existed in the epilimnion. Lower water temperature, oxygen depletion and lack of currents were typical of the hypolimnion. During formation of the dark laminae large quantities of organic matter preserved as alginite B were produced. The epilimnion was possibly inhabited by planktonic algae and cyanobacteria, which formed the base of the food chain. The microbial remains were sedimented into the dark laminae of the lake floor and their degradation produced an oxygen-free hypolimnion. A reducing environment occurred irregularly and is represented by pyrite-rich layers (yellow layers). Oxygen-free times are also documented by an undisturbed lamination, which is not influenced by bioturbation. The lack of endobenthic and epibenthic organisms is a result of these environmental conditions.

Stratification of the lake broke down during the formation of the light laminae, when complete circulation occurred. This was probably caused by heavy rainfall, resulting in the sudden influx of colder water and increased input of plant detritus and macro-plant remains from the delta area. The complete circulation caused an upwelling of H_2S from the lake floor, which killed the fish fauna (*Elonichthys*) and enriched the fossils in the light laminae (Claus-

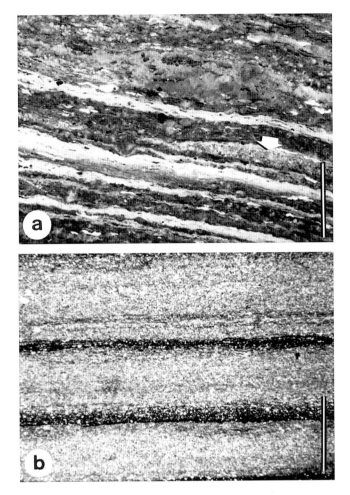

Fig. 8. Microphotographs of laminites. (**a**) Lake Enspel laminae with a succession of bright alginite B and pale chrysophyte–diatom layers (sample T2, reflected fluorescence). A distinct chrysophyte layer is indicated by an arrow (scale bar represents 100 μm). (**b**) Profundal facies of Lake Meisenheim with a succession of bright and dark layers (sample ME11A, transmitted light). The dark layers are rich in organic matter and so-called calcareous algae species 'type ME' (bright spots) (scale bar represents 400 μm).

ing *et al.* 1992). After this, the lake floor quickly returned to its anoxic state.

Cyclicity was caused by changes between dry periods, when the lake was stratified and there was only minor input of organic material, and rainy periods, when there was complete circulation with increased organic input.

The rhythmic development of the phytoplankton during annual wet and dry climatic phases in modern lakes has been studied by John (1986). According to his investigations, the dry season is the time of maximum phytoplankton growth. A number of factors are responsible: good light conditions because of little cloud cover and low turbidity, high concentration of nutrients, less

competition by the periphyton and aquatic macrophytes, and an increased stability of the water column. At the end of the dry and beginning of the wet season, increasing productivity of the phytoplankton was observed. One cause is overnight vertical water mixing, which returns nutrients from the hypolimnion to the epilimnion. Additional nutrient supplies are delivered by the rivers during rainfall and first floodings. In large lakes the water body does not rise at the beginning of the wet period, but is stable for months. The additional water and suspension input may lead to enhanced sedimentation rates and in shallower lakes the currents may reach the lake floor.

During the flood season, the phytoplankton and phytobenthos growth are drastically reduced. The high suspended sediment content of the incoming river water prevents light penetration and this shifts the euphotic zone towards the lake surface. The flood water contains abundant organic and inorganic matter in suspension, but only after decay are the nutrients released. Rapid decay of phytoplankton material is typical for modern tropical lakes with high water temperatures (John 1986). The originally well-preserved organisms are degraded and the nutrients released. The opportunity for the preservation of complete specimens is rather small, and homogenized organic matter is primarily accumulated in the sediment. This would correspond to the observations from most Autunian lake horizons, where only alginite B is present and identifiable phytoplankton organisms are rare (Clausing 1996a). At the beginning of the next dry season abundant nutrients are, therefore available to support new phytoplankton blooms.

John (1986) did not describe lamination formation, but the rhythmical productivity should also be reflected in the sediments. As a result of decomposition, a phytoplankton succession comparable with that of modern lakes from the Northern Hemisphere (Schwoerbel 1993) may not be preserved. The massive input of organic material and the following decay, on the other hand, should at least partially remain in the sediment as homogenized organic matter (e.g. as alginite). The difference in the organic matter content in the sediments of the wet and the dry seasons is thus assumed to distinguish light and dark laminae.

Climatic models for the Autunian age

Palaeogeographical research has led to the development of models for the continent configuration in Late Palaeozoic time. In Stephanian time, Central Europe was located at a latitude of about 10°N (Schäfer 1989; Schäfer & Stamm 1989). The continental plates moved northwards during Autunian time (Matte 1986) and probably reached a latitude of 20°N in Late Saxonian times (Witzke 1990).

The present climatic model is based on the assumption of a world-wide glacial phase as proposed by Patzkowski et al. (1991). In Late Palaeozoic time several glacial and interglacial events occurred (Veevers & Powell 1987), which are difficult to correlate chronostratigraphically (Kozur 1994). Evidence for major climatic changes are found at the Stephanian–Autunian boundary. In Autunian and Early Saxonian

times a distinct cold phase seems probable, whereas data on Stephanian time are contradictory. Glaciation ended during Middle Saxonian time (Crowell 1995).

The Saar–Nahe Basin (SNB) was part of the Variscan orogenic belt (Fig. 9). The northern margin was formed by the Rhenish Slate Mountains whereas the location of the southern margin is not yet known. Information on the topographic height of the surrounding mountains is also lacking. Phytogeographical and meteorological modelling suggests that Central Europe was part of a humid tropical belt during Westphalian time. Beginning in Stephanian time, this changed to a summer-wet climate (Ziegler 1990; Patzkowski et al. 1991). For Stephanian–Autunian times temperature gradients similar to to those of today, and a distinct monsoonal climate, have been proposed (Parrish 1995). During the summer, winds from the Tethys area to the south delivered humidity to the southern part of the Variscan chain. In its position within that chain the SNB could be reached by rainfall. In winter, dry winds from the east to north reached the SNB after travelling over the continental mass. This represented the dry season in the area.

Kutzbach & Ziegler (1994) published a simulation for the Late Permian climate. They proposed climatic models with and without extensive inland lakes and seas. They showed that the climate was sensitive to the presence of larger lakes and seas. The effects of temperature damping, however, are probably negligible for the short-lived lakes of the Saar–Nahe Basin. The simulation with and without inland lakes and seas seemed to have only minor effects on the climate in the research area on the northern continent. It is more important that topographic features were also included in the modelling, even if the knowledge of Central Europe is relative poor. The presence of the highlands intensifies the monsoons and allows precipitation farther into the continent.

From Early to Late Permian time the continental mass moved northward, whereas the continent arrangement remained identical. The distribution of the mountain ranges was also similar. According to Crowell (1995), glaciation ended in Late Permian time. From this a general rise in temperature is assumed (Kutzbach & Ziegler 1994), which would have emphasized the seasons but not changed the seasonal pattern or wind distribution itself. The possible wind direction during Early Permian time in Central Europe should have been similar to that of Late Permian time (Fig. 9).

A monsoonal climate for Early Permian time

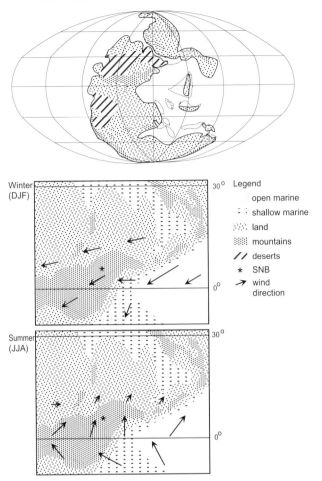

Fig. 9. Early Permian palaeogeography of the Northern Hemisphere after Kutzbach & Ziegler (1994), and the position of the Saar–Nahe Basin (SNB) in the Variscan orogenic belt. Wind direction in summer and winter according to Kutzbach & Ziegler (1994), palaeogeography based on Scotese & Langford (1995). DJF, December–January–February; JJA, June–July–August.

is also implied by the model by Kutzbach & Ziegler (1994). The addition of topography to the simulation leads to an increase in precipitation in the area of the Variscan mountains during the summer. This indicates that a clear differentiation between summer and winter is present in the Variscan mountains and the Saar–Nahe Basin. Assuming these ideas are correct, then these seasonal patterns coincide with the observations made on the laminae development in the lakes of the Saar–Nahe Basin.

For the area of the Variscan mountains in Central Europe Kutzbach & Ziegler (1994) determined a tropical to warm temperate, humid climate. This would generate a tropical rainforest to a temperate evergreen forest for these mountains. This seems to be an appropriate

description of the Early Permian situation around the Saar–Nahe Basin.

Summary

The usefulness of laminated lake sediments of Pleistocene and Holocene origin from various environments for the reconstruction of annual recurring events has been shown in numerous studies. The morphological features and composition of the Paleogene laminites studied is also too similar to recent observations to be caused by factors other than climatic changes alone. Variation is clearly induced by taphonomical reductions of the preserved organisms. For Permian time, laminite composition is more difficult to interpret. As a result of the lack of identical organisms and the generally poor preservation, the features observed are not as clear as in the younger examples. Nevertheless, the morphologically distinct lamination configuration suggests a similar process of formation. Additionally, the interpreted data show a correlation with the currently available climate models for the Carboniferous to Permian age. An interpretation of laminites as annually produced laminae (i.e. varves) in the lakes of the Saar–Nahe Basin fits the current climatic models. Laminites that are non-annually produced also exist and can be distinguished by their sedimentological criteria.

The database is currently small and the results presented are first indication of the possible application of laminite studies in Permian climate modelling based on the inclusion of organic components in the analyses.

The authors wish to thank many colleagues for numerous stimulating discussions on the topics of lamination and climatic reconstructions. The Deutsche Forschungs-Gemeinschaft is acknowledged for funding the studies during the Research Projects BO553/6 'Rotliegend-Paläoökologie' and CL154/1-1 'Enspel-Primärproduktion'. Critical suggestions for improvement of the text were given by M. Schudack (Berlin), I. Lerche (Columbia–SC) and two anonymous reviewers.

References

ALLEN, P. A. & COLLINSON, J. D. 1986. Lakes. *In*: READING, H. G. (ed.) *Sedimentary Environments and Facies*, 2nd edn. Blackwell, Oxford, 63–94.

ANDERSON, R. Y. & DEAN, W. E. 1988. Lacustrine varve formation through time. *Palaeogeography Palaeoclimatology, Palaeoecology*, **62**, 215–235.

——, ——, BRADBURY, J. P. & LOVE, D. 1985. Meromictic lakes and varved lake sediments in North America. *US Geological Survey Bulletin*, **1607**, 1–19.

BOY, J. A. 1995. Über die Micromelerpetontidae (Amphibia: Temnospondyli). 1. Morphologie und Paläoökologie des *Micromelerpeton credneri* (Unter-Perm; SW-Deutschland). *Paläontologische Zeitschrift*, **69**, 429–457.

CLAUSING, A. 1992. Fluorescence microscopy of lake sediments from the Saar–Nahe Basin (Upper Carboniferous–Lower Permian; SW-Germany). *Leica Scientific and Technical Information*, **10**, 72–79. (Corrigenda, 1993: **10**, 188.)

—— 1996a. Organische Produktion und Degradation in Seen des europäischen Rotliegend. *Hallesches Jahrbuch für Geowissenschaften*, **B18**, 89–108.

—— 1996b. Palynofacies studies of lake sediments with examples from the Saar–Nahe Basin (Permocarboniferous; SW-Germany). *Mainzer geowissenschaftliche Mitteilungen*, **25**, 27–48.

—— 1997. Remarks on acritarchs and taphonomy: studies for palaeoenvironmental conclusions. *Acta Universitatis Carolinae, Geologica*, **40**, 361–374.

—— 1998. Mikro-organofazielle Studien an Sedimenten des Enspel-Sees (Oberoligozän, Westerwald, Deutschland). *Hallesches Jahrbuch für Geowissenschaften*, **B20**, 119–133.

——, SCHMIDT, D. & SCHINDLER, T. 1992. Sedimentation und Paläoökologie unterpermischer Seen in Mitteleuropa. 1. Meisenheim-See (Rotliegend; Saar–Nahe-Becken). *Mainzer geowissenschaftliche Mitteilungen*, **21**, 159–198.

CROWELL, J. C. 1995. The ending of the Late paleozoic ice age during the Permian period. *In*: SCHOLLE, P. A., PERYT, T. M. & ULMER-SCHOLLE, D. S. (eds) *The Permian of Northern Pangea. 1. Palaeogeography, Paleoclimates, Stratigraphy*. Springer, Berlin, 62–74.

CROWLEY, K. D., DUCHON, C. E. & RHI, J. 1986. Climate record in varved sediments of the Eocene Green River Formation. *Journal of Geophysical Research*, **91**, 8637–8647.

GAUPP, R. & WILKE, A. 1998. Zur Sedimentologie der oberoligozänen Seesedimente von Enspel/Westerwald. *Hallesches Jahrbuch für Geowissenschaften*, **B20**, 97–118.

GERDES, G., KRUMBEIN, W. E. & REINECK, H.-E. 1991. Biolaminations—Ecological versus depositional dynamics. *In*: EINSELE, G., RICKEN, W. & SEILACHER, A. (eds) *Cycles and Events in Stratigraphy*. Springer, Berlin, 592–607.

GLENN, C. R. & KELTS, K. 1991. Sedimentary rhythms in lake deposits. *In*: EINSELE, G., RICKEN, W. & SEILACHER, A. (eds) *Cycles and Events in Stratigraphy*, Springer, Berlin, 188–221.

GOTH, K. 1990. Der Messeler Ölschiefer - Ein Algenlaminit. *Courier Forschungs-Institut Senckenberg*, **131**.

——, DELEEUW, J. W., PüTTMANN, W. & TEGELAAR, E. W. 1988. Origin of Messel Oil Shale kerogen. *Nature*, **336**, 759–761.

HASZELDINE, R. S. 1984. Muddy deltas in fresh water lakes, and tectonism in the Upper Carboniferous coalfield of England. *Sedimentology*, **31**, 811–822.

HENK, A. 1993a. Subsidenz und Tektonik des Saar–Nahe-Beckens (SW-Deutschland). *Geologische Rundschau*, **82**, 3–19.

—— 1993b. Late orogenic evolution in the Variscan Internides: the Saar–Nahe Basin, south-west Germany. *Tectonophysics*, **223**, 273–290.

HUC, A. Y., LE FOURNIER, J. & VANDENBROUCKE, M. 1990. Northern Lake Tanganyika—An example of organic sedimentation in an anoxic rift lake. *In*: KATZ, B. J. (ed. *Lacustrine Bsin Exploration: Case Sudies and Modern Analogs*. AAPG Memoir, **50**, 169–185.

JOHN, D. M. 1986. *The Inland Waters of Tropical West Africa. An Introduction and Botanical Review*. Ergebnisse der Limnology—Advances in Limnology, **23**.

JUNG, G. 1990. *Seen werden, Seen vergehen: Entstehung, Geologie, Geomorphologie, Altersfrage, Limnologie und Ökologie. Eine Landschaftsgeschichte der Seen allgemein*. Ott, Thun.

JUNGE, G. 1998. *Die Bändertone Mitteldeutschlands und*

angrenzender Gebiete. Ein regionaler Beitrag zur quartären Stausee-Entwicklung im Randbereich des elsterglazialen skandinavischen Inlandeises. Altenburger Naturwissenschaftliche Forschungen, **9**.

KEMP, A. E. S. (ed.) 1996. *Palaeclimatology and Palaeoceanography from Laminated Sediments.* Geological Society, London, Special Publications, **116**.

KOZUR, H. 1994. Application of the 'parastratigraphic' and international scale for the continental Permian of Europe. *In*: HEIDTKE, U. (compiler) *New Research on Permo-Carboniferous Faunas*, Pollichia Buch, **29**, 9–22.

KUTZBACH, J. E. & ZIEGLER, A. M. 1994. Simulation of Late Permian climate and biomes with an atmosphere–ocean model: comparisons with observations. *In*: ALLEN, J. R. L., HOSKINS, B. J., SELWOOD, B. L., SPICER, R. A. & VALDES, P. J. (eds) *Palaeoclimates and their Modelling. With Special Reference to the Mesozoic Era.* Chapman & Hall, London, 119–132.

LAMBERT, A. & HSÜ, K. J. 1979. Non-annual cycles of varve-like sedimentation in Walensee, Switzerland. *Sedimentology*, **26**, 453–461.

MATTE, P. 1986. La chaîne varisque parmie les chaînes paléozoiques péri atlantiques, modèle d'évolution et position des grands blocs continentaux au Permo-Carbonifère. *Bulletin de la Société géologique de France*, Série 8, **2**, 9–24.

MINGRAM, J. 1994. Sedimentologie und Zyklizität laminierter eozäner Ölschiefer von Eckfeld/Eifel. *Mainzer Naturwissenschaftliches Archiv Beiheft*, **16**, 55–85.

NEGENDANK, J. F. W., BRAUER, A. & ZOLITSCHKA, B. 1990. Die Eifelmaare als erdgeschichtliche Fallen und Quellen zur Rekonstruktion des Paläoenvironments. *Mainzer geowissenschaftliche Mitteilungen*, **19**, 235–262.

PARRISH, J. T. 1995. Geologic evidence of Permian climate. *In*: SCHOLLE, P. A., PERYT, T. M. & ULMER-SCHOLLE, D. S. (eds) *The Permian of Northern Pangea. 1. Palaeogeography, Paleoclimates, Stratigraphy.* Springer, Berlin, 53–61.

PATZKOWSKI, M. E., SMITH, L. W., MARKWICK, P. J., ENGBERTS, C. J. & GYLLENHAAL, E. D. 1991. Application of the Fujita-Ziegler paleoclimate model: Early Permian and Late Cretaceous examples. *Palaeogeography, Palaeoclimatology, Palaeoecology*, **86**, 67–85.

SAGEMAN, B. B. & BIGA, C. R. 1997. Diversity and species abundance patterns in late Cenomanian black shale biofacies, western Interior, U.S. *Palaios*, **12**, 449–466.

SCHÄFER, A. 1989. Variscan molasse in the Saar–Nahe Basin (W-Germany), Upper Carboniferous and Lower Permian. *Geologische Rundschau*, **78**, 499–524.

—— & KORSCH, R. J. 1998. Formation and sediment fill of the Saar–Nahe Basin (Permo-Carboniferous, Germany). *Zeitschrift der deutschen Geologischen Gesellschaft*, **149**, 233–269.

—— & SNEH, A. 1983. Lower Rotliegend fluviolacustrine sequences in the Saar–Nahe-Basin. *Geologische Rundschau*, **72**, 1135–1146.

—— & STAMM, R. 1989. Lakustrine Sedimente im Permokarbon des Saar–Nahe-Beckens. *Zeitschrift der deutschen Geologischen Gesellschaft*, **140**, 259–276.

——, RAST, U. & STAMM, R. 1990. Lacustrine paper shales in the Permocarboniferous Saar–Nahe-Basin (West Germany)—Depositional environment and chemical characterization. *In*: HELING, D., ROTHE, P., FÖRSTNER, U. & STOFFERS, P. (eds) *Sediments and Environmental Geochemistry.* Springer, Berlin, 220–238.

SCHMIDT, E. 1978. Ökosystem See. Das Beziehungsgefüge der Lebensgemeinschaft im eutrophen See und die Gefährdung durch zivilisatorische Eingriffe. *Biologische Arbeitsbücher*, **12**, 3rd edn. Quelle & Meyer, Heidelberg.

—— 1996. Ökosystem See. Der Uferbereich des Sees. *Biologische Arbeitsbücher*, **12/1**, 5th edn. Quelle & Meyer, Wiesbaden.

SCHULZ, U., LEYTHAEUSER, D. & SCHWARK, L. 1997. High-resolution geochemical investigation of an Oligocene lacustrine paleoenvironment—Impact of volcanic activity on biological communities of the fossil Enspel maar lake. *Würzburger geographische Manuskripte*, **41**, 195–196.

SCHWOERBEL, J. 1993. *Einführung in die Limnologie*, *Uni-Taschenbücher*, **31**, 7th edn. Fischer, Stuttgart.

SCOTESE, C. R. & LANGFORD, R. P. 1995. Pangea and the paleogeography of the Permian. *In*: SCHOLLE, P. A., PERYT, T. M. & ULMER-SCHOLLE, D. S. (eds) *The Permian of Northern Pangea. 1. Palaeogeography, Paleoclimates, Stratigraphy.* Springer, Berlin, 3–19.

STOLLHOFEN, H. 1994. Vulkaniklastika und Siliziklastika des basalen Oberrotliegend im Saar–Nahe-Becken (SW-Deutschland): Terminologie und Ablagerungsprozesse. *Mainzer geowissenschaftliche Mitteilungen*, **23**, 95–138.

—— & STANISTREET, I. G. 1994. Interaction between bimodal volcanism, fluvial sedimentation and basin development in the Permo-Carboniferous Saar–Nahe Basin (south-west Germany). *Basin Research*, **6**, 245–267.

STORCH, G., ENGESSER, B. & WUTTKE, M. 1996. Oldest fossil record of gliding in rodents *Nature*, **379**, 439–441.

TALBOT, M. R. & ALLEN, P. A. 1996. Lakes. *In*: READING, H. G. (ed.) *Sedimentary Environments: Processes, Facies and Stratigraphy*, 3rd edn. Blackwell, Oxford, 83–124.

TYSON, R. V. 1995. *Sedimentary Organic Matter, Organic Facies and Palynofacies.* Chapman & Hall, London.

VEEVERS, J. J. & POWELL, C. M. 1987. Late Paleozoic glacial episodes in Gondwanaland reflected in transgressive depositional sequences in Euramerica. *Geological Society of America Bulletin*, **98**, 475–487.

WITZKE, B. J. 1990. Palaeoclimatic constraints for Paleozoic palaeolatitudes of Laurentia and Euramerica. *In*: MCKERROW, W. S. & SCOTESE, C. R. (eds) *Palaeozoic Palaeogeography and Biogeography*, Geological Society, London, Memoir, **12**,

57–73.

WUTTKE, M. & RADTKE, G. 1993. Agglutinierende Mikrobenmatten im Profundal des mitteleozänen Eckfelder Maar-Sees bei Manderscheid/Eifel (Bundesrepublik Deutschland). *Mainzer Naturwissenschaftliches Archiv*, **31**, 115–126.

ZIEGLER, A. M. 1990. Phytogeographic patterns and continental configuration during the Permian period. *In* : MCKERROW, W. S. & SCOTESE, C. R. (eds) *Palaeozoic Palaeogeography and Biogeography*, Geological Society, London, Memoir, **12**, 363–379.

ZOLITSCHKA, B. 1989. Jahreszeitlich geschichtete Sedimente aus dem Holzmaar und dem Meerfelder Maar. *Zeitschrift der deutschen Geologischen Gesellschaft*, **140**, 25–33.

—— 1990. *Spätquartäre jahreszeitlich geschichtete Seesedimente ausgewählter Eifelmaare. Paläolimnologische Untersuchungen als Beitrag zur spät- und postglazialen Klima- und Besiedlungsgeschichte.* Documenta Naturae, **60**.

—— 1998. Paläoklimatische Bedeutung laminierter Sedimente. Holzmaar (Eifel, Deutschland), Lake C2 (Nordwest-Territorien, Kanada) und Lago Grande di Monticchio (Basilicata, Italien). *Relief Boden Paläoklima*, **13**, Bornträger, Berlin.

Global terrestrial productivity in the Mesozoic era

D. J. BEERLING

Department of Animal and Plant Sciences, University of Sheffield, Sheffield S10 2TN, UK

Abstract: This paper describes a global-scale modelling approach for investigating the structure and productivity of terrestrial vegetation in the Mesozoic era. General circulation model (GCM) palaeoclimate simulations (by the University of Reading) for Kimmeridgian and Cenomanian time are described as driving datasets for the University of Sheffield Dynamic Global Vegetation Model (SDGVM). The validity of these GCM palaeoclimates is reviewed. Global patterns of terrestrial net primary productivity (NPP) and leaf area index (LAI) have been simulated with the SDGVM, which models plant physiological processes and the biogeochemical cycling of carbon and nitrogen in soils and vegetation. An important feature of this modelling approach is that it requires no underlying map of soils or vegetation type. Global NPP in Kimmeridgian and Cenomanian time was high (117.6 and 106.8 Gt C a^{-1}, respectively) compared with the present-day level (57.0 Gt C a^{-1}). The high concentrations of atmospheric CO_2 at each interval significantly influenced the NPP and LAI of Mesozoic vegetation, to an extent dependent on climate. CO_2 effects on the structure of vegetation in Kimmeridgian and Cenomanian time were sufficient to markedly influence key land surface variables required by GCMs for climate predictions, and point to the need to include direct CO_2–vegetation interactions in climate models. Two approaches to testing the modelled NPP patterns were devised, one using palaeobotanical information on tree growth, and the other based on a comparison of measurements and predictions of the stable carbon isotope ratios of fossil plants. Both approaches provided some support for the global-scale simulations, indicating the feasibility of modelling vegetation activity in ancient climates from a knowledge of present-day processes.

General trends in world-wide patterns of terrestrial productivity in the Mesozoic era can be inferred non-quantitatively from floral boundaries based on the plant fossil record (Vakhrameev 1991). The positioning of such boundaries is, however, subjective, because of the limited number of sites from which fossils have been recovered. An alternative method of grouping information from the plant fossil record is to use multivariate analyses of palaeofloristic data to provide a non-subjective means of grouping phytogeographical units and their distribution at distinct intervals of geological time (Spicer et al. 1993) although, again when considered at the global scale, the spatial coverage of fossil floras is incomplete. Therefore investigation of shifts in productive zones of the Earth's surface through the Mesozoic era requires a different approach. Previous attempts to gain the 'global picture' in Mesozoic time have been based on correlations between climate and vegetation, the distribution of coals (Parrish et al. 1982; Ziegler et al. 1987, 1993; Lottes & Ziegler 1994), and inferences of productivity based on modern vegetation analogues of those found in fossil plant beds (Ziegler et al. 1987, 1993). These types of correlative approaches assume a uniformitarian relationship between plants and climate irrespective of atmospheric composition, yet experimental work on the direct effects of atmospheric CO_2 concentrations on plants suggest this may not be the case (e.g. Hättenschwiler et al. 1996, 1997; Briffa et al. 1997; Beerling 1998).

An emerging approach to characterizing plant productivity in the distant past is based on palaeoclimate (Allen et al. 1993) and vegetation modelling (Beerling et al. 1998). Such an approach has the advantage that it is largely based on a mechanistic understanding of the systems involved, and so is less prone to errors of some correlative methods. In addition, it offers greater spatial detail than the fossil record, and provides quantitative information for comparison with the other methods of reconstructing climate and plant productivity. This paper describes global climate and vegetation modelling simulations for the Mesozoic era, and more fully describes earlier work on the mid-Cretaceous period (Beerling et al. 1999), but with an extension back to Late Jurassic time.

From: HART, M. B. (ed.) *Climates: Past and Present.* Geological Society, London, Special Publications, **181**, 17–32, 1-86239-075-4/$15.00

Consideration is also given to CO_2 sensitivity analyses and the likely potential for feedback between vegetation and climate at these times.

According to the distribution of climatically sensitive sediments, fossils and oxygen isotope measurements, the climate of the Mesozoic era was appreciably more equable than today (Hallam 1985), in the sense of a flatter pole-to-equator temperature gradient, probably through a combined increased CO_2 and water vapour 'greenhouse effect' (Hallam 1985; Berner 1997; Sellwood & Valdes 1997). It therefore represents a useful interval for evaluating the performance of climate and vegetation models under high CO_2 concentrations, an issue critical if they are to be used to predict future global change.

Global patterns of climate in the Mesozoic era have been derived by the UK Universities Global Atmospheric Modelling Program (UGAMP) general circulation model (GCM) of the Earth's climate system for the Late Jurassic Kimmeridgian age and the mid-Cretaceous Cenomanian age (Valdes et al. 1995, 1996). Each climate has been evaluated against sedimentary data to test model strengths and weaknesses and gain an impression of how realistically the GCM replicated past climates (Price et al. 1995, 1997a). Each global palaeoclimate simulation (monthly temperature, precipitation and humidity) was used to drive the process-based University of Sheffield Dynamic Global Vegetation Model (SDGVM), which calculates the net primary productivity (NPP) and leaf area index (LAI) of terrestrial vegetation, from largely mechanistic descriptions of plant ecophysiological processes (Woodward et al. 1995). The SDGVM is dynamically coupled with the CENTURY biogeochemistry model of soil carbon and nitrogen cycling in vegetation and soils (Parton et al. 1993). The coupling allows litter formation and decomposition, and so the vegetation and soil carbon cycle is completed. An important consequence of this approach is that predictions of the characteristics of terrestrial vegetation are made without any underlying (or prescribed) map of vegetation or soil type (Beerling et al. 1999).

Here, the SDGVM has been used to predict global patterns of NPP and LAI in Kimmeridgian and Cenomanian time. A quantitative assessment is also made of the direct effects of the high-CO_2 environment of the Mesozoic era on several of the key vegetation characteristics required by GCMs to model past and future climates. The final section considers the use of tree rings and stable carbon isotope analyses of plant fossils as two approaches to testing the model predictions.

GCM palaeoclimate simulations for the Mesozoic era

The palaeoclimate simulations for Kimmeridgian and Cenomanian time used to drive the SDGVM were performed at the University of Reading and used the UGAMP GCM, described in detail elsewhere (Valdes et al. 1995, 1996), which is based on Newton's laws of motion, and the first and second laws of thermodynamics (Valdes et al. 1995). It has c. 50% higher resolution than most other GCMs used for palaeoclimate simulations with a horizontal grid of $3.75° \times 3.75°$, and its 19 levels in the vertical and the extra resolution allow better representation of the coastlines and mid-latitude depressions; representation of the latter is important for a good simulation of climate in these regions. Simulations for both intervals do not explicitly model the oceans and these runs were either with prescribed sea surface temperatures or using a mixed-layer (slab) ocean model. These sea surface temperatures are energetically consistent with the choice of CO_2 and solar constant, and with data derived from the Cretaceous oceans by Sellwood et al. (1994). For each interval, model results were integrated for 5 years and the last 2 years were averaged to produce the Kimmeridgian and Cenomanian 'climate' consisting of monthly values of precipitation, temperature and humidity. The UGAMP model includes a range of parameterizations, including a full radiation scheme (with seasonal and diurnal variations), a relative humidity based cloud scheme, and a simple three-layer soil model. Vegetation is not explicitly represented but its effects on the albedo and surface roughness are included.

Mean annual temperature (MAT) and mean monthly precipitation (MMP) for Kimmeridgian and Cenomanian time are given in Fig. 1. Both model climates are very warm in equatorial regions, with MATs of 30°C. Over large regions of the land masses at high latitudes of the Southern Hemisphere in both Kimmeridgian and Cenomanian time MATs fall to 0 to –5°C, whereas high latitudes (> 50°N) in the Northern Hemisphere are warmer, c. 10–15°C. Precipitation is simulated to be higher in Kimmeridgian than in Cenomanian time (Fig. 1 b and d), with the highest values of both model climates occurring in equatorial regions. Strong precipitation gradients are evident across continental regions, with regions of low precipitation extensive, particularly in the Cenomanian results. Quantitative comparisons with reconstructions based on analyses of the physiognomy of fossil leaves, although potentially biased by the

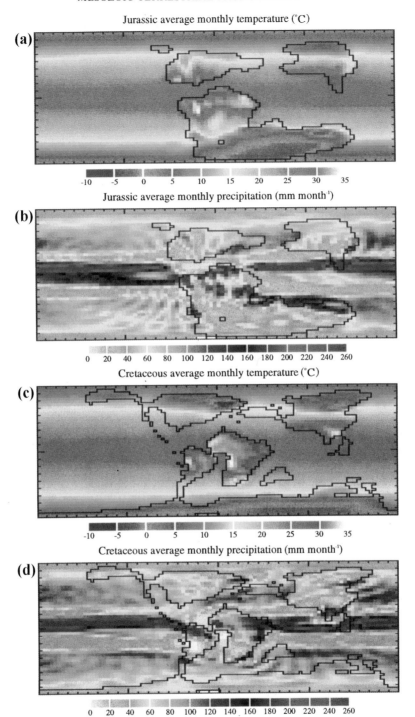

Fig. 1. Global patterns of mean temperature and precipitation simulated for the Kimmeridgian (**a**, **b**) and Cenomanian time (**c**, **d**) by the UGAMP GCM. Latitudinal and longitudinal ticks are, 10° intervals. (For details of the simulations, see text.) Data provided by P. J. Valdes (University of Reading) (pers. comm.).

D. J. BEERLING

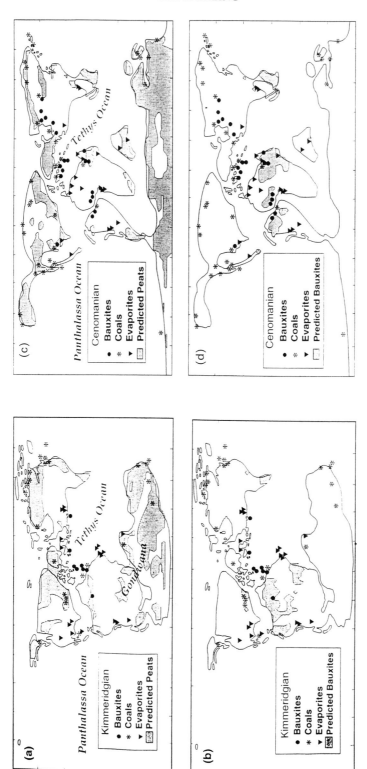

Fig. 2. Observed global distribution of bauxites, coals and evaporites, and the predicted distribution of peats and bauxites for Kimmeridgian (**a, b**) and Cenomanian time (**c, d**). Predicted distributions were made from the UGAMP GCM output and the empirical proxy climate models defining the occurrence of peats and bauxites for the present data. From Price *et al.* (1997*a*).

(a)

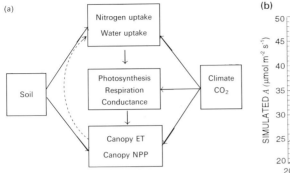

(b)

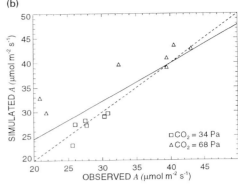

(c)

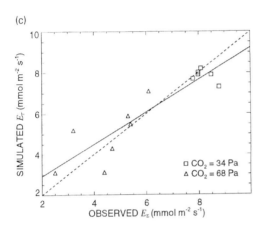

(d)

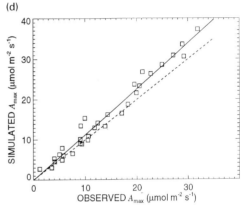

(e)

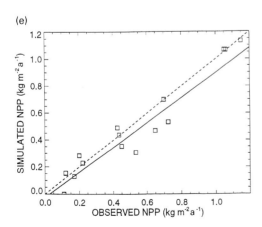

Fig. 3. (**a**) Schematic diagram of the University of Sheffield Dynamic Global Vegetation Model (SDGVM). The diagram illustrates the interaction between soil nutrient status (determined by the biogeochemistry model), water-holding capacity, climate, CO_2 concentration and leaf gas exchange. These responses are scaled to the canopy to predict canopy transpiration and net primary productivity (NPP). The broken arrow shows the feedback between water loss via canopy transpiration and plant water uptake. The lower set of panels illustrate the vegetation model performance against point measurements. Comparison of model predictions and observations for (**b**) canopy photosynthesis and (**c**) canopy transpiration for a mixed forest grown at two partial pressures of CO_2, 34 Pa and 65 Pa, and a comparison of (**d**) maximum rates of leaf photosynthesis (A_{max}) and (**e**) NPP for a worldwide range of vegetation types under current climate and CO_2. In each case (**b**–**e**) the broken line indicates perfect fit between predictions and measurements, and the continuous line indicates the regression line fitted to the data. From Woodward *et al.* (1995). Reproduced with permission from the American Geophysical Union.

22   D. J. BEERLING

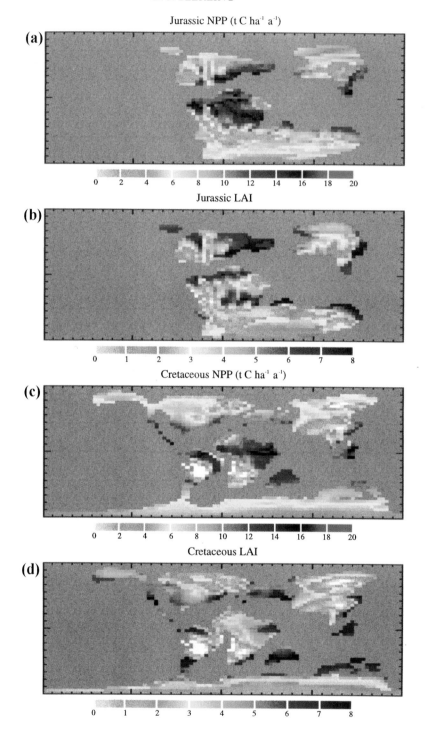

Fig. 4. Equilibrium SDGVM solutions for net primary productivity (NPP) and leaf area index (LAI) in Kimmeridgian (**a**, **b**) and Cenomanian time (**c**, **d**).

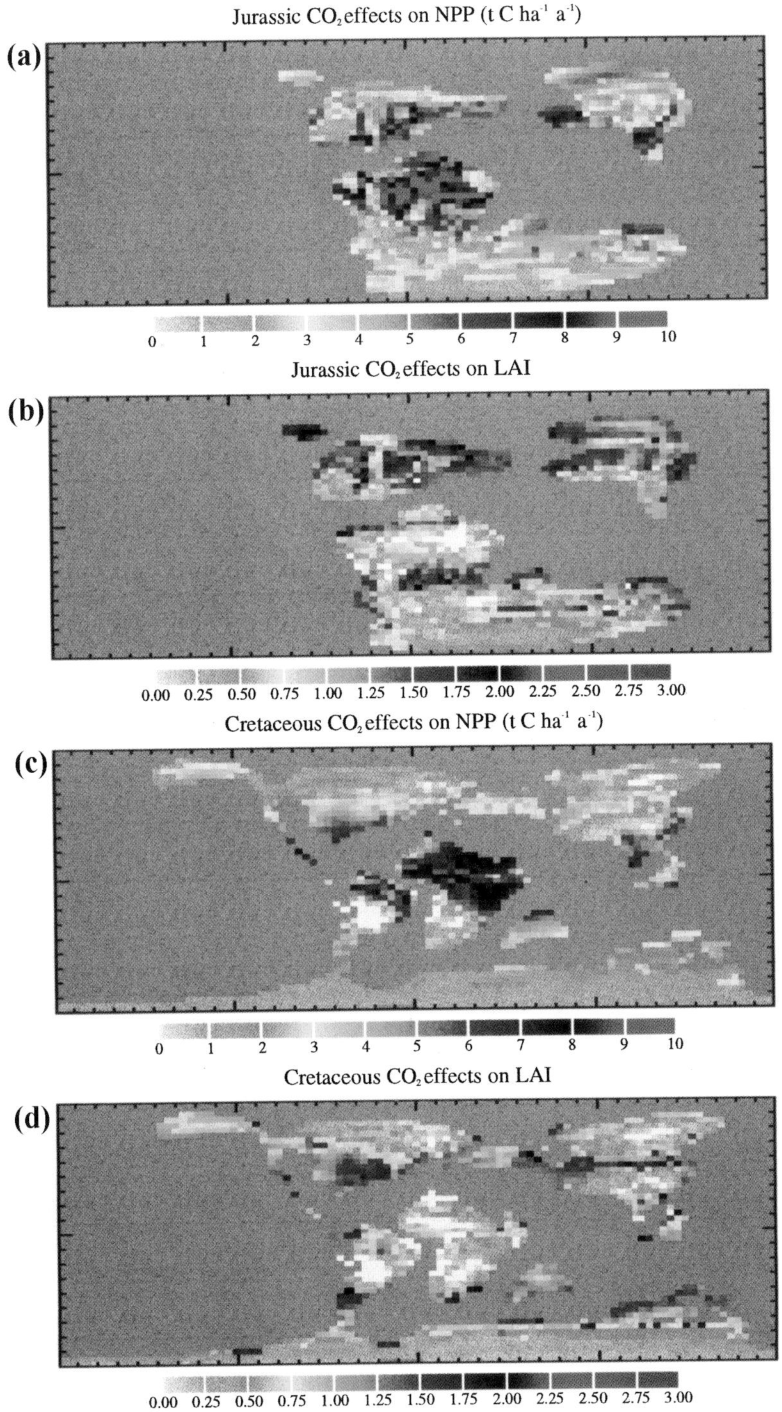

Jurassic CO_2 effects on NPP (t C ha^{-1} a^{-1})

(a)

0   1   2   3   4   5   6   7   8   9   10

Jurassic CO_2 effects on LAI

(b)

0.00 0.25 0.50 0.75 1.00 1.25 1.50 1.75 2.00 2.25 2.50 2.75 3.00

Cretaceous CO_2 effects on NPP (t C ha^{-1} a^{-1})

(c)

0   1   2   3   4   5   6   7   8   9   10

Cretaceous CO_2 effects on LAI

(d)

0.00 0.25 0.50 0.75 1.00 1.25 1.50 1.75 2.00 2.25 2.50 2.75 3.00

Fig. 5. The influence of CO_2 on NPP and LAI predictions for Kimmeridgian (**a**, **b**) and Cenomanian time (**c**, **d**). The maps show the difference between the model runs at elevated and ambient CO_2 concentrations, i.e. a positive value indicates an increase under the higher CO_2 concentration. (See text for choice and rationale of CO_2 concentrations.)

influence of high CO_2 concentrations (Beerling 1998), show similar values: Spicer & Parrish (1990) and Herman & Spicer (1996) reconstructed MATs for Cenomanian time at 75°N of 10–12.5°C and an MMP of 122 mm.

Validation of the two Mesozoic global palaeoclimate simulations is an important step before their use for vegetation modelling. One approach to testing has been to predict the global distribution of the bauxites and coals for comparison with their observed distribution (Price *et al.* 1997*a,b*). Each climatic indicator is predicted from GCM output using quantitative proxy models; temperature, precipitation and evaporation constraints of each have been given elsewhere (Price *et al.* 1997*a,b*). Comparison of the predicted and observed distributions for Kimmeridgian and Cenomanian time (Fig. 2) shows that 70% and 65% of the observed coal occurrences respectively were within the areas predicted for coal formation by the GCM (Fig. 2a and c). Predictions of bauxite formation for the land area used by the GCM show general agreement with observations (Fig. 2b and d). In Cenomanian time, for which there are most observations of bauxite distribution, 45% of the observed points occurred within the predicted regions. Some discrepancy may relate to erosion of bauxite deposits and weaknesses in the GCM hydrological cycle. Climatic controls on evaporites are less certain and preclude prediction of their distribution. The observed evaporite distribution in Kimmeridgian time probably reflects arid conditions around the Gulf of Mexico, Southern Eurasia, South America and Saudi Arabia, whereas for Cenomanian time the main arid regions were northern and eastern Africa. For both intervals, these evaporite sites typically occur in regions of low MMP (Fig. 1). Overall, the comparison of the GCM outputs with geological data suggests that sufficient regional detail is captured for their use in vegetation modelling.

Modelling productivity of terrestrial vegetation

The current simulations use the SDGVM, described in detail elsewhere (Woodward *et al.* 1995). The SDGVM first predicts the temperature-dependent uptake of nitrogen and water from the soil. Nitrogen uptake determines rates of photosynthesis, dark respiration and stomatal conductance of leaves, and is assigned to leaf layers in proportion to irradiance. Rates of photosynthesis are modified according to variations in temperature, CO_2, humidity, irradiance and stomatal conductance, and depend on soil

water content and temperature. Canopy conductance controls soil water loss by evapotranspiration. Rates of each process are then integrated to predict canopy evapotranspiration and net CO_2 assimilation (photosynthesis–respiration) (Fig. 3a). LAI is defined by the annual long-term hydrological and carbon budgets. The hydrological budget constrains leaf mass, and related structural variables, so that the average values maintained over the year do not deplete soil moisture. The carbon budget constrains the LAI by checking that the respiration required to synthesize the tissues does not exceed net assimilation. The model then adjusts LAI to a maximum value that satisfies both criteria using monthly values of climate. A value of zero indicates desert, whereas a near-maximal value of 10 is more indicative of a tropical forest.

The NPP computed by the vegetation model is then decomposed by the various routines within the CENTURY biogeochemistry model (Parton *et al.* 1993). CENTURY is interactively coupled to the vegetation model on a monthly basis. CENTURY simulates the carbon (C) and nitrogen (N) dynamics of vegetation and soils from NPP of any given climate and atmospheric CO_2 concentration to predict soil carbon and nitrogen concentrations. The newly derived soil nutrient status is used to drive the SDGVM, and the iterations are continued until equilibrium values of LAI and NPP are achieved. The outputs of the full model, of concern here, are the functional attribute net primary productivity (NPP, g C m^{-2} a^{-1}) and the structural attribute leaf area index (LAI, amount of leaf area per unit area of ground) for vegetation at any given terrestrial point on the globe.

The responses of each sub-model of the vegetation model have been tested against observational data from experiments (Fig. 3b–e). Predictions of canopy photosynthesis and evapotranspiration for mature *Eucalyptus* and *Acacia* trees exposed to ambient and elevated partial pressures of CO_2 show good agreement with measurements (Fig. 3b and c). Under the current CO_2 concentration, leaf-scale predictions of the maximum photosynthetic rates (A_{max}) and ecosystem NPP for a global range of different vegetation types closely match observations (Fig. 3d and e). The coupled SDGVM has also been successfully tested against observations of gas exchange, catchment runoff and soil nutrient concentrations from an entire boreal ecosystem experimentally exposed *in situ* to increased CO_2 and raised temperature for the past 3 years (Beerling *et al.* 1997). These tests indicate adequate representation of plant and edaphic processes in the SDGVM for it to be of use when applied to past climates and atmospheres.

Global terrestrial productivity in the Mesozoic era

The equilibrium solutions of the SDGVM for NPP and LAI in Kimmeridgian and Cenomanian time indicate that the climate and CO_2 concentrations of the Mesozoic era were suitable for a highly productive land surface (Fig. 4), assuming no major down regulation of the photosynthetic proteins (Drake *et al.* 1997). The global NPP in Kimmeridgian and Cenomanian time has been estimated at 117.6 and 106.8 Gt $C\,a^{-1}$ respectively. The high productivity of the Mesozoic terrestrial biosphere is emphasized when these figures are compared with a model estimate of global NPP for current CO_2 and climate of 57.0 Gt $C\,a^{-1}$ and for a GCM-simulated climate with a doubling of the present CO_2 concentration (69.6 Gt $C\,a^{-1}$) (Cao & Woodward 1997). Presumably, such high productivities were in keeping with, or even required by, large herbivorous and carnivorous dinosaurs and reptiles that depended directly or indirectly on plant growth for support (Fischer 1981).

Both the Kimmeridgian and Cenomanian results indicate the potential for forest growth at very high latitudes, even under the prevailing light regime (Creber & Chaloner 1985), confirming earlier modelling efforts (Beerling 1997). In Kimmeridgian time, regions of highest NPP occur around the eastern tip and northern regions of Gondwana (the main landmass in the Southern Hemisphere), and the southeastern most edges of Eurasia and North America (Fig. 4a), each of which corresponds to the occurrence of coals (Fig. 2a and b), suggesting a connection between coal formation and high productivity. The NPP of vegetation near the equator in Kimmeridgian time is very high, *c.* 20 t $C\,ha^{-1}a^{-1}$, over double that of tropical evergreen forests in Brazil under today's climate and CO_2 (6.8–10.5 t C $ha^{-1}a^{-1}$) (McGuire *et al.* 1992; Lloyd & Farquhar 1996), but is in keeping with theoretical expectations for these forests to be strong responders to CO_2 and climate change (Lloyd & Farquhar 1996). Phytogeographical analyses of the Late Jurassic plant fossil record indicate that the modelled high-productivity areas around the southeastern edges of Eurasia were dominated by cool temperate and dry subtropical floras (Ziegler *et al.* 1993).

Regions of high NPP in Cenomanian time are confined to central Africa and northern parts of South America (Fig. 4c), and these areas too show some correspondence to Cenomanian coal occurrence (Fig. 2b and c). Spicer *et al.* (1993) suggested that plant productivity in Late Cretaceous time was concentrated in mid- to high latitudes, and although no quantitative data were offered, this patterning agrees with the model results (Fig. 4). The modelled low productivity (2–6 t C $ha^{-1}a^{-1}$) in Late Cretaceous time of large areas of Eurasia corresponds, according to the phytogeographical data of plant fossils, to the domination of these regions by xeromorphic taxa and compares with the productivity of savannah vegetation in Minnesota of 7.2 t C $ha^{-1}a^{-1}$ and xeromorphic forests in Puerto Rico of 5.5 t C $ha^{-1\,-1}$ (McGuire *et al.* 1992). Further north in Eurasia, where NPP values are higher, the phytogeographical data indicate a switch in dominance to broad-leaved deciduous genera (Spicer *et al.* 1993).

The world-wide patterns of the structure of terrestrial vegetation canopies, measured as LAI, reflect most strongly the geographical distribution of land surface precipitation (Fig. 1a and c) and NPP, areas of high LAI coinciding with high mean monthly precipitation and high NPP (Fig. 4b and d). It is interesting to note that the global distribution of evaporites in Kimmeridgian and Cenomanian time shows close correspondence to regions of low LAI (Fig. 2). Although this is due in part to the driving climatic datasets, LAI is controlled by the hydrological and carbon budgets of a given site, as well as by many different and interacting plant physiological and soil processes. In this sense, LAI might be a more effective predictor of evaporite distribution than simple proxy models using rainfall patterns and temperature. Low LAI values occur in arid areas of the Kimmeridgian and Cenomanian simulations because even in high-CO_2 environments, where stomata closure is induced and plant water-use efficiency increased, no additional layers in the canopy can be added if this leads to annual transpiration exceeding annual precipitation.

Quantifying CO_2 effects on terrestrial vegetation

The high-CO_2 environment of the Mesozoic era (Berner 1997) has frequently been suggested to contribute conditions suitable for substantial forest growth (Creber & Chaloner 1985; Ziegler *et al.* 1987). If this is the case, inclusion of the direct effects of CO_2 in vegetation modelling is required (Beerling 1994, 1997), particularly when used to define land surface boundary conditions in GCMs (Pollard & Thompson 1995; Sellers *et al.* 1996; Betts *et al.* 1997; Otto-

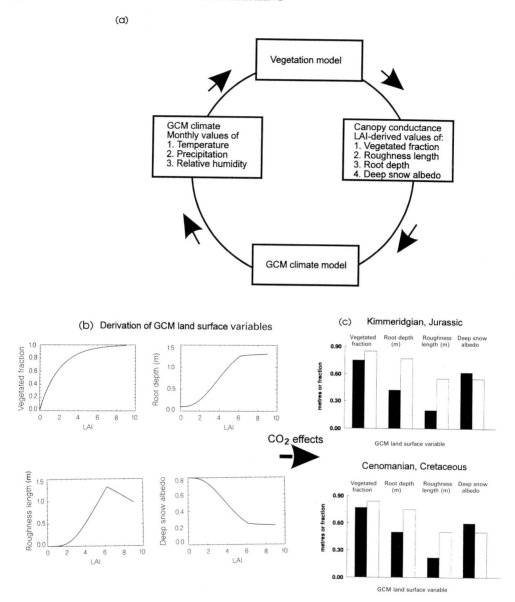

Fig. 6. (a) Scheme illustrating how general circulation models of the Earth's climate (GCMs) can be iteratively coupled with the vegetation model to refine climate predictions. It should be noted that the vegetation model is required to define the boundary-layer conditions of canopy conductance and the four land surface factors described. (b) Empirical functions for deriving four land surface factors required by GCMs to calculate climate from leaf area index (LAI). (c) Calculation of each land surface variable at ambient CO_2 (■) and elevated CO_2 (□) concentrations for each period. It should be noted that the effect of elevated CO_2 is marked on three out the four factors in Kimmeridgian and Cenomanian time, with substantial implications for vegetation feedback on climate. Reproduced by permission of *Nature* [Betts *et al.* (1997)]. Copyright (1997) Macmillan Magazines Ltd.

Bliesner & Upchurch 1997). Therefore, to quantify the potential for CO_2 impacts on the NPP and LAI of vegetation, two sensitivity analyses have been performed. For Kimmerid-

gian and Cenomanian time, the model was run with an ambient pre-industrial CO_2 concentration of 300 ppmv and an elevated 'best guess' CO_2 concentration estimated from Berner's

Jurassic plant $^{13}CO_2$ discrimination

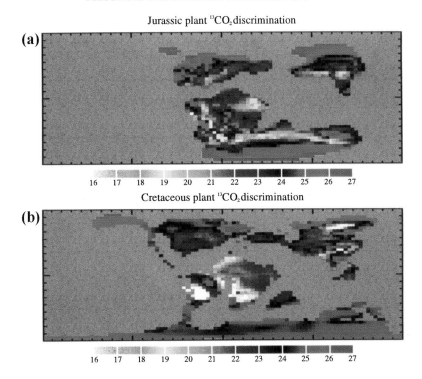

Cretaceous plant $^{13}CO_2$ discrimination

Fig. 7. Modelled global patterns of discrimination against the heavy isotope of carbon, $^{13}CO_2$, by the terrestrial biosphere in (**a**) Kimmeridgian and (**b**) Cenomanian time.

(1997) long-term carbon cycle model (1500 ppmv for Kimmeridgian time, 900 ppmv for Cenomanian time) with a constant atmospheric O_2 content of 25% for all model runs (Berner & Canfield 1989). Each ambient and elevated CO_2 run used the same Jurassic or Cretaceous palaeoclimate simulation. The difference between elevated and ambient CO_2 runs represents the direct effect of CO_2 on NPP and LAI. The pre-industrial concentration was used to represent the value under which current vegetation characteristics were obtained and assigned to palaeo-vegetation in modelling exercises (e.g. Pollard & Thompson 1995; Otto-Bliesner & Upchurch 1997).

It emerges from the sensitivity analyses that global NPP is markedly affected by the different atmospheric CO_2 concentrations (Fig. 5). In Kimmeridgian time, CO_2 effects account for 127% increase in global NPP, from 51.9 to 117.6 Gt C a^{-1}, and for Cenomanian time an 87% increase from 56.6 to 106.8 Gt C a^{-1}. The absolute global pattern of NPP and LAI remains unchanged by CO_2, but the two variables show responses that are geographically distinct. Increases in NPP occur at higher CO_2 concentra-

tions at low latitudes, particularly around the central equatorial belt, whereas the increases in LAI occur mainly at mid-latitudes in both hemispheres. These two different responses arise because in equatorial regions the high temperatures and high rates of precipitation allow NPP to increase under elevated CO_2; this effectively represents an increase in the water-use efficiency of plant growth (i.e. same climate but increased CO_2).

Increases in LAI resulting from CO_2 enrichment are not evident in these regions because tropical vegetation typically has a high LAI anyway, and rapid light attenuation through a dense canopy means it is effectively uneconomical for the plant to add further leaf layers operating at very low irradiances (Beerling 1999). This situation holds even though the light compensation point for photosynthesis decreases at high CO_2 (Long & Drake 1991). Instead, therefore, LAI increases only where the trade-off between photosynthesis and maintenance respiration stimulates NPP, a situation that typically occurs at mid-latitude sites with adequate temperature and precipitation (Beerling 1999).

(a)

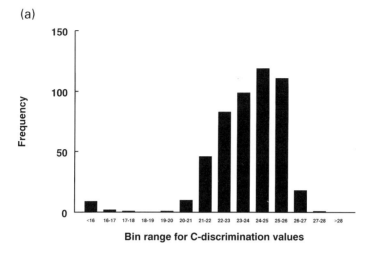

(b)

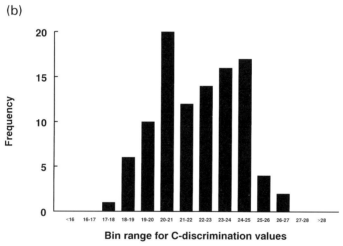

Fig. 8. Frequency histograms of (**a**) modelled $^{13}CO_2$ discrimination values in Cenomanian time from Fig. 7b ($n = 1517$ points) and, for comparison, (**b**) measurements of $^{13}CO_2$ discrimination from Cretaceous plant fossils (n = 103). Data for (**b**) from Bocherens *et al.* (1994).

Implications of CO₂ effects for the feedback between vegetation and climate

Terrestrial vegetation influences climate through its modification of the radiative balance and energy partitioning between the Earth's surface and the atmosphere, both of which can substantially alter land surface temperatures (Henderson-Sellers *et al.* 1995; Sellers *et al.* 1996; Betts *et al.* 1997). Vegetation influences climate through its structural features, e.g. LAI, albedo, canopy height and roughness, and through its physiological effects, e.g. canopy stomatal conductance to water vapour (Henderson-Sellers *et*

al. 1995; Sellers *et al.* 1996; Betts *et al.* 1997; Otto-Bliesner & Upchurch 1997). Iterative coupling of vegetation and climate models therefore (Fig. 6) provides one means of incorporating structural and physiological effects of vegetation in climate predictions. In many case however, the structural and physiological variables are taken from current vegetation and are fixed and unresponsive to the environment, especially CO_2, thus introducing uncertainties into the model results.

Therefore, this section considers a preliminary assessment of the importance of CO_2 on four GCM land surface variables (vegetated fraction, root depth, roughness length and deep snow albedo). Each variable has been estimated from empirical functions, using the global mean LAI of the Kimmeridgian and Cenomanian simulations at pre-industrial and 'best guess' CO_2 concentrations. The results indicate substantial CO_2 effects on three of the four factors (Fig. 6). The higher mean global LAI for Kimmerdgian and Cenomanian time at the higher CO_2 concentration would have resulted in a greater vegetated fraction of sites, increased aerodynamic roughness length of the vegetation and deeper rooting depths (by up to 0.5 m). These comparisons are calculated for global average LAIs and indicate the potential for erroneous predictions of climate when CO_2–vegetation interactions are ignored in climate models (Henderson-Sellers et al. 1995; Sellers et al. 1996; Otto-Bliesner & Upchurch 1997). Atmospheric CO_2 effects on LAI varied geographically (Fig. 5), but were most marked at mid-latitudes, where the LAI differences as a result of CO_2 effects were up to three in Kimmeridgian and two in Cenomanian time (Fig. 5). From the relationships in Fig. 6, this suggests greater fractional cover of vegetation and increased roughness length in these regions, resulting in large feedbacks to climate. The net effect of this feedback on land surface air temperatures is not, however, easy to predict without coupling the SDGVM and GCM (Betts et al. 1997).

Validating global predictions of vegetation function in the Mesozoic era

The model simulations of NPP and LAI received limited support from the sedimentary and palaeoecological data discussed above. Additional testing of the SDGVM predictions can be made by comparison with productivity estimates from fossil tree ring widths (e.g. Beerling 1997) and measurements of plant stable carbon isotope composition, as an integrator of physiological activity through the growing season. Creber & Chaloner (1985) assembled data on the global distribution of fossil tree ring widths of Mesozoic woods and showed wide ring widths in high latitudes for Late Jurassic and mid-Cretaceous woods greatly exceeding those of modern woods from mid-latitudes, the implication being that Mesozoic climates were warmer and CO_2 concentrations higher than now. Experimental evidence confirming the relationship between wide ring widths and elevated CO_2 is equivocal (Hättenschwiler et al. 1996, 1997;

Briffa et al. 1997; Beerling 1998), although, clearly, the existence of forests at high latitudes in Mesozoic time argues for a more equable climate and supports the SDGVM predictions of NPP at this time (Fig. 4). From detailed tree ring analyses, and assumptions on the conversion of incident irradiance to wood, Creber & Chaloner (1985) calculated a productivity of forests in Antarctica for late Cretaceous time of 4.6 t C $ha^{-1} a^{-1}$. This value is remarkably close to the model predictions for this region and at this time (Fig. 4) of 4–5 t C $ha^{-1} a^{-1}$, and therefore provides support for the modelling at this particular site.

The second test of the model predictions is more mechanistic and compares measured and predicted stable carbon isotope ratios of plants in Cretaceous time, although model results are presented for Kimmeridgian time as well, as values of isotopic discrimination provide revealing information on plant metabolism during growth. The stable carbon isotope ratio ($\delta^{13}C$) of plants is related to leaf gas exchange through the well-validated theoretical model of Farquhar et al. (1982):

$$\delta^{13}C = \delta^{13}C_a - a - (b - a) \times \frac{p_i}{p_a} \quad (1)$$

where $\delta^{13}C_a$ is the isotopic composition of the atmospheric CO_2 utilized by the plant, a is the discrimination against $^{13}CO_2$ by diffusion in free air and through the stomatal pores (4.4‰), b is the fractionation associated with Ribulose-1,5-carboxylase–oxygenase (RuBisCo, 27‰) and p_i/p_a is the ratio of CO_2 concentration inside the leaf (p_i) relative to that in the atmosphere (p_a). When the $\delta^{13}C_a$ is known, the extent of plant discrimination against $^{13}CO_2$ (Δ) can be related to leaf gas exchange by

$$\Delta = a + (b - a) \times \frac{p_i}{p_a} \quad (2)$$

Equation 2 shows that Δ is largely controlled by the p_i/p_a ratio, which in turn is influenced by the photosynthetic and stomatal activity of the plants over the entire growing season. When photosynthetic rates are low and/or stomatal conductance is high, p_i increases and Δ approaches 27, i.e. the value for RuBisCo. Conversely, when photosynthetic rates are high and/or stomatal conductance is low, p_i increases and Δ approaches four. In this way, plant Δ represents an integrator of stomatal and photosynthetic activity over the lifetime of the tissue, and measurements on plant material can be compared with predicted values as a means of testing the modelling approach (Beerling, 1997; Beerling & Woodward 1997).

Taking an estimate of $\delta^{13}C_a$ for Kimmeridgian and Cenomanian time from the ocean carbonate record (Frakes et al. 1992), plant Δ has been predicted from the NPP-weighted estimates of p_i/p_a (equation 2), calculated by the SDGVM and averaged through the canopy. Global maps of $^{13}CO_2$ discrimination by plants in Kimmeridgian and Cenomanian time (Fig. 7) show similar patterns; Δ values approach 27 at high latitudes and decrease moving towards lower latitudes. Similar global patterns in terrestrial plant Δ have been modelled for the present-day climate and CO_2 concentration (Lloyd & Farquhar 1994). The reason for this patterning is related to variations in surface temperature, precipitation and leaf-to-air vapour pressure deficit (VPD) acting in concert on photosynthetic rates and stomatal conductance. At high latitudes, lower temperatures reduce photosynthetic rates so that growing season p_i values are high. Presumably, lower temperatures also lower the evaporative demand from the leaf, and lower leaf-to-air VPDs, allowing higher stomatal conductances, which also contribute to higher Δ values. At lower latitudes, reductions in Δ reflect lower p_i values caused by higher rates of photosynthesis and possibly some effects of seasonally high leaf-to-air VPD reducing stomatal conductance.

The paucity of carbon isotope measurements on Jurassic plant fossil carbon precludes testing of Fig. 7a. Comparison can, however, be made between modelled and observed (Bocherens et al. 1994) Δ values for Cretaceous time (Fig. 7b). Frequency histograms of the modelled and observed datasets show remarkably similar trends (Fig. 8), in terms of both absolute values and their spread, thereby providing some support for the approaches to modelling Mesozoic climates and vegetation described here.

Conclusion

The most important conclusion of this work is that it appears possible to model the activities of the global climate system and plants in the geological past, and at the global scale, based on a knowledge of present-day processes. Sensitivity analyses indicated the importance of including the direct effects of CO_2 within vegetation models, and the potential for large feedbacks between vegetation function and climate, which have not yet been adequately included in the GCM simulations of past climates. Future work in Sheffield now aims to address this need.

I thank P. Valdes for providing the UGAMP GCM palaeoclimate fields for late Jurassic and mid-Cretaceous time, I. Woodward for access to the SDGVM, and M. Lomas for expert technical support. I gratefully acknowledge funding of this work by a Royal Society University Research Fellowship.

References

ALLEN, J. R. L., HOSKINS, B. J., SELLWOOD, B., SPICER, R. A. & VALDES, P. J. (eds) 1993. Discussion meeting. Palaeoclimates and their modelling with special reference to the Mesozoic era. *Philosophical Transactions of the Royal Society of London*, **341**, 1–343.

BEERLING, D. J. 1994. Modelling palaeophotosynthesis: late Cretaceous to present. *Philosophical Transactions of the Royal Society of London, Series B*, **346**, 421–432.

—— 1997. The net primary productivity and water use of forests in the geological past. *Advances in Botanical Research*, **26**, 193–227.

—— 1998. The future as the key to the past for palaeobotany? *Trends in Ecology and Evolution*, **13**, 311–316.

—— 1999. Atmospheric carbon dioxide, past climates and the plant fossil record. *Botanical Journal of Scotland*, **51**, 46–68.

—— & WOODWARD, F. I. 1997. Changes in land plant function over the Phanerozoic: reconstructions based on the fossil record. *Botanical Journal of the Linnean Society*, **124**, 137–153.

——, CHALONER, W. G. & WOODWARD, F. I. (eds) 1998. Discussion meeting. Vegetation–climate interactions: past, present and future. *Philosophical Transactions of the Royal Society* of London, **353**, 1–171.

——, WOODWARD, F. I., LOMAS, M. & JENKINS, A. J. 1997. Testing the responses of a dynamic global vegetation model to environmental change: a comparison of observations and predictions. *Global Ecology and Biogeography Letters*, **6**, 439–450.

——, —— & VALDES, P. J. 1999. Global terrestrial productivity in the mid-Cretaceous (100 Ma): model simulations and data. *Geological Society of America Special Paper*, **332**, 385–390.

BERNER, R. A. 1997. The rise of plants and their effect on weathering and atmospheric CO_2. *Science*, **276**, 544–546.

—— & CANFIELD, D. E. 1989. A new model of atmospheric oxygen over Phanerozoic time. *American Journal of Science*, **289**, 333–361.

BETTS, R. A., COX, P. M., LEE, S. E. & WOODWARD, F. I. 1997. Contrasting physiological and structural feedbacks in climate change simulations. *Nature*, **387**, 796–799.

BOCHERENS, H., FRIIS, E. M., MARIOTTI, A., RAUNSGAARD, K. & PEDERSEN, G. 1994. Carbon isotopic abundances in Mesozoic and Cenozoic fossil plants: palaeoecological implications. *Lethaia*, **26**, 347–358.

BRIFFA, K. R., SCHWEINGRUBER, F. H., JONES, P. D., OSBORN, T. J., SHIYATOV, S. G. & VAGANOV, E. A. 1997. Reduced sensitivity of recent tree growth to temperature at high northern latitudes. *Nature*, **391**,

678–682.

CAO, M. & WOODWARD, F. I. 1997. Net primary and ecosystem production and carbon stocks of terrestrial ecosystems and their responses to climate change. *Global Change Biology*, **4**, 185–198.

CREBER, G. T. & CHALONER, W. G. 1985. Tree growth in the Mesozoic and early Tertiary and the reconstruction of palaeoclimates. *Palaeogeography, Palaeoclimatology, Palaeoecology*, **52**, 35–60.

DRAKE, B. G., GONZÁLEZ-MELER, M. A. & LONG, S. P. 1997. More efficient plants: a consequence of rising atmospheric CO_2? *Annual Review of Plant Physiology and Plant Molecular Biology*, **48**, 607–637.

FARQUHAR, G. D., O'LEARY, M. H. & BERRY, J. A. 1982. On the relationship between carbon isotope discrimination and the intercellular carbon dioxide concentration in leaves. *Australian Journal of Plant Physiology*, **9**, 121–137.

FISCHER, A. G. 1981. Climatic oscillations in the biosphere. *In*: NITECKI, M. H. (ed.) *Biotic Crises in Ecological and Evolutionary Time*. Academic Press, New York, 103–131.

FRAKES, L. A., FRANCIS, J. E. & SYKTUS, J. L. 1992. *Climate Modes of the Phanerozoic*. Cambridge University Press, Cambridge.

HALLAM, A. 1985. A review of Mesozoic climates. *Journal of the Geological Society, London*, **142**, 433–445.

HÄTTENSCHWILER, S., MIGLIETTA, F., RASCHI, A. & KÖRNER, C. 1997. Thirty years of *in situ* tree growth under elevated CO_2: a model for future forest responses? *Global Change Biology*, **3**, 463–471.

——, SCHWEINGRUBER, F. H. & KÖRNER, C. 1996. Tree ring responses to elevated CO_2 and increased N deposition in *Picea abies*. *Plant, Cell and Environment*, **19** 1369–1378.

HENDERSON-SELLERS, A., MCGUFFIE, K. & GROSS, C. 1995. Sensitivity of global climate model simulations to increased stomatal resistance and CO_2 increases. *Journal of Climate*, **8**, 1738–1743.

HERMAN, A. B. & SPICER, R. A. 1996. Palaeobotanical evidence for a warm Cretaceous Arctic Ocean. *Nature*, **380** 330–333.

LLOYD, J. & FARQUHAR, G. D. 1994. ^{13}C discrimination during CO_2 assimilation by the terrestrial biosphere. *Oecologia*, **99**, 201–215.

——,—— 1996. The CO_2 dependence of photosynthesis, plant growth responses to elevated atmospheric CO_2 concentrations and their interactions with soil nutrient status. I. General principles and forest ecosystems. *Functional Ecology*, **10**, 4–32.

LONG, S. P. & DRAKE, B. G. 1991. Effect of long-term elevation of CO_2 concentration in the field on the quantum yields of photosynthesis in the C_3 sedge, *Scirpus olneyi*. *Plant Physiology*, **96**, 221–226.

LOTTES, A. L. & ZIEGLER, A. M. 1994. World peat occurrence and the seasonality of climate and vegetation. *Palaeogeography, Palaeoclimatology, Palaeoecology*, **106**, 23–37.

MCGUIRE, A. D., MELLILO, J. M., JOYCE, L. A., KICKLIGHTER, D. W., GRACE, A. L., MOORE, B. & VOROSMARTY, C.J. 1992. Interactions between carbon and nitrogen dynamics in estimating net primary productivity for potential vegetation in North America. *Global Biogeochemical Cycles*, **6**, 101–124.

OTTO-BLIESNER, B. L. & UPCHURCH, G. R. 1997. Vegetation-induced warming of high-latitude regions during the Late Cretaceous period. *Nature*, **385**, 804–807.

PARRISH, J. T., ZIEGLER, A. M. & SCOTESE, C. R. 1982. Rainfall patterns and the distribution of coals and evaporites in the Mesozoic and Cenozoic. *Palaeogeography, Palaeoclimatology, Palaeoecology*, **40**, 67–101.

PARTON, W. J., SCURLOCK, J. M. O., OJIMA, D. S. *et al.* 1993. Observations and modeling of biomass and soil organic carbon matter dynamics for the grassland biome worldwide. *Global Biogeochemical Cycles*, **7**, 785–809.

POLLARD, D. & THOMPSON, S. L. 1995. Use of a land surface transfer scheme (LSX) in a global climate model: response to doubling stomatal resistance. *Palaeogeography, Palaeoclimatology, Palaeoecology (Global and Planetary Change Section)*, **10**, 129–161.

PRICE, G. D., SELLWOOD, B. W. & VALDES, P. J. 1995. Sedimentological evaluation of general circulation model simulations for 'greenhouse' Earth: Cretaceous and Jurassic case studies. *Sedimentary Geology*, **100**, 159–180.

——, VALDES, P. J. & SELLWOOD, B. W. 1997a. Quantitative palaeoclimate GCM validation: Late Jurassic and mid-Cretaceous case studies. *Journal of the Geological Society, London*, **154**, 769–772.

——, —— & —— 1997b. Predictions of modern bauxite occurrence: implications for climate reconstruction. *Palaeogeography, Palaeoclimatology, Palaeoecology*, **131**, 1–13.

SELLERS, P. J., BOUNOUA, L., COLLATZ, G. J. *et al.* 1996. Comparison of radiative and physiological effects of doubled atmospheric CO_2 on climate. *Science*, **271**, 1402–1406.

SELLWOOD, B. W. & VALDES, P. J. 1997. Geological evaluation of climate General Circulation Models and model implications for cloud cover. *Terra Nova*, **9**, 75–78.

——, PRICE, G. D. & VALDES, P. J. 1994. Cooler estimates of Cretaceous temperatures. *Nature*, **370**, 453–455.

SPICER, R. A. & PARRISH, J. T. 1990. Late Cretaceous–early Tertiary palaeoclimates of northern high latitudes: a quantitative view. *Journal of the Geological Society, London*, **147**, 329–341.

——, REES, P. A. & CHAPMAN, J. L. 1993. Cretaceous phytogeography and climate signals. *Philosophical Transactions of the Royal Society of London, Series B*, **325**, 291–305.

VAKHRAMEEV, V. A. 1991. *Jurassic and Cretaceous Floras and Climates of the Earth*. Cambridge University Press, Cambridge.

VALDES, P. J., SELLWOOD, B. W., PRICE, G. D. 1995. Modelling Jurassic Milankovitch climate variations. *In*: HOUSE, M. R. & GALE, A. S. (eds) *Orbital Forcing Timescales and Cyclostratigraphy*. Geological Society, London, Special Publications, **85**, 115–132.

——, —— & —— 1996. Evaluating concepts of Cretaceous equability. *Palaeoclimates, Data and*

Modelling, 1, 139–158.

WOODWARD, F. I., SMITH, T. M. & EMANUEL, W. R. 1995. A global land primary productivity and phytogeography model. Global Biogeochemical Cycles, 9, 471–490.

———, RAYMOND, A. L., GIERLOWSKI, T. C., HORREL, M. A., ROWLEY, D. B. & LOTTES, A. L. 1987. Coal, climate and terrestrial productivity: the present and early Cretaceous compared. In: SCOTT, A. C. (ed.) Coal and Coal-bearing Strata: Recent Advances. Geological Society, London, Special Publications, 32, 25–49.

ZIEGLER, A. M., PARRISH, J. M., GYLLENHAAL, E. D. et al. 1993. Early Mesozoic phytogeography and climate. Philosophical Transactions of the Royal Society of London Series B, 341, 297–305.

Climatic modelling in the Cretaceous using the distribution of planktonic Foraminiferida

MALCOLM B. HART

Plymouth Environmental Research Centre and Department of Geological Sciences, University of Plymouth, Drake Circus, Plymouth PL4 8AA, UK

Abstract: During the Cretaceous, planktonic Foraminiferida underwent a major diversification, colonizing a wider range of water depths and geographical regions. Orville Bandy, over 30 years ago, was the first to recognize the potential of using this regional distribution to reconstruct different palaeoceanographic regimes. Using new data from the South Atlantic Ocean, Antarctic Ocean and the Indian Ocean it is possible to show the poleward migration of warm-water taxa during the Cretaceous and compare it with data already available from the Northern Hemisphere. Instead of the present-day nine latitudinal zones based on planktonic taxa it is only possible to identify five with any degree of reliability. These are from north to south, the Boreal, Transition, Tropical (= Tethyan), Transition and Austral. In some of the developing oceans during the Cretaceous (e.g. the Eastern Indian Ocean, South Atlantic Ocean) there are local biogeoprovinces with quite distinctive local assemblages and morphotypes. While these do not reflect on the climatic zonation they provide an interesting insight into the development of the plankton. The foraminiferal distributions are compared with climatic maps produced by the assessment of other data sources.

During the Cretaceous it is generally accepted (Merritts *et al.* 1998, fig. 13-2) that the global climate was very different to that of the present day. Some models (e.g. Larson *et al.* 1993) indicate that the mid-Cretaceous (~Turonian) recorded the warmest temperatures and the highest sea-levels of the Phanerozoic. In recent years Spicer *et al.* (1994), Sellwood *et al.* (1994), Price *et al.* (1995, 1997) and Valdes *et al.* (1996), using Cretaceous global circulation models (GCMs), have attempted to predict climatic patterns using basic physical data and the location of shorelines, oceans, mountain relief, etc., rather than simplistic palaeogeographical reconstructions. These GCMs have been validated, or at least tested, using vegetation patterns and the presence/absence of evaporites, bauxites and other geological features.

Many of these models, although they predict (see Price *et al.* 1995, fig. 3b) possible oceanic circulation patterns (related to the winds in the models), use very little palaeobiological data from the marine realm. Marine microfossils do, however, provide a significant amount of data in the way of isotopic sea surface temperatures (ssts) and, in some cases, information on sea-floor temperatures (e.g. Corfield *et al.* 1990; Huber *et al.* 1995; Norris & Wilson 1998) by use of both planktonic and benthonic Foraminiferida and the enclosing nanno-chalks. The application of foraminiferal faunal provinces to the interpretation of Cretaceous palaeoclimates has only been undertaken to a very limited extent (Sliter 1977; Bailey & Hart 1979) and it is this work that is being extended in this paper.

Globotruncana/Rotalipora lines

One of the first attempts at the reconstruction of Cretaceous planktonic foraminiferal provinces and their application to palaeoceanography was that of Bandy (1967). In a wide-ranging review of both the Cretaceous fauna and the environmental controls on its distribution (latitude, bathymetry, salinity, water masses, life cycle, etc.), Bandy effectively summarized the information available at that time (mainly taxonomic and biostratigraphical) on the distribution of both modern and Cretaceous taxa. There were two fundamental problems with Bandy's (1967) analysis, although they must not be taken as criticisms.

1. There was a significant lack of palaeotemperature data in his analysis. Indeed, his (Bandy 1967, text-fig. 2) temperature curve

From: HART, M. B. (ed.) *Climates: Past and Present.* Geological Society, London, Special Publications, **181**, 33–41, 1-86239-075-4/$15.00

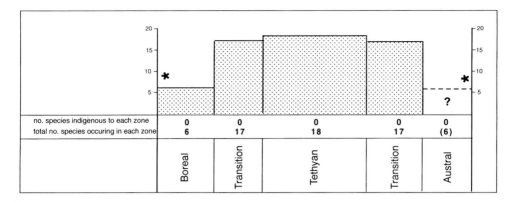

(a)

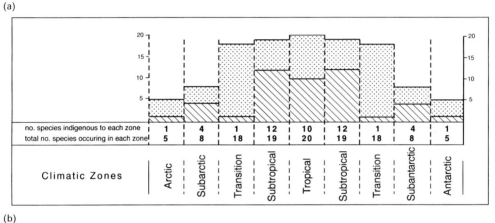

(b)

Fig. 1. (**a**) Distribution of latest Albian planktonic Foraminiferida. The * marks the occurrence of high latitude assemblages that are dominated by agglutinated Foraminiferida (Chamney 1976; Hart unpublished data). (**b**) Distribution of Recent planktonic Foraminiferida based on data in Be (1977).

for the whole of the Cretaceous was based on only four data points.

2. His estimations of the palaeo-latitudinal distribution of faunas was (obviously for 1967) based on a pre-plate tectonic world using present-day maps for the assessment of latitude. For any taxon this inevitably amalgamated locality information and, in some cases, distorted latitudinal ranges.

Despite these limitations, Bandy (1967, text-fig.1) showed that in the late Cretaceous (Turonian–Campanian) keeled taxa (*Rotalipora* spp., *Praeglobotruncana* spp. and *Globotruncana* spp.) appeared to migrate further north than in the Maastrichtian and pre-Turonian. In NW Europe the database being used was largely that of Williams–Mitchell (1948) and Barr (1962), with no information for the North Sea Basin and very little from Scandinavia. Bandy (1967, text-fig.13) noted that, at the species level, certain taxa (e.g. '*Globotruncana concavata*' and

'*Globotruncana calcarata*') were more restricted by latitude than some other species. The development of these 'lines' depicting the northern limits of keeled taxa were a natural development of Bandy's (1960) earlier paper on palaeoclimate zonation and Bandy & Arnal's (1960) work on foraminiferal ecology.

The work of Bandy was used, and extended, by Frerichs (1970, 1971) who was the first micropalaeontologist to introduce the concept of morphotypes into the analysis of planktonic Foraminiferida. Using descriptors (e.g. keels, apertures, clavate chambers, spines, etc.) Frerichs (1971, text-fig.3) considered the repeated evolution of characters over the Jurassic to Recent interval, an exercise extended in greater detail by Hart & Bailey (1979), Hart (1980), Caron & Homewood (1983), Caron (1983) and Leckie (1989). The concept of a water depth stratified Cretaceous fauna has been (rightly) challenged by the palaeotemperature data (e.g. Norris & Wilson 1998) although the data are, as

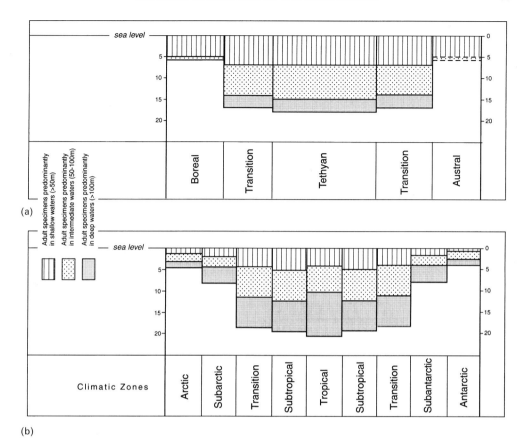

Fig. 2. Distribution of (**a**) planktonic morphotypes in the latest Albian. (**b**) Recent planktonic Foraminiferida based on data in Be (1977).

yet, equivocal. In Cretaceous oceans the temperature gradient in the water column appears to have been much less than the present day and this causes the palaeotemperature data to cluster, rather than providing what would appear to be a full range of temperatures from the surface to deeper water. Critics of the depth zonation model cannot, however, question the areal distribution evidence (see Hart & Ball 1986) which shows keeled assemblages present in faunas from known deeper water environments (oceanic basins and deeper parts of the shelves as against more near-shore environments). This debate will continue.

Distribution of the modern fauna

At the time of Bandy's (1967) review there was relatively little information available on the distribution of the modern fauna (Be 1960, 1965; Be & Ericson 1963). Some of the work on the ecology of the modern fauna had been concentrated on the variations in coiling direction (Bolli 1950, 1951) with relatively little experimental work with cultures. During the 1960s and 1970s, however, there was considerable interest in the biology and areal distribution of the fauna and this was summarized in Be (1977, and the extensive reference list therein). This major review provided data on: (i) the vertical distribution of the modern taxa in the water column; and (ii) the geographical distribution of the modern species.

Be (1977, fig.8) summarized the latitudinal ranges of the modern taxa in a series of nine latitudinal zones (Be & Toderlund 1971; Be 1977, fig. 7) and these are presented in a slightly different way in Fig. 1(b). If one amalgamates the data from Be (1977, fig. 8 and table 4) one can assess the latitudinal variation coupled with the preferred position in the water column (see Fig. 2(b)). This distribution shows a marked diversity drop at the northern and southern edges of the 'Transition' zones between 40° and

50° N/S (except in the North Atlantic Ocean where the distribution is distorted by the Gulf Stream).

Figure 1(b) shows quite clearly how very few species are indigenous to the north/south 'Transition' zones, while Fig. 2(b) indicates that the ecology of the taxa follows the general pattern.

The areal distribution of the modern taxa (Be 1977, fig. 7) shows that the great majority of the faunal province boundaries are parallel or near-parallel to latitudes except in the areas where major ocean current systems create a distorting effect. This is particularly marked in the North Atlantic Ocean where 'Transition' and 'Sub-Arctic' zone faunas are displaced northwards by the Gulf Stream. In the Southern Hemisphere the faunal boundaries are almost sub-parallel across the Pacific, Indian and, to a lesser extent, South Atlantic oceans.

In the Cretaceous succession of the UK, Hart & Carter (1975) drew attention to the Cenomanian/Turonian planktonic faunas, indicating thet they contained many keeled taxa that are normally associated with Tethyan assemblages. The Turonian fauna of *Helvetoglobotruncana helvetica* (Bolli, 1945), *Marginotruncana sigali* (Reichel, 1950), *M. pseudolinneiana* (Reuss, 1945) and *M. coronata* (Bolli, 1945) clearly represents a warm-water assemblage although, at the time of writing, the authors were unaware that the Early to Middle Turonian was probably the warmest part of the Phanerozoic succession. In the higher parts of the Cretaceous succession, however, keeled planktonic species – though present – are quite rare and include only more cosmopolitan species. Following DSDP Leg 80 (Goban Spur, NE Atlantic Ocean) it was discovered that warm-water faunas of Campanian to Maastrichtian age were present at the palaeolatitude of the UK, although many of the species were not recorded from the onshore successions (Hart & Ball 1986). This, effectively, pushed the palaeozoogeographic boundaries much further north than even Bailey & Hart (1979) had indicated. These faunas did not extend much further north and the principal diversity change appeared to be around 45°N palaeolatitude. The results were the same as those reported from the South Atlantic Ocean (Falkland Plateau) by Sliter (1977). Sliter (1977, fig. 12) had indicated that, following a comprehensive review of Southern Ocean faunas in the Cretaceous, he could only identify a 'Tethyan' zone, a 'Transition' zone and an 'Austral' zone in the assemblages. This was mirrored in the Northern Hemisphere (Bailey & Hart 1979, fig. 4) with an almost identical distribution. Using data from the latest Albian (*P. buxtorfi* zone or

equivalent) it is possible to generate profiles comparable to those of Be (1977). Figures 1(a) and 2(a) show the resulting five-fold subdivision of the fauna and this can be compared to the nine-fold profiles for the modern fauna (Figs 1(b) and 2(b)). The sudden drop in diversity, both north and south, appears to be slightly pole-wards of the present boundaries, perhaps reflecting the flatter temperature profile (Sellwood *et al.* 1994, fig.2) of the Cretaceous. Two time slices of the Cretaceous are now presented to highlight some of the issues/problems.

The latest Albian

In the latest Albian the planktonic foraminiferal zonation of Robaszynski & Caron (1995) uses the *Rotalipora appenninica* Interval Zone or, where appropriate, the *Planomalina buxtorfi* Taxon Range Zone. The assemblage at this level, immediately below the first appearance of *Rotalipora globotruncanoides*, is quite distinctive and includes *Biticinella breggiensis* (Gandolfi, 1942), *Favusella washitensis* (Carsey, 1926), *Globigerinelloides bentonensis* (Morrow, 1934), *G. ultramicra* (Subbotina, 1949), *Guemelitria cenomana* (Keller, 1935), *Hedbergella delrioensis* (Carsey, 1934), *H. planispira* (Tappan, 1940), *H. simplex* (Morrow, 1934), *Heterohelix moremani* (Cushman, 1938), *Planomalina buxtorfi* (Gandolfi, 1942), *P. praebuxtorfi* (Wonders, 1980), *Praeglobotruncana delrioensis* (Plummer, 1931), *P. stephani* (Gandolfi, 1942), *Rotalipora appenninica* (Renz, 1936), *R. ticinensis* (Gandolfi, 1942), *Ticinella roberti* (Gandolfi, 1942) and *T. madecassina* Sigal, 1966.

In southern England there is a flood of quite large *G. bentonensis* associated with (Magniez-Jannin 1981) the northernmost occurrence of *P. buxtorfi* and this flood can be traced northwards into Cambridgeshire and the North Sea Basin. This abundance of large *G. bentonensis* (associated with rare *P. buxtorfi* and primitive *R. appenninica*) appears to mark the Maximum Flooding Surface (MFS) of the 3rd Order Sequence Al 10 (Hart 2000). Using data from Europe, North Africa, West Africa, the South Atlantic Ocean, the Brazilian Margin, southern USA, the Western Interior Seaway (USA), Alaska, the Pacific Ocean, Australasia, India, East Africa and the peri-Antarctic region, the distribution of the latest Albian fauna has been assessed. In the latest Albian it is reasonably clear that diverse planktonic assemblages, dominated by ticinellids and early rotaliporids, are 'Tethyan' and that to the north (in particular) there is a 'Transition' zone with only the more cosmopolitan ticinellids and very rare rotalipor-

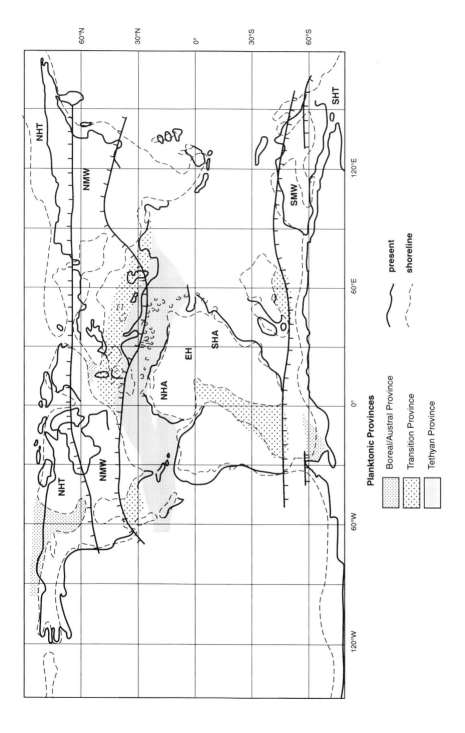

Fig. 3. Latest Albian palaeogeography based on a map provided by Prof. R. A. Spicer (Open University). The Tethyan, Transitional and Boreal Provinces of planktonic Foraminiferida are indicated. The climatic belts of Spicer are identified by the following codes: NHT, Northern High Latitude Temperate; NMW, Northern mid-Latitude Wet; NHA, Northern Hot Arid; EH, Equatorial Humid; SHA, Southern Hot Arid; SMW, Southern mid-Latitude Wet; SHT, Southern High Latitude Temperate.

ids. This passes northwards into an assemblage dominated by only hedbergellids. The 'Transition' zone fauna just reaches southern England where rare *P. buxtorfi* can be found with very rare *R. appenninica* in the uppermost Gault Clay Formation of Copt Point, Folkestone (Magniez-Jannin 1981). This assemblage is not recorded anywhere to the north of this palaeolatitude, although a similar assemblage is recorded from the Goban Spur west of the UK (Carroll, pers. comm.). In the rest of the UK and in the North Sea Basin, North Germany and Poland the latest Albian fauna is dominated by *Hedbergella s.l.* and rare *Heterohelix moremani, Guembelitria cenomana* and *Globigerinelloides bentonensis*. Further to the north, in the northern North Sea Basin and Alaska (Tappan 1962; Chamney 1976) faunas are dominated by agglutinated taxa. A similar association of agglutinated foraminifera is also present in the Antarctic Peninsula (Hart unpublished data). These agglutinated assemblages are indicated in Fig. 1(a) by the Austral and Boreal asterisks.

In the region of the developing Indian Ocean, *P. buxtorfi* Zone sediments are present on the Exmouth Plateau (Wonders 1992) and in the Cauvery Basin (Tewari 1996). In the South Atlantic Ocean the Upper Albian succession has been investigated by Koutsoukos (1989) at various locations on the Brazilian Continental Margin. In these sediments, the fauna is missing some of the important 'Tethyan' markers, and this probably indicates a poor circulation system at that time. As indicated in Fig. 3, the South Atlantic Ocean was not fully connected through to the North Atlantic Ocean and a high latitude circulation was probably blocked by both India and the developing Kerguelan Plateau. Only in low latitudes can a distribution sub-parallel to lines of palaeolatitude be seen. The climatic boundaries based on the vegetation maps of Spicer are shown for comparison. Within the 'Tethyan' zone shallow-water areas (e.g. Oman; see Simmons & Hart 1987) are characterized by the presence of rudist assemblages and large Foraminifera (e.g. *Orbitolina* spp.). In such successions the planktonic faunas that are recorded are generally restricted to the shallow-water hedbergellids. This distribution was also recorded by Koutsoukos & Hart (1990) on the Brazilian Continental Margin where the hedbergellids and favusellids are recorded in the shallow-water, near-shore environments. Offshore, in deeper water, a more diverse planktonic fauna is recorded, although it is not fully characteristic of the warmer waters of the Tethyan belt. In their work on Cretaceous climates, Price *et al.* (1995) modelled the

Cenomanian interval (Price *et al.* 1995, figs 1(b), 2(a),(b)) and while this time slice is some 2 million years after the latest Albian interval that is being considered here, their maps can be used to consider the distribution of the planktonic Foraminifera. The 'Tethyan' faunal assemblage mirrors the general east–west wind pattern through the palaeo-tropics. The prevailing wind direction in the Southern Ocean is west–east, and this would ensure that migration from the Indian Ocean towards the Southern Atlantic Ocean would not take place. This might explain the detailed differences of these Southern Hemisphere faunas, and further work on their discrimination is in progress. The high latitude agglutinated faunas of the Antarctic Peninsula and Alaska are also isolated from open ocean circulation, especially as the South America/Antarctic isthmus was almost certainly closed at this time.

The Late Santonian

By the Late Santonian (*Dicarinella asymetrica* Zone) the Cretaceous planktonic fauna was more diverse (see Hart 1999), sea-levels were probably much higher and the continents (Fig. 4) were more dispersed. The *Dicarinella asymetrica* Zone fauna includes *Archaeoglobigerina blowi* Pessagno, 1967, *A. bosquensis* Pessagno, 1967, *A. cretacea* (d'Orbigny, 1840), *Contusotruncana fornicata* (Plummer, 1931), *Dicarinella asymetrica* (Sigal, 1952), *D. concavata* (Brotzen, 1934), *Globigerinelloides ultramicra* (Subbotina, 1949), *Globotruncana arca* (Cushman, 1926), *Hastigerinoides alexanderi* (Cushman, 1931), *H. subdigitata* (Carman, 1929), *H. watersi* (Cushman, 1931), *Hedbergella flandrini* Porthault, 1970, *H. holmdelensis* Olsson, 1964, *Heterohelix globulosa* (Ehrenberg, 1840), *H. reussi* Cushman, 1938), *Marginotruncana canaliculata* (Reuss, 1854), *M. coronata* (Bolli, 1945), *M. marginata* (Reuss, 1845), *M. pseudolinneiana* Pessagno, 1967, *M. schneegansi* (Sigal, 1952), *Pseudoguembelina costulata* (Cushman, 1938), *Shackoina multispinata* (Cushman & Wickenden, 1930) and *Whiteinella baltica* Douglas & Rankin, 1969.

In this reconstruction (Fig. 4) the 'Tethyan' zone is much larger and includes both the bulk of the Indian Ocean and the South Atlantic Ocean. The latter is now fully open between West Africa and the NE corner of South America. In the Northern Hemisphere the UK and NW Europe are a part of the 'Transition' zone as indicated by Bailey & Hart (1979). The only area with a slightly different assemblage is that of Western Australia (Wright & Apthorpe

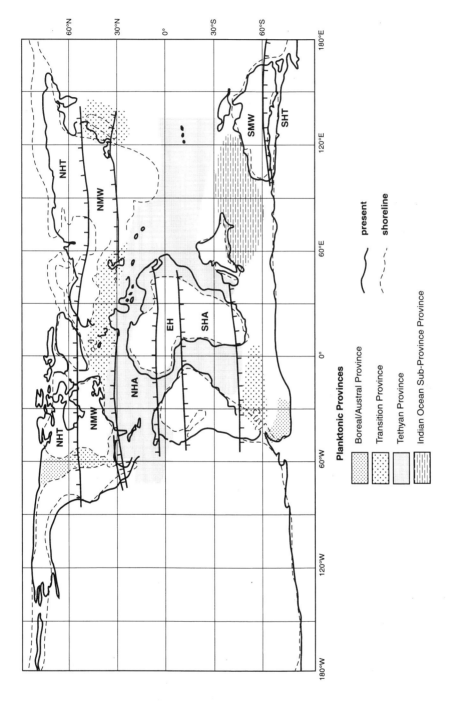

Fig. 4. Late Santonian palaeogeography based on a map provided by Prof. R. A. Spicer (Open University). The Tethyan, Transitional and Boreal Provinces of planktonic Foraminiferida are indicated, together with the Indian Ocean Sub-Province. The climatic belts of Spicer are identified by the same codes as Fig. 3. The Tethyan Province is almost exactly the same area as that covered by the 'framework fossil assemblages' of Sohl (1987). These assemblages are algal/coral or coral/rudist or rudist-dominated, and were reproduced in the palaeogeographical maps of Hallam (1994). In these shallow water carbonate-dominated environments there are assemblages of larger Foraminiferida.

1976, 1995; Apthorpe 1979), the Exmouth Plateau (Wonders 1992) and SE India (Tewari 1996). In this area the planktonic fauna is unusual with slightly atypical ranges and with many of the well-known species having unusual ornamentation. The reasons for this 'isolation' are unknown because, by this time, the Eastern Indian Ocean was quite large and open to both the north and the south. Circulation to the south was probably still restricted by the mass of the Kerguelan Plateau and might not be helped by the west–east circulation described above.

Summary

The palaeolatitudinal distribution of the planktonic Foraminifera in the Cretaceous has been described in terms of simple diversity and compared to the modern distribution. A five-fold subdivision can be recognized, as compared to the nine-fold subdivision of the modern oceans. The diversity pattern supports the concept of a flatter pole–equator temperature gradient for the Cretaceous. In two reconstructions for the latest Albian and Late Santonian the distribution of the fauna is compared to the circulation models of Price et al. (1995). This work is now being extended in much greater detail using Multi-Dimensional Scaling (MDS) and other statistical techniques that are being used to compare the faunas from as wide a range of localities as possible. Although the number of data points for the Cretaceous are still limited in some parts of the world it is suggested that this methodology will provide useful palaeoclimatic information that can be built into current models for this interesting period in the development of the modern oceanic system.

The author wishes to thank all those who have provided reprints of their work and entered into lengthy discussions of the subject. Particular thanks go to Greg Price (University of Plymouth) and Bob Spicer (Open University). John Abraham is thanked for the production of the final versions of the diagrams used in this paper.

References

APTHORPE, M. C. 1979. Depositional history of the Upper Cretaceous of the Northwest Shelf, based upon Foraminifera. *Australian Petroleum Exploration Association Journal*, **19**, 74–89.

BAILEY, H. W. & HART, M. B. 1979. The correlation of the Early Senonian in Western Europe using Foraminiferida. *Aspekte der Kreide Europas, IUGS, Series A*, **6**, 159–169.

BANDY, O. L. 1960. Planktonic foraminiferal criteria for paleoclimatic zonation. *Tohoku University, Scientific Reports, Series 2 (Geology), Special Volume No. 4*, 1–8.

—— 1967. Cretaceous planktonic foraminiferal zonation. *Micropaleontology*, **13**, 1–31.

—— & ARNAL, R. E. 1960. Concepts of foraminiferal paleoecology. *Bulletin of the American Association of Petroleum Geologists*, **44**, 1921–1932.

BARR, F. T. 1962. Upper Cretaceous planktonic foraminifera from the Isle of Wight, England. *Palaeontology*, **4**, 552–580.

BE, A. W. H. 1960. Some observations on Arctic planktonic Foraminifera. *Contributions to the Cushman Foundation for Foraminiferal Research*, **11**, 64–68.

—— 1965. The influence of depth on shell growth in *Globigerinoides sacculifer* (Brady). *Micropaleontology*, **11**, 81–97.

—— 1977. An ecological, zoogeographic and taxonomic review of Recent planktonic Foraminifera. *In*: RAMSAY, A. T. S. (ed), *Oceanic Micropalaeontology*, Acdemic Press, London, Volume 1, 1–100.

—— & ERICSON, D. B. 1963. Aspects of calcification in planktonic Foraminifera. *In*: Comparative Biology of Calcified Tissues. *Annals of the New York Academy of Sciences*, **109**, 65–81.

—— & TODERLUND, D. S. 1971. Distribution and ecology of living planktonic Foraminifera in surface waters of the Atlantic and Indian Oceans. *In*: FUNNELL, B. M. & RIEDEL, W. R. (eds), *The Micropalaeontology of Oceans*. Cambridge University Press, Cambridge, 105–149.

BOLLI, H. M. 1950. The direction of coiling in the evolution of some Globorotaliidae. *Cushman Foundation for Foraminiferal Research, Contributions*, **1**, 82–89.

—— 1951. Notes on the direction of coiling of rotalid foraminifera. *Cushman Foundation for Foraminiferal Research, Contributions*, **2**, 139–143.

CARON, M. 1983. La spéciation chez les foraminifères planctoniques: une réponse adaptée aux contraintes de l'environnement. *Zitteliana*, **10**, 671–676.

—— & HOMEWOOD, P. 1983. Evolution of early planktic foraminifers. *Marine Micropaleontology*, **7**, 453–462.

CHAMNEY, T. P. 1976. Foraminiferal morphogroup symbol for paleoenvironmental interpretation of drill cutting samples: Arctic America, Albian Continental Margin. *Maritime Sediments, Special Publication*, **1B**, 585–624.

CORFIELD, R. M., HALL, M. A. & BRASIER, M. D. 1990. Stable isotope development for foraminiferal habitats during the development of the Cenomanian/Turonian oceanic anoxic event. *Geology*, **18**, 175–178.

FRERICHS, W. E. 1970. Paleobathymetry, paleotemperature and tectonism. *Geological Society of America, Bulletin*, **81**, 3445–3452.

—— 1971. Evolution of planktonic Foraminifera and paleotemperatures. *Journal of Paleontology*, **45**, 963–968.

HALLAM, A. 1994. *An Outline of Phanerozoic Biogeography*. Oxford Biogeography Series, No.10, Oxford University Press, Oxford.

HART, M. B. 1980. A water depth model for the evolution of the planktonic Foraminiferida. *Nature*, **286**, 252–254.

—— 1999. The evolution and biodiversity of Cretaceous planktonic Foraminiferida. *Géobios*, **32**, 247–255.

—— 2000. Foraminifera, sequence stratigraphy and regional correlation; an example from the uppermost Albian of Southern England. *Revue de Micropaléontologie*, **43**, 27–45.

—— & BAILEY, H. W. 1979. The distribution of planktonic Foraminiferida in the mid-Cretaceous of NW Europe. *Aspekte der Kreide Europas, IUGS, Series A*, **6**, 527–542.

—— & BALL, K. C. 1986. Late Cretaceous anoxic events, sea-level changes and the evolution of the planktonic Foraminifera. *In*: SUMMERHAYES, C. P. & SHACKLETON, N. J. (eds), *North Atlantic Palaeoceanography*. Geological Society, London, Special Publications, **21**, 67–78.

—— & CARTER, D. J. 1975. Some observations on the Cretaceous Foraminiferida of SE England. *Journal of Foraminiferal Research*, **5**, 114–126.

HUBER, B. T., HODELL, D. A. & HAMILTON, C. P. 1995. Middle–Late Cretaceous climate of the southern high latitudes: Stable isotopic evidence for minimal equator-to-pole thermal gradients. *Geological Society of America, Bulletin*, **107**, 1164–1191.

KOUTSOUKOS, E. A. M. 1989. *Mid- to Late Cretaceous microbiostratigraphy, palaeo-ecology and palaeogeography of the Sergipe Basin, Northeastern Brazil*. PhD thesis, Polytechnic Southwest (now University of Plymouth).

—— & HART, M. B. 1990. Cretaceous foraminiferal morphogroup distribution patterns, palaeocommunities and trophic structures: a case study from the Sergipe Basin, Brazil. *Transactions of the Royal Society of Edinburgh, Earth Sciences*, **81**, 221–246.

LARSON, R. L., FISCHER, A. G., ERBA, E. & PREMOLI-SILVA, I. 1993. Summary of workshop results. *In*: LARSON, R. L. *et al. Apticore – Albicore: A workshop on global events and rhythms of the mid-Cretaceous*, Perugia, October 1992.

LECKIE, R. M. 1989. A paleoceanographic model for the early evolutionary history of planktonic foraminifera. *Palaeogeography, Palaeoclimatology, Palaeoecology*, **73**, 107–138.

MAGNIEZ-JANNIN, F. 1981. Découverte de *Planomalina buxtorfi* (Gandolfi) et d'autres Foraminifères planctoniques inattendus dans l'Albien Supérieur d'Abbots Cliff (Kent, Angleterre): consequences paléogeographiques et biostratigraphiques. *Géobios*, **14**, 91–97.

MERRITTS, D., DE WET, A. & MENKING, K. 1998. *Environmental Geology*. W.H. Freeman and Company, New York.

NORRIS, R. D. & WILSON, P. A. 1998. Low-latitude sea-surface temperatures for the mid-Cretaceous and the evolution of planktic foraminifera. *Geology*, **26**, 823–826.

PRICE, G. D., SELLWOOD, B. W. & VALDES, P. J. 1995. Sedimentological evaluation of general circulation model simulations for the 'greenhouse' Earth: Cretaceous and Jurassic case studies. *Sedimentary Geology*, **100**, 159–180.

——, VALDES, P. J. & SELLWOOD, B. W. 1997. Quantitative palaeoclimate GCM validation: Late Jurassic and mid-Cretaceous case studies. *Journal of the Geological Society, London*, **154**, 769–772.

ROBASZYNSKI, F. & CARON, M. 1995. Foraminifères planctoniques du Crétacé: commentaire de la zonation Europe-Méditerranée. *Bulletin de le Société Géologique de France*, **6**, 681–692.

SELLWOOD, B. W., PRICE, D. G. & VALDES, P. J. 1994. Cooler estimates of Cretaceous temperatures. *Nature*, **370**, 453–455.

SIMMONS, M. D. & HART, M. B. 1987. The biostratigraphy and microfacies of the early to mid-Cretaceous carbonates of Wadi Mi'aidin, Central Oman Mountains. *In*: HART, M. B. (Ed.), *The Micropalaeontology of Carbonate Environments*, Ellis Horwood Ltd., Chichester, 176–207.

SLITER, W. V. 1977. Cretaceous foraminifera from the southwestern Atlantic Ocean, Leg 36, Deep Sea Drilling Project. *Initial Reports of the Deep Sea Drilling Project*, **36**, 519–545.

SOHL, N. F. 1987. Cretaceous gastropods: contrasts between Tethys and the temperate Provinces. *Journal of Paleontology*, **61**, 1085–1111.

SPICER, R. A., REES, P.McA. & CHAPMAN, J. L. 1994. Cretaceous phytogeography and climate signals. *In*: ALLEN, J. R. L., HOSKINS, B. J., SELLWOOD, B. W., SPICER, R. A. & VALDES, P. J. (Eds), *Palaeoclimates and Their Modelling: With Special Reference to the Mesozoic Era*, Chapman & Hall, London, 69–78.

TAPPAN, H. 1962. Foraminifera from the Arctic Slope of Alaska; pt. 3, Cretaceous Foraminifera. *US Geological Survey, Professional Paper*, **236-C**, 91–209.

TEWARI, A. 1996. *The Middle to Late Cretaceous microbiostratigraphy (foraminifera) and lithostratigraphy of the Cauvery Basin, Southeast India*. PhD thesis, University of Plymouth.

VALDES, P. J., SELLWOOD, B. W. & PRICE, G. D. 1996. Evaluating concepts of Cretaceous equability. *Palaeoclimates*, **1**, 139–158.

WILLIAMS-MITCHELL, E. 1948. The Zonal value of Foraminifera in the Chalk of England. *Proceedings of the Geologists' Association*, **59**, 91–112.

WONDERS, A. A. H. 1992. Cretaceous planktonic foraminiferal biostratigraphy, Leg 122, Exmouth Plateau, Australia. *Proceedings of the Ocean Drilling Program, Scientific Results*, **122**, 587–599.

WRIGHT, C. A. & APTHORPE, M. 1976. Planktonic foraminiferids from the Maastrichtian of the Northwest Shelf, Western Australia. *Journal of Foraminiferal Research*, **6**, 228–240.

—— & —— 1995. C. A. Wright's Cretaceous planktonic foraminiferal zonation for the Northwest Shelf, Australia [Abstract]. *Abstract Volume of the Second International Symposium on Cretaceous Stage Boundaries, Brussels, 8–16 September 1995*.

The Maastrichtian (Late Cretaceous) climate in the Northern Hemisphere

LENA B. GOLOVNEVA

Komarov Botanical Institute of the Russian Academy of Sciences, 2 Popov str., St Petersburg 197376, Russia (e-mail: lena@LG1575.spb.edu)

Abstract: This investigation of the Maastrichtian climate in the Northern Hemisphere is based on taxonomic and ecological studies of fossil floras, leaf physiognomy and the distribution of dinosaurian faunas. Fossil plant evidence indicates that the climate during Maastrichtian time was warm temperate at high and middle latitudes, and subtropical south of 40°N. Precipitation was relatively high (about 700–800 mm) and evenly distributed over the year. The annual range of temperatures was similar to that of modern maritime climates, but the latitudinal gradient was lower than at present. At high latitudes cold-month mean temperatures were about 3–4°C and probably never dropped below 0°C for extended periods. It seems that these comparatively mild winter temperatures in polar regions were a result of the heating of these areas by warm oceanic upwelling.

The Maastrichtian stage is the youngest stage of the Cretaceous Period, immediately preceding the Cretaceous–Tertiary boundary. At that time there was a great turnover in the Earth's biota and the development of terrestrial ecosystems was significantly affected by changes in climate. Thus, the reconstruction of the Maastrichtian palaeoclimate is of particular importance for the evaluation of climatic change near this boundary. To determine the climate, the taxonomic and ecological characters of fossil land-plant assemblages and dinosaur faunas were analysed.

The conclusions presented in this paper are based largely on the study of 12 fossil floral assemblages from northeastern Asia and the western part of North America from 35°N to 70°N (Fig. 1). These floral assemblages were chosen because they are accurately dated and well described. In Europe there are no rich Maastrichtian macrofloral assemblages that can be used for investigation of palaeoclimates. The floras from northeast Russia and Kazakhstan were studied on the basis of collections in the Komarov Botanical Institute, and other floras were analysed using published data. As a rule, literature-based climatic inferences do not significantly differ from evaluations based on original collections.

The northeastern Asia Maastrichtian floras are known from the Koryak Upland, the island of Sakhalin, Central Asia and Mongolia. In the Koryak Upland, rich floral assemblages come from deposits of the Rarytkin (Late Maastrich-

tian) and the Kakanaut (Middle Maastrichtian) Formations (Golovneva 1994, 1995). The Maastrichtian flora of the Avgustovka River from Sakhalin was studied by Krassilov (1979). This assemblage is from the upper part of the Krasnoyarka Formation and includes 16 species. In Kazakhstan, Maastrichtian floras have been found in the lower part of the Severozaysan Formation (Zaysan Depression), at the localities of Zhuvankara and Tajzhusgen (Shilin & Romanova 1978; Akhmetiev & Shevyreva 1989), and at Ulken-Kalkan in the Ili River Basin (Makulbekov 1974). The Ulken-Kalkan flora is rather poor and contains only nine species. It is considered to be of Maastrichtian age; this age is suggested by the presence of dinosaurian bones in the underlying conglomerates although the flora could equally be Early Danian in age. In Mongolia the Maastrichtian flora comes from deposits of the Nemegt Formation, which also contains a rich vertebrate fauna. The Nemegt flora includes about 10 species although it has been studied only in a preliminary way (Krassilov & Makulbekov 1995; Makulbekov 1995). Kogosukruk is the northernmost Maastrichtian locality (70°N) in Alaska (Spicer & Parrish 1987). It is dated to Late Campanian–Early Maastrichtian time by marine fossils and includes only five or six species. Numerous Early and Late Maastrichtian leaf assemblages are known in the Western Interior of North America from the deposits of the Edmonton, Brazeau, Hell Creek, Lance,

From: HART, M. B. (ed.) *Climates: Past and Present.* Geological Society, London, Special Publications, **181**, 43–54, 1-86239-075-4/$15.00

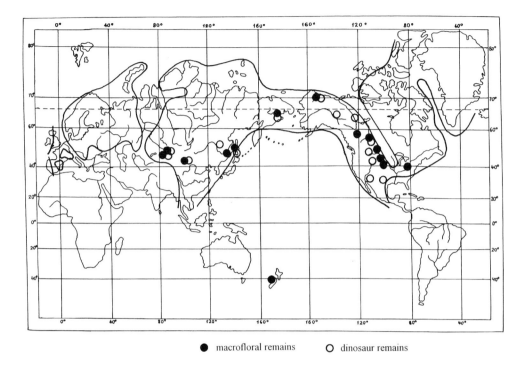

● macrofloral remains   ○ dinosaur remains

Fig. 1. Distribution of the Maastrichtian floral and dinosaur remains in the Northern Hemisphere (continuous line indicates the approximate geographical extent of the sea and land masses, after Gorodnitskiy *et al.* (1978).

Medicine Bow, Denver, Raton and Vermejo Formations (Berry 1935; Dorf 1942*a,b*; Bell 1949; Shoemaker 1966; Rouse, 1967; Johnson 1992). Many localities contain rich and well-described floral remains. The flora of the Ripley Formation (Early Maastrichtian) at 35°N is the southeastern-most occurrence considered (Berry 1925).

Several characters of the fossil floras were examined, including their general diversity, the relative abundance of different taxonomic and ecological groups, and the ratio between evergreen and deciduous gymnosperms (Table 1). Palaeoclimatic analysis of the structure of plant assemblages and putative tolerances of living relatives has produced only a generalized interpretation of the Maastrichtian climate. Morphological characters of angiosperm leaves correlate especially well with environmental factors (Richards 1952; Wolfe 1979). Physiognomic analysis of leaves of dicotyledonous woody plants can provide not only general inferences on climate but also rather precise quantitative estimations of climatic factors (Bailey & Sinnot 1915; Dilcher 1973; Wolfe 1979; Wolfe & Upchurch 1987). In this study I applied a recently developed physiognomic

analytical procedure called CLAMP (climate leaf analysis multivariate program; Wolfe 1993). This program uses 31 leaf morphological characters to obtain eight climate variables: mean annual temperature, cold-month mean temperature, warm-month mean temperature, mean annual precipitation, mean month growing season precipitation, mean growing season precipitation, precipitation during the three consecutive driest months and length of the growing season. The application of this method has been described in detail elsewhere (Wolfe 1993; Herman & Spicer 1996; Kovach & Spicer 1996). Only floras that included more than 15–20 species of woody dicotyledonous species were scored for CLAMP analysis.

For a more accurate evaluation of the Maastrichtian climate and verification of the CLAMP data, the distribution of the vertebrate fauna, primarily dinosaurs, was analysed and oxygen-isotope data were surveyed. These data were compiled from various publications (e.g. Teis & Nidin 1973; Savin 1977; Yasamanov 1978; Jerzykiewicz & Russell 1991; Weishampel 1991; Nessov 1995). By comparing evidence from independent sources, more precise inferences of the climate can be produced.

Table 1. *Taxonomic and ecological diversity of the Maastrichtian floral assemblages*

Fossil floras	Latitude (°N)	Number of species						Morphological characters of woody angiosperm leaves								Gymnosperm groups		Reference used for calculations
		General	Woody-angiosperms	Aquatic-angiosperms	Conifers	Cycadophytes	Ferns	Entire-margined leaves		Leaf size (Wolfe, 1993)						Evergreen	Deciduous	
								n	%	micro1	micro2	micro3	meso1	meso2	meso3			
Kogosukruk	70	10	1	1	1	–	2	–	–	–	100	–	–	–	–	–	1	Spicer & Parrish (1997)
Rarytkin	63	68	30	7	10	–	5	8	30	3	14	24	18	24	16	5	9	Golovneva (1994)
Kakanaut	63	43	23	6	7	2	1	3	15	9	34	37	9	8	3	4	6	Golovneva (1994)
Edmonton	53	35	17	8	5	1	2	5	32	–	24	47	22	5	2	4	3	Bell (1949)
Sakhalin	49	17	5	2	3	3	4	2	40	–	6	36	33	17	6	2	4	Krassilov (1979)
Zaisan	48	22	11	1	8	–	1	2	22	12	38	21	23	11	5	5	3	Shilin & Romanova (1978)
Hell Creek	46	33	30	–	2	–	–	11	31	1	11	28	31	26	3	–	3	Shoemaker (1996)
Ulken-Kalkan	44	9	5	1	1	–	–	–	–	–	–	50	50	–	–	–	1	Makulbekov (1974)
Nemegt	43	7	2	3	2	–	–	–	–	100	–	–	–	1	–	1	2	Makulbekov (1995)
Lance	43	50	29	7	2	–	4	15	53	3	44	36	21	–	–	2	1	Dorf (1942b)
Medicine Bow	40	39	31	2	2	–	3	20	66	–	19	61	13	3	4	2	–	Dorf (1942a)
Ripley	35	54	36	2	6	–	5	22	63	16	36	22	10	3	4	6	–	Berry (1925)

Maastrichtian floral assemblages

The Maastrichtian floras from various parts of
the Northern Hemisphere are characterized by
widely varying species composition, despite the
general similarity of the gymnosperms. The
taxonomic composition of the angiosperms
differs considerably on both generic and specific
levels. During Maastrichtian time, the floras of
each area seemed to have developed independently and belonged to different phytogeographical units (Golovneva 1994).

In all floras, except the Alaskan, angiosperm
taxa prevail over ferns, conifers and cycadophytes. The relative abundance of ferns is low in
comparison with other Mesozoic floras. In many
Cretaceous floras ferns amounted to 30% of the
total diversity and apparently formed non-forest
communities on swampy or poor, disturbed
soils. Increased abundance of ferns in floras is
usually correlated with epochs of climatic
warming (Samylina 1974). During Maastrichtian time the reduced number of fern species and
their lower relative abundance (Table 1) were the
consequence of the evolutionary replacement of
fern communities by angiosperms, rather than a
result of climatic cooling. However, the rare
occurrence of ferns along with a small number of
thin coal layers may indicate an insufficiently
moist climate.

The remains of cycadophytes and ginkgos in
the Maastrichtian deposits occur only at high
latitudes, where they are rather abundant in
some localities. These plants had already become
extinct by Late Cretaceous time in mid–low
latitudes. On the whole, extant cycadophytes are
thermophilic plants. Throughout Cretaceous
time they were typical elements of tropical and
subtropical floras. In the Cretaceous temperate
floras an increased ratio of cycadophytes to
other taxa has always been considered to
indicate climatic warming. Krassilov (1985)
proposed a cycadophyte index to evaluate
climatic changes during Early and mid-Cretaceous time. By Late Cretaceous time cycadophyte
diversity had gradually decreased at all latitudes
and, strange as it may seem, northern temperate
regions became one of their last refuges.
Vakhrameev (1991) ascribed this phenomenon
to a milder and more humid climate in these
regions. Thus, for Maastrichtian time the
latitudinal distribution of cycadophytes cannot
be the climatic indicator, but within northern
territories their relative abundance can be used
to evaluate climates of different Cretaceous ages
(Herman & Lebedev 1991; Golovneva 1994).

Conifers in the high-latitude Maastrichtian
floras of the Northern Hemisphere are usually
represented by both evergreen and deciduous
taxa. The number of conifer species generally
decreases from north to south. This is accompanied by an increased percentage of evergreen
conifers. In the northern regions deciduous
Taxodiaceae dominated the conifers. In northeastern Asia they prevailed in floras up to
approximately 45°N, including those at the
Zaysan and Ulken-Kalkan localities in eastern
Central Asia. Evergreen Araucariaceae appeared
only in southern Gobi (Nemegt Fm.). In North
America the distribution of deciduous Taxodiaceae extended as far south as Wyoming. In more
southern areas conifers were not abundant and
were represented mainly by evergreen Taxodiaceae, Cupressaceae and Araucariaceae. The
distribution of conifers in the Northern Hemisphere can, therefore, be interpreted as an
indicator of a rise in temperature and a decrease
in rainfall from north to south.

Most Maastrichtian angiosperms are represented by extinct taxa. This impedes use of
nearest living relatives for climatic determinations. Leaves of angiosperms in the Maastrichtian floras of the Northern Hemisphere are on
average of moderate size and gradually became
smaller southwards. This may be related to more
favourable moist conditions at high–mid-latitudes. At the taxonomic level this reduction is
reflected by the slow disappearance of plata-
noids and trochodendroids. The range of leaf
sizes and scarcity of drip-tips indicate relatively
moderate precipitation (Wolfe 1990). In northern regions leaves of angiosperms are characterized mainly by their dentate margins and thin
texture. Lobed and simple pinnately veined
leaves with entire margins and thick texture
become more abundant to the south. These data
indicate warmer and drier climates in southern
areas by comparison with more northern ones
(Wolfe 1979; Wolfe & Upchurch 1987). The
percentage of entire-margined species varies
from 15% at 63°N to 65% at 40°N. Accordingly, leaf margin analysis (Wolfe 1979) allows
us to conclude that the latitudinal range of mean
annual temperatures ranged from 5°C to 20°C.

Thus, the Maastrichtian climate can be
evaluated as generally warm temperate in northern regions and subtropical at mid–low latitudes, with moderate humidity decreasing
slightly southwards.

CLAMP palaeoclimatic inferences

The results of the CLAMP analysis are in
accordance with evidence derived from the
general analysis of fossil floral assemblages
(Table 2). CLAMP predictions suggest that the

Table 2. Climate data for Maastrichtian time obtained from CLAMP analyses

Fossil floras	Latitue (°N)	Mean annual temperature (°C)	Cold-month mean temperature (°C)	Warm month mean temperature (°C)	Mean annual precipation (mm)	Mean monthly growing season precipation (mm)	Mean growing season precipation (mm)	Three consecutive driest months precipation (mm)	Length of the growing season (months)
Rarytkin	63	11	4	19	1722	129	1410	305	6.9
Kakanaut	63	10	3	19	1414	98	948	181	6.3
Edmonton	53	12	5	19	1804	141	1586	335	7.1
Sakhalin	49	14	8	20	1892	136	1214	293	8
Zaisan	48	11	4	19	1515	106	1065	230	7
Hell Creek	46	12	6	19	1899	131	1094	302	7
Lance	43	14	8	21	1574	119	1216	242	8
Medicine Bow	40	17	13	23	1933	172	1946	348	9.7
Ripley	35	17	11	23	1498	115	1126	204	9.2
SD		1.8	3.3	3.1	430	23	280	70	1.1

climate in Maastrichtian time was warm temperate (*sensu* Wolfe 1979, mean annual temperature (MAT) about 10–14°C) at high and mid-latitudes and subtropical south of about 40°N (MAT about 15–17°C). At high and mid-latitudes cold-month mean temperatures were apparently about 3–8°C, and warm-month mean temperatures about 19–20°C. To the south of 40°N, winter temperatures increased to 10–13°C and summer temperatures to 20–23°C. The estimated annual range of temperatures is about 12–16°C and, therefore, similar to that of modern maritime climates. This is consistent with the proximity of many of the fossil localities to sea coasts during Maastrichtian time, although today they are situated in the continental interior (Fig. 1). Predicted precipitation is relatively high in all observed areas. It averaged about 1500–1700 mm and was fairly evenly distributed over the year. Periods of drought are not detectable by CLAMP, but the three driest months were slightly less moist than the mean monthly precipitation. It seems that the drier period corresponded to winter, because the mean monthly precipitation of the growing season was higher than that of the drier period. On the whole, the Maastrichtian climate of northern latitudes was characterized by low seasonal variation in precipitation and moderate seasonality in temperature. Above the Arctic Circle, the light regime was an additional seasonal factor that controlled the structure of vegetation. However, the influence of this factor is uncertain. According to the relationship between temperature and precipitation, the Maastrichtian climate is close to the modern warm temperate and subtropical climates of the eastern coast of North America from Maryland to northern Florida.

Discussion

Temperature

CLAMP temperature predictions are in good agreement with general climate inferences and data based on leaf margin percentages. The evaluation of temperatures in polar areas is of particular importance because the Maastrichtian climate of these regions differed significantly from modern conditions. Palaeobotanical evidence suggests that the winters were mild. Sometimes temperatures could drop to 0°C at mid-latitudes and even below 0°C at high latitudes, but the magnitude of minimum winter temperatures and duration of frost periods are uncertain.

The most northerly known Maastrichtian macrofloral localities are Kakanaut from the Koryak Upland and Kogosukruk in Alaska. The Maastrichtian position of the North Pole is assumed to be significantly nearer the Pacific Ocean than it is today, and not far from the coast of northeastern Siberia (Smith & Briden 1978; Howarth 1981). Accordingly, these localities were farther north during Maastrichtian time and well within the Arctic Circle. It follows from different continental reconstructions that Kakanaut was at about 75–80°N and Kogosukruk at about 75–85°N. Of these two floras the Kakanaut flora is more diverse, and includes more than 40 species. This number of species is sufficient for valid estimation of climate by CLAMP. Mean annual temperature is determined as 10°C and cold-month mean temperature about 3°C. The Kogosukruk locality contains a poorer floral assemblage, which is characterized by strong dominance of deciduous conifers and rare occurrence of angiosperms, represented only by 3–4 species. The depauperation of this flora by comparison with other Cretaceous Alaskan floras was previously interpreted as a dramatic drop in diversity during Maastrichtian time. The mean annual temperature for this area was estimated to be about 5°C (Parrish *et al.* 1987; Spicer & Parrish 1987), implying a frequent occurrence of winter temperatures below 0°C. However, pollen assemblages from the North Slope of Alaska (Frederiksen 1991) and from other localities in Alaska and Arctic Canada (Tabbert 1967; Rouse & Srivastava 1972; Felix & Burbridge 1973) usually involve more than 30 species of angiosperms, represented by approximately 15 genera. In taxonomic composition and diversity these assemblages are very similar to pollen assemblages from localities further south, in Alberta and Saskatchewan, and also to assemblages of the Koryak Upland and Yenisei depression in northern Siberia (Samoylovich 1977). Palynofloras of these regions are characterized by the dominance of *Aquilapollenites* pollen and by the presence of other characteristic genera from the *Triprojectacites* group (Samoylovich 1977; Batten 1984). Close resemblance of spore and pollen assemblages implies that macrofloras were also similar in composition and diversity. Hence, the depauperation of the Maastrichtian Kogosukruk flora may not reflect a change in climate. Perhaps, it could be attributable to a taphonomic factor instead. This leads to the conclusion that the climate in polar areas during Maastrichtian time was similar to that indicated by the flora of the Kakanaut locality, i.e. cold-month mean temperatures were not lower than 3–4°C. This assumption is confirmed by the

Table 3. *Diversity of vertebrate faunas of the North Pacific region during Maastrichtian time (on the basis of data from Spicer & Parrish (1987), Nessov & Golovneva (1990), Wheishampel (1991) and Nessov (1997)*

Area or locality	Latitude (°N)	Number of dinosaur genera	Main vertebrate groups
Alaska (Kogosukruk)	70	5	fish, dinosaurs, mammals
Yukon Territory, Canada	66	1	dinosaurs
Northwest Territories, Canada	64	2	dinosaurs
Koryak Upland (Kakanaut)	63	2	sharks, plesiosaurs, dinosaurs
Amur region	49–50	4	dinosaurs, turtles, crocodiles
Kazakhstan (Zaysan)	48	8	dinosaur egg shells
Sakhalin	47	1	dinosaurs
Kazakhstan (Ulken-Kalkan)	44	5	dinosaurs
Southern Gobi (Nemegt)	43	13	crocodiles, turtles, dinosaurs
Southwestern Canada and northwestern USA	40–53	≈20	fish, amphibians, turtles, crocodiles, lizards, dinosaurs, mammals
Southern USA	< 40	≈26	fish, amphibians, turtles, crocodiles, lizards, dinosaurs, mammals

presence of taxa from the subtropical families Loranthaceae, Nyssaceae, Proteaceae and Olacaceae in pollen assemblages at high latitudes (Samoylovich 1977). The disappearance of subtropical thermophilic elements is observed only in the Khatanga–Lena palynological province in northeastern Siberia. It is not yet clear if this depauperation of palynofloras is associated with proximity to the North Pole or with more severe winters in the Siberian continental interior.

Comparison of the predicted Maastrichtian polar climate with the database developed by Wolfe (1993) shows that climatic characteristics of the Kakanaut locality are closest to those of several sites in Maryland, including Battle Creek Cypress Swamp State Park. Forests consisting of deciduous Taxodiaceae, which were similar to the modern genera *Taxodium* and *Metasequoia*, were very common plant communities of high latitudes in Maastrichtian time, and sometimes they were dominant in the vegetation. Thus, the predicted polar climatic values do not contradict the putative temperature tolerance of the Maastrichtian plants.

However, the discovery of dinosaurian remains in polar regions was not expected, because dinosaurs have been previously regarded as animals adapted to warm climates. In the North Pacific region dinosaurian and other vertebrate faunas are associated with most of the same localities as the floral remains (Table 3). Maastrichtian high-latitude dinosaurs have been reported from the North Slope of Alaska (Spicer & Parrish 1987), the Northwest Territories and

Yukon Territories of Canada (Weishampel 1991), as well as from the Koryak Upland of northeast Russia (Nessov & Golovneva 1990). The diversity of the dinosaur fauna decreases northwards, and is presumed to be linked not only to diminishing temperatures, but also to a decreasing number of localities at high latitudes. However, a drop in diversity to the north of *c.* 55–60°N (from 5–10 genera to 1–3 genera) can be probably related to the former position of the Arctic Circle and long dark polar nights. Taxonomically, only one biogeographical border in the distribution of dinosaurs in North America has been identified. This is marked by the appearance of sauropods south of 35–40°N (Lehman 1987). In Maastrichtian time, sauropods were also distributed in tropical regions of South America, India and southern Europe (Weishampel 1991). This boundary corresponds to that between warm temperate and subtropical climates, which is inferred from floral data, and also corresponds to the boundary between the *Aquilapollenites* and *Normapolles* palynological provinces (Samoylovich 1977; Batten 1981, 1984). The fauna of mid- and high latitudes is taxonomically very similar, but high-latitude assemblages were more depauperate. The decreased diversity reflects, in particular, the absence (or rarity) of amphibians and nondinosaurian reptiles, which were abundant and taxonomically diverse further south (Clemens & Nelms 1993). Among the large reptiles that lived within the Arctic Circle during Maastrichtian time were plesiosaurs and dinosaurs represented by the herbivorous hadrosaurs and ceratopsians, as well as by carnivorous carnosaurs and small terapods. This rich fauna of reptiles could not survive long severe winters. Small modern reptiles of the temperate zone can hibernate in shelters during the winter, but large dinosaurs could not do the same because of their size (Parrish et al. 1987). Hence, polar dinosaurs were probably active throughout the winter. Most recent reptiles need average winter temperatures to be not less than 10–12°C to sustain the necessary level of activity, although they can endure slight frosts for short periods (McKenna 1980; Hutchison 1982). If dinosaurs were simple ectothermic animals and had temperature tolerances similar to those of other reptiles, we should assume that mean winter temperatures during Maastrichtian time were also about 10–12°C or above. These estimates of cold period temperatures are much higher than the estimations inferred from plant physiognomy. Wolfe (1987) suggested that the deciduousness of high-latitude forests resulted from the polar light regime rather than from lower in temperature. He estimated mean annual temperatures of polar regions to be about 10–15°C. However, the gradual change in abundance of species with entire-margined leaves along a latitudinal gradient, which is not accompanied by an abrupt change close to the former Arctic Circle, and the presence of a relatively large numbers of these species (20–30%) at high latitudes, indicates that deciduousness was determined primarily by seasonality in temperature, rather than by the light regime. It seems that some entire-margined species could have been evergreen and survived the polar winter in a dormant state.

Some investigators have supposed that polar dinosaurs, like caribou today, could have migrated southwards to escape cold winters (Hotton 1980). However, because of the lower thermal gradient in Maastrichtian time by comparison with the present, the dinosaurs would have had to cover a vast distance of 2500–4000 km to reach significantly warmer environments. The predominance of juvenile individuals in dinosaurian assemblages contradicts this concept of southward migration (Nessov & Golovneva 1990; Clemens & Nelms 1993). It is highly improbable that hatchlings and young dinosaurs were involved in such long-distance annual migrations.

These conclusions indicate that dinosaurs were able to remain in polar areas all year round and tolerated a cool, dark winter. It is possible that they differed metabolically from modern reptiles. Smaller turtles and crocodiles did not live so far north, although they had the

capacity to hibernate (Clemens & Nelms 1993). In this case the occurrence of dinosaurs cannot be used as a climatic indicator because their temperature tolerance is uncertain. It is most likely that they were not fully endothermic and used some other physiological mechanisms (e.g. gigantothermy) to maintain their body temperature (Paladino & Spotila 1994). In any case, the presence of reptiles with naked skin indicates that winter temperatures in polar regions were generally above freezing and never dropped below 0°C for long periods.

The summarized data obtained from oxygen-isotope analyses indicate that mean annual temperatures were about 18–22°C (Douglas & Savin 1973; Saito & Van Donk 1974) or 20–25°C (Frakes et al. 1994) at tropical latitudes; about 11–16°C in mid-latitudes (Teis & Nidin 1973; Yasamanov 1978; Frakes et al. 1994); and about 10°C at high latitudes (Teis & Nidin 1973; Boersma 1984; Frakes et al. 1994). The Arctic Ocean is generally believed to have lacked a permanent ice cap in Maastrichtian time (Clark 1974).

Evidence coming from oxygen-isotope data for high and mid-latitudes is therefore in good agreement with climate predictions derived from plant physiognomy.

Precipitation

CLAMP predictions for precipitation in Maastrichtian time are high (about 1500–1700 mm) and indicate an even distribution throughout the year. No latitudinal differences in rainfall were revealed by CLAMP. This amount of precipitation corresponds to a humid climate and closed-canopy forest vegetation. However, general physiognomic analysis of fossil floral assemblages, scarcity of drip-tips and range of leaf size suggest only moderate humidity that decreased slightly southwards. Coal layers are insignificant (usually several centimetres) in the Maastrichtian deposits at high latitudes and almost disappear in the mid-latitudes. The existence of large dinosaurs is consistent with open vegetation and, therefore, semi-arid conditions (Wing & Tiffney 1987). In its turn, disturbance of vegetation by large herbivores must have favoured the formation of open landscapes. It is probable that the precipitation values derived from CLAMP analysis are somewhat overestimated. This overestimation can be caused by taphonomic factors because fossil assemblages are primarily formed by plants from river valleys, lake shores and other lowland environments. In these conditions the magnitude of moisture content is always much greater than for

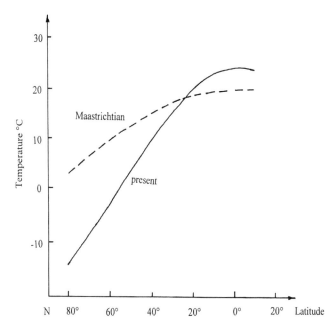

Fig. 2. Temperature gradients for present (continuous line, after Barron (1983)) and the Maastrichtian stage (dotted line).

upland areas. It seems that the vegetation during Maastrichtian time would be a mosaic of open woodland on uplands and closed-canopy forests along river valleys. Wolfe (1990) estimated the mean annual precipitation at mid-latitudes to be about 700–900 mm, deriving this value from the estimated precipitation for the driest months. This approach seems more appropriate for estimating precipitation during Maastrichtian time.

Latitudinal temperature gradient

On the whole, the Maastrichtian climate was characterized by high average temperatures and high equability both in precipitation and seasonal range of temperature. Oxygen-isotopic and palaeobotanical data indicate that temperatures were relatively lower in the tropics and significantly higher at high latitudes by comparison with conditions today. The latitudinal thermal gradient was less pronounced (Fig. 2). The difference between mean annual temperatures of polar and tropical areas during Maastrichtian time was about 10°C. This is 4–5 times less than today.

The greatest differences between the Maastrichtian and modern climates are observed at high latitudes. The much higher Arctic temperatures during Maastrichtian time may have resulted from heating of polar areas by warm oceanic upwellings (Nessov & Golovneva 1990; Nessov 1997). The temperature of the bulk of deep oceanic waters is now about 2°C. These water masses are formed as the result of the descent of cold, dense water at high latitudes. Later they appear near the western margins of continents at low latitudes as upwellings of cold waters containing a large amount of nutrients. In tropical and subtropical areas only a thin surface layer of water is warmed. Thus, the modern vertical ocean circulation is determined by cold high-latitude downwellings.

However, during Maastrichian time the picture was different. The temperatures of deep ocean waters reached 10–14°C (Boersma 1987). These waters were probably heated at low latitudes in shallow epicontinental seas. As a result of surface evaporation the salinity and density of these waters increased, leading to their descent to the bottom and into the deeper parts of the ocean. These warm downwellings transferred the heat from the sea surface to the depths of the ocean. At present, similar downwellings are observed in the Mediterranean and Red Seas, but the influence of this type of downwelling on the heat balance of the oceans is not great, because the areal extent of these seas is comparatively small. In Cretaceous time, epicontinental seas in subtropical arid regions were widespread. As a result, vertical oceanic circulation was determined not by cold downwellings at high latitudes, but by warm downwellings at low latitudes. At high latitudes the warm deep waters rose to the surface. During Campanian time, these upwellings are well marked by findings of hesperornidids, because the distribution of these birds in the Cretaceous period was related to the occurrence of water masses of high productivity (Nessov & Yarkov 1993).

Warm upwellings at high latitudes heated both the surface waters and the atmosphere. At that time the ocean could have been an accumulator of heat, storing it at depth for a long time and then redistributing it from the mid-latitudes to the poles. This model could explain the low temperature gradients during Maastrichtian time and the general equability of the climate.

The research described in this paper was made possible in part by Grant N 96-04-50335 and in part by Grant N 96-05-65843 from the Russian Foundation of Fundamental Research. I would like to thank R. A. Spicer for his help in obtaining CLAMP data, and I. A. Mershchikova for her helpful editorial advice.

References

AKHMETIEV, M. A. & SHEVYREVA, N. S. 1989. The flora of the Tsagaian-type near Zaysan Lake (eastern Kazakhstan). *Izvestiya Akademii Nauk SSSR, Seriya Geologicheskaya*, 1989(6), 80–90. [in Russian].

BAILEY, I. W. & SINNOT, E. W. 1915. A botanical index of Cretaceous and Tertiary climates. *Science*, **41**, 831–834.

BARRON, E. J. 1983. A warm, equable Cretaceous: the nature of the problem. *Earth-Science Reviews*, **18**, 305–338.

BATTEN, D. J. 1981. Stratigraphic, palaeogeographic and evolutionary significance of Late Cretaceous and early Tertiary Normapolles pollen. *Review of Palaeobotany and Palynology* **35**, 125–137.

—— 1984. Palynology, climate, and the development of Late Cretaceous floral provinces in the Northern Hemisphere; a review. *Geological Journal, special issue*, **11**, 127–164.

BELL, W. A. 1949. Uppermost Cretaceous and Paleocene floras of western Alberta. *Canada Geological Survey Bulletin* , **13**, 1–231.

BERRY, E. W. 1925. The flora of the Ripley Formation. US Geological Survey, Professional Paper, **136**, 1–94.

—— 1935. A preliminary contribution to the floras of the Whitemud and Ravenscrag Formations. Canada Geological Survey Memoirs, **182**, 1–65.

BOERSMA, A. 1984. Campanian through Paleocene paleotemperature and carbon isotope sequence and the Cretaceous–Tertiary boundary in the Atlantic Ocean. *In*: BERGGREN, W. A. & VAN

COUVERING, J. A. (eds). *Catastrophes and Earth History. The New Uniformitarianism*. Princeton University Press, Princeton, NJ, 247–278.

CLARK, D. L. 1974. Late Mesozoic and Early Cenozoic sediment cores from the Arctic Ocean. *Geology*, **2**, 41–44

CLEMENS, W. A. & NELMS, L. G. 1993. Paleoecological implications of Alaskan terrestrial vertebrate fauna in latest Cretaceous time at high paleolatitudes. *Geology*, **21**, 503–506.

DILCHER, D. L. 1973. The Eocene floras of southeastern North America. IN: GRAHAM, A. (ed.) *Vegetation and Vegetational History of Northern Latin America*, Elsevier, Amsterdam, 39–59.

DORF, E. 1942a. Stratigraphy and palaeontology of the Fox Hills and Lower Medicine Bow Formations of southern Wyoming and northeastern Colorado. Carnegie Institution of Washington Publication, **580**, 1–78.

—— 1942b. Flora of the Lance Formation in its type locality, Niobrara County, Wyoming. Carnegie Institution of Washington Publication, **580**, 79–159.

DOUGLAS, J. G. & SAVIN, S. M. 1973. Oxygen and carbon isotope analyses of the Cretaceous and Tertiary foraminifera from the central North Pacific. *Initial Reports of the Deep Sea Drilling Project*, 17. US Government Printing Office, Washington, DC, 591–605.

FELIX, CH. J. & BURBRIDGE, P. P. 1973. A Maastrichtian age microflora from Arctic Canada. *Geoscience and Man*, **7**, 31–38.

FRAKES, L. A., PROBST, J. & LUDWIG, W. 1994. Latitudinal distribution of paleotemperature on land and sea from early Cretaceous to middle Miocene. *Comptes Rendus de l'Academie des Sciences, Serie II*, **318**, 1209–1218.

FREDERIKSEN, N. O. 1991. Pollen zonation and correlation of Maastrichtian marine beds and associated strata, Ocean Point dinosaur locality, North Slope, Alaska. *US Geological Survey Bulletin*, **1990-E**, 1–24

GOLOVNEVA, L. B. 1994. Maastrichtian–Danian floras of Koryak Upland. *Proceedings of Komarov Botanical Institute*, **13**, 1–148. [in Russian].

—— 1995. Environmental changes and patterns of floral evolution during the Cretaceous–Tertiary transition in Northeastern Asia. *Paleontological Journal*, **29**(2a), 36–49.

GORODNITSKIY, A. M., ZONENSHAIN, L. P. & MIRLIN, E. G. 1978. *Reconstruction of Positions of the Continents during the Phanerozoic (on the Basis of Palaeomagnetic and Geological Data)*. Nauka, Moscow [in Russian].

HERMAN, A. B. & LEBEDEV, E. L. 1991. *Stratigraphy and Flora of the Cretaceous Deposits of Northeastern Kamchatka*. Nauka, Moscow [in Russian].

—— & SPICER, R. A. 1996 Palaeobotanical evidence for a warm Cretaceous Arctic Ocean. *Nature*, **380**, 330–333.

HOTTON, N. 1980. An alternative to dinosaur endothermy: the happy wanderers. *In*: THOMAS, R. D. K. & OLSON, E. C. (eds) *A Cold Look at the Warm-blooded Dinosaurs*. American Association for the Advancement of Science Selected Symposium, **28**, 311–350.

HOWARTH, M. K. 1981. Palaeogeography of the Mesozoic. *In*: Greenwood, P. H. (ed.) *The Evolving Earth*. Cambridge University Press, Cambridge. 197–220.

HUTCHISON, J. H. 1982. Turtle, crocodilian and champsosaur diversity changes in the Cenozoic of the north–central region of western United States. *Palaeogeography, Palaeoecology, Palaeoclimatology*, **37**, 149–164.

JERZYKIEWICZ, T. & RUSSELL, D. A. 1991. Late Mesozoic stratigraphy and vertebrates of the Gobi Basin. *Cretaceous Research*, **12**, 345–377.

JOHNSON, K. R. 1992. Leaf-fossil evidence for extensive floral extinction at the Cretaceous–Tertiary boundary, North Dakota, USA. *Cretaceous Research*, **13**, 91–117.

KOVACH, W. L. & Spicer, R. A. 1996. Canonical correspondence analysis of leaf physiognomy: a contribution to the development of a new palaeoclimatological tool. *Palaeoclimates*, **2**, 125–138.

KRASSILOV, V. A. 1979. *Cretaceous Flora of Sakhalin*. Nauka, Moscow [in Russian].

——— 1985. *The Cretaceous Period. Evolution of the Earth's Crust and Biosphere*. Nauka, Moscow [in Russian]

——— & MAKULBEKOV, N. M. 1995. Maastrichtian aquatic plants from Mongolia. *Paleontological Journal*, **29**(2a), 119–140.

LEHMAN, T. M. 1987. Late Maastrichtian paleoenvironments and dinosaur biogeography in the Western Interior of North America. *Palaeogeography, Palaeoecology, Palaeoclimatology*, **60**, 189–217.

MAKULBEKOV, N. M. 1974. The Late Cretaceous flora of Ulken-Kalkan (the Ilijskaja depression). *In*: KOJAMKULOVA, B. S. *et al.* (eds.) *The Fauna and Flora from Mesocenozoic of South Kazakhstan*. Materials on the History of Fauna and Flora of Kazakhstan, **6**, 108–120 [in Russian].

——— 1995. Florogenesis in central Asia at the Cretaceous–Palaeogene boundary. *International Conference: Evolution of Ecosystems, Abstracts*, PIN RAN, Moscow, 57.

MCKENNA, M. C. 1980. Eocene paleolatitude, climate and mammals of Ellesmere Island. *Palaeogeography, Palaeoecology, Palaeoclimatology*, **30**, 349–362.

NESSOV, L. A. 1995. *Dinosaurs of Northern Eurasia: New Data about Assemblages, Ecology and Palaeobiography*. St Petersburg University, St Petersburg [in Russian].

——— 1997. *Cretaceous Nonmarine Vertebrates of Northern Eurasia*. St Petersburg [in Russian].

——— & GOLOVNEVA, L. B. 1990. History of the flora, vertebrates and climate in the late Senonian of the north-eastern Koryak Uplands. In: Krassilov, V. A. (ed.) *Continental Cretaceous of the USSR*. Vladivostok, 191–212 [in Russian].

——— & YARKOV, A. A. 1993. Hesperornitids in Russia. *Russian Ornithological Journal*, **2**, 37–54.

PALADINO, V. F. & SPOTILA, J. S. 1994. Does the physiology of large living reptiles provide insights into the evolution of endothermy and paleophysiology of extinct dinosaurs? Paleontological Society Special Publication, **7**, 261–274.

PARRISH, J. M., PARRISH, J. T., HUTCHISON, J. H. & SPICER, R. A. 1987. Late Cretaceous vertebrate fossils from the North Slope of Alaska and implication for dinosaur ecology. *Palaios*, **2**, 377–389.

RICHARDS, P. W. 1952. *The Tropical Rain Forest*. Cambridge University Press, Cambridge.

ROUSE, G. E. 1967. A late Cretaceous plants assemblage from east–central British Columbia. *Canadian Journal of Earth Sciences*, **4**, 1185–1197.

——— & SRIVASTAVA, S. K. 1972. Palynological zonation of Cretaceous and Early Tertiary rocks of the Bonnet Plume formation Northeastern Yukon. *Canadian Journal of Earth Sciences*, **9**, 1163–1179.

SAITO, T. & VAN DONK, J. 1974. Oxygen and carbon isotope measurements of Late Cretaceous and Early Tertiary Foraminifera. *Micropaleontology*, **20**, 152–177.

SAMOYLOVICH, S. P. 1977. New scheme of floristic differentiation of Northern Hemisphere during Late Senonian. *Palaeontological Journal*, **1977**(3), 118–127 [in Russian]

SAMYLINA, V. A. 1974. The Early Cretaceous floras of the Northeast of USSR (on the problem of development of Cenophyte floras). *Komarovskie chtenija*, **27**, 1–56. [in Russia]

SAVIN, S. M. 1977. The history of the earth's surface temperature during the past 100 million years. *Annual Review of Earth and Planetary Sciences*, **5**, 319–355.

SHILIN, P. V. & ROMANOVA, E. V. 1978. *The Senonian flora of Kazakhstan*. Nauka, Alma-Ata [in Russian]

SHOEMAKER, R. E. 1966. Fossil leaves of the Hell Creek and Tullock Formations of Eastern Montana. *Palaeontographica*, **119B**, 54–75.

SMITH, A. G. & BRIDEN, J. C. 1978. *Mesozoic and Cenozoic Paleocontinental Maps*. Cambridge University Press, Cambridge.

SPICER, R. A. & PARRISH, J. T. 1987. Plant megafossils, vertebrate remains, and paleoclimate of the Kogosukruk Tongue (Late Cretaceous), North Slope, Alaska. *US Geological Survey Circular*, **998**, 47–48.

TABBERT, R. L. 1967. Cretaceous pollen and spores from the Ivishak river area arctic Alaska. *Review of Palaeobotany and Palynology*, **2**, 8–9.

TEIS, R. V. & NIDIN, D. P. 1973. *Palaeothermometry and Oxygen-isotopic Composition of Organogenous Carbonate Rocks*. Nauka, Moscow [in Russian].

VAKHRAMEEV, V. A. 1991. *Jurassic and Cretaceous Floras and Climates of the Earth*. Cambridge University Press, Cambridge.

WEISHAMPEL, D. B. 1991. Dinosaur distribution. *In*: WEISHAMPEL, D. B., *et al.* (eds.) *The Dinosauria*. University of California Press, Berkeley, 63–140.

WING, S. L. & TIFFNEY, B. H. 1987. Interactions of angiosperms and herbivorous tetrapods. *In*: FRIIS, E. M. *et al.* (eds) *The Origins of Angiosperms and their Biological Consequence*. Cambridge University Press, Cambridge. 203–224.

WOLFE, J. A. 1979. Temperature parameters of humid to mesic forests of eastern Asia and their relation to forests of other regions of the Northern Hemisphere and Australasia. *US Geological Survey, Professional Paper*, **1106**, 1–37.

——— 1987. Late Cretaceous–Cenozoic history of deciduousness and terminal Cretaceous event. *Paleobiology*, **13**, 215–226.

——— 1990. Palaeobotanocal evidence for a marked temperature increase following the Cretaceous/Tertiary boundary. *Nature*, **343**, 153–156.

——— 1993. A method of obtaining climatic parameters from leaf assemblages. *US Geological Survey Bulletin*, **2040**, 1–71.

——— & UPCHURCH, G. R. 1987. North American nonmarine climates and vegetation during the Late Cretaceous. *Palaeogeography, Palaeoecology, Palaeoclimatology*, **61**, 33–77.

YASAMANOV, N. A. 1978. *Landscape and Climatic Conditions during the Jurassic, Cretaceous and Palaeogene in Southern USSR*. Nedra, Moscow [in Russian].

Palaeoclimatic evolution in the Miocene from the Transylvanian Depression reflected in the fossil record

CARMEN CHIRA, SORIN FILIPESCU & VLAD CODREA

Babes - Bolyai University, Geology Department
Str. Kogalniceanu 1, 3400 Cluj-Napoca, Romania

Abstract: The palaeoclimate during the Miocene of the Transylvanian Depression (Romania) is interpreted mainly by the response of calcareous nannoplankton, microfauna (foraminifera), molluscs and mammals (rhinoceros) assemblages.

The first significant palaeoclimatic event is recorded during the early Miocene. The climatic warming at the level of the Vima Formation was followed by a more important warming during the Eggenburgian, which is preserved in the fossil record of the Corus Formation. During the Ottnangian, a cooling episode is recorded in the Hida Formation, probably related to Atlantic and Boreal influences. The climate on land was probably sub-tropical, humid, but much cooler than the Eggenburgian, as the remains of rhinoceros from this formation indicate.

The Middle Miocene marine sub-tropical assemblages are present in the Dej Formation (early Badenian), but towards its upper part sub-tropical species become scarce. The endemic elements, with boreal influences, are recognized in Kossovian strata. On land, the Moravian rhinoceros also suggest a sub-tropical climate, with marshy, densely afforested areas. Another rhinoceros found in the upper part of the Moravian, indicates a tendency to the continentalization of the climate. The early Sarmatian began with short period of warming, followed again by a cooling episode (Feleac Formation).

The last significant warming is recorded in the Late Miocene, during the Pannonian (Lopadea Formation). The small aceratheres from the Pannonian suggest marshy areas with a relatively warm climate.

The purpose of this paper is to compare the most recent biostratigraphical data on the Miocene of the Central Paratethys (Steininger *et al.* 1990; Roegl 1996) with the data provided, especially for the Middle Miocene, by the calcareous nannoplankton and molluscs (Chira 1992, 1995, 1996), foraminifera (Filipescu & Garbacea 1994, 1997; Filipescu 1996) and mammals (Codrea 1991, 1992, 1996*a,b*) from the Transylvanian Basin (Figs 1, 2). The data are added to those of other researchers who have dealt with this specific interval (Vancea 1960; Ciupagea *et al.* 1970; Meszaros *et al.* 1975, 1976, 1977; Popescu & Marinescu 1978; Moisescu & Popescu 1980; Meszaros 1985, 1992; Marunteanu 1992; Nicorici & Meszaros 1994).

Early Miocene

In the Transylvanian Depression, the first nannoplankton and foraminifera with Miocene affinities are found in the Vima Formation. The Mediterranean character of the assemblages suggests a warming of the climate at this level.

The **Egerian** stage is defined by the *Crassostrea gryphoides aginensis* Biozone. Other characteristic molluscs are *Crassostrea gryphoides crassissima* (Lamarck), *Crassostrea gingensis* Defrance and *Ostrea edulis lamellosa* Brocchi. This stage is also correlated with the lower part of the NN1 nannoplankton zone (*Triquetrorhabdulus carinatus* Biozone, Moisescu & Popescu 1980). It is generally thought that the NN1 zone occurs within the 'Zimbor Beds' (Meszaros *et al.* 1975) and 'Vima Group' (Meszaros 1992); the middle part of the Vima Formation is thought to belong to either the Aquitanian or late Egerian stages (Marunteanu 1992).

During the early Miocene, significant changes in the fossil assemblages are recorded in the Corus Formation and, more obviously, in the

From: HART, M. B. (ed.) *Climates: Past and Present.* Geological Society, London, Special Publications, **181**, 55–64, 1-86239-075-4/$15.00

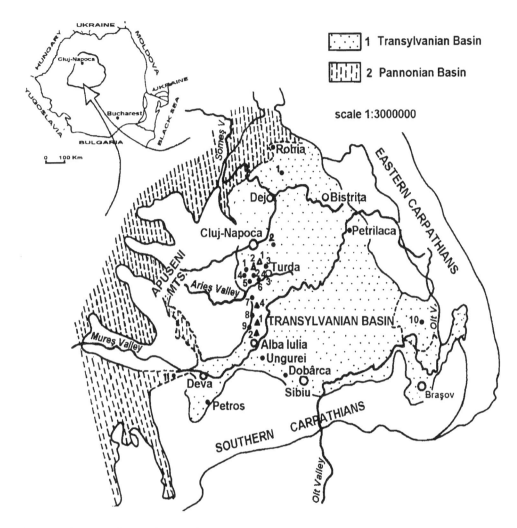

Fig. 1. Location map of the study area of the Transylvanian Depression
○ Eggenburgian sites: 1, Deleni; 2, Petrestii de Sus.
● Badenian sites: 1, Ciceu Corabia; 2, Cojocna; 3, Tureni–Copaceni; 4, Livada; 5, Borzesti; 6, Cheia; 7, Moldovenesti–Podeni; 8, Lopadea Veche; 9, Garbova de Sus; 10, Lueta–Meresti; 11, Lapugiu de Sus.
△ Sarmatian sites: 1, Aiton; 2, Ceanu Mic; 3, Turda; 4, Mahaceni–Podeni.
▲ Pannonian sites: 1, Lopadea Veche; 2, Garbovita.

Chechis Formation. The few Mediterranean elements, and the lack of low latitude planktonic foraminifera, were mentioned by Popescu *et al.* (1995).

The **Eggenburgian** stage is generally characterized by a marine sub-tropical Indo-Pacific mollusc fauna (Roegl 1996). This stage is characterized in Transylvania by marine sub-tropical molluscan faunas belonging to the *Chlamys gigas* Biozone. The most representative assemblages can be found in the Corus Forma-

tion with its typical sections near Cluj–Napoca, at Coasta cea Mare (Nicorici *et al.* 1979) and at Corus. Occurring with *Chlamys (Macrochlamys) gigas* Schlotheim are *Anadara fichteli* (Deshayes), *Glycymeris fichteli* (Deshayes), *Chlamys multistriata* (Poli), *Pecten pseudobeudanti* Deperet & Roman and *Laevicardium kuebecki* (Hauer). The calcareous nannoplankton is typical of the NN2 to NN3 zones (Moisescu & Popescu 1980).

The terminal part of the Vima Formation

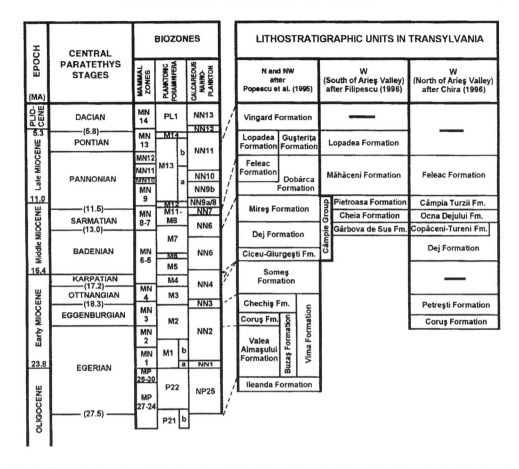

Fig. 2. Chronostratigraphy and biochronology of the Miocene from the Central Paratethys (partly after Roegl 1996) and correlation with the stratigraphic units from the Transylvanian Depression.

belongs to the NN2 zone (*Discoaster druggi* Biozone) which is also present almost throughout the Chechis Formation (Marunteanu 1992, 1993). The NN3 zone (*Sphenolithus belemnos* Biozone) is recorded in the Chechis Formation (Meszaros *et al.* 1976; Marunteanu 1992) and the Hida Formation (Meszaros *et al.* 1976). The Corus Formation is difficult to define using nannofossil assemblages (Meszaros 1992) because of the poor conditions of preservation.

The next warming interval might be correlated with the deposition of the Chechis Formation (Eggenburgian), characterized by the *Parvamussium duodecimlamellatum–Pecten hornensis* Biozone, nannoplankton zones NN2 (*Discoaster druggi* Biozone) and NN3 (*Sphenolithus belemnos* Biozone) and also by the M1–M2 planktonic foraminifera zone. The typical molluscan assemblages were described by Suraru (1968) from the northwestern part of the

depression and include *Pecten* cf. *hornensis* Deperet & Roman, *Pecten pseudobeudanti* Deperet & Roman and *Pecten pseudobeudanti rotundata* Schaffer.

The recently investigated Eggenburgian occurrences from the western Transylvanian Depression contain molluscan assemblages with oysters, *Teredo* sp. and *Turritella (Haustator)* cf. *vermicularis* (Brocchi) at Petrestii de Sus (Chira 1994) and Borzesti (Ghergari & Chira 1994), but only at Deleni has *Chlamys gigas* been recorded.

The **Ottnangian** stage is defined by marine molluscs of Atlantic and Boreal origin and includes the *Pecten beudanti strictocostata* Biozone within the Hida Formation, and part of the NN3 and NN4 – *Helicosphaera ampliaperta* Biozone (Moisescu & Popescu 1980). The NN4 zone was recorded by Meszaros *et al.* (1977) within the Hida Formation.

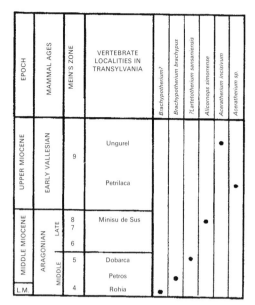

Fig. 3. Stratigraphic distribution of Miocene rhinoceros in Transylvania (after Codrea 1996).

As a consequence of the break of the connections between the Central Paratethys and the Mediterranean, the marine faunas are replaced by brackish or fresh-water faunas at the level of the Hida Formation (Somes Formation of Cioflica & Popescu 1971).

Compared to what it is already known from the extra-Carpathian area and in the Romanian sector of the Pannonian Depression (including the 'gulf-basins' situated east of it, inside the Apuseni Mountains) data on the fossil mammals from the Transylvanian Depression are extremely scarce. The only information is that referring to large mammals, although this is not very conclusive. Up to the present, any information concerning micro-mammals is completely lacking, with the exception of the marine mammals (Codrea 1996a) and the only discoveries worth taking into consideration are those referring to the rhinoceros (Fig. 3).

From the Lower Miocene there is a record of an isolated worn-out tooth from the Hida Formation (Ottnangian – NN4, Nicorici & Meszaros 1994; MN4, Codrea 1996b), which was found in the Rohia conglomerates (Salaj county). Despite its poor preservation, it was assigned to ?*Brachipotherium*, most probably *Brachipotherium brachipus* (Lartet). As indicated by Guerin (1980) and Cerdeno & Nieto (1995) this rhinoceros, which had a hippopotamus-like appearance was characteristic of afforested, marshy areas. During the deposition of the

conglomerates bearing the mentioned tooth the climate was warm, sub-tropical and sufficiently humid, even though the annual average precipitation (1600 mm) was much lower than that of the Eggenburgian from northwestern Transylvania (Petrescu 1990).

Middle Miocene

The connection with the Tethys was re-established at the beginning of the **Badenian** (Ciceu–Giurgesti Formation, Popescu 1970). The marine faunas reached their maximum proliferation in the late early Badenian, as in the well-known fossiliferous site at Lapugiu de Sus (Popescu *et al.* 1995). The cause of this proliferation was a global warming episode – the most important in the Miocene – which occured at this level (Dej Formation). The sub-tropical marine assemblages can be assigned to a part of the NN4 and NN5 zones (*Sphenolithus heteromorphus* Biozone) and *Neopycnodonte navicularis – Clio falauxi* Biozone (Lubenescu *et al.* 1978; Moisescu & Popescu 1980). The NN5 zone, with *S. heteromorphus, Discoaster exilis* and *D. variabilis* as dominant species (Marunteanu 1992), was identified in the 'Ciceu–Giurgesti Beds', 'Dej Tuff' (Popescu & Gheta 1984) and in the tuffaceous sequence with lenses of gypsum (Meszaros *et al.* 1977; Meszaros 1992). Dumitrica *et al.* (1975) assumed the presence of the NN4 Zone at the base of the 'Ciceu Giurgesti Beds', despite the absence of *Helicosphaera ampliaperta*. The absence of this species in the upper NN4 may be explained by Indo-Pacific influences (Marunteanu 1992).

The warming episode at the beginning of the Middle Miocene coincides with the beginning of the Indo-Mediterranean faunal invasion, characterized by a marked increase in the planktonic component of the fauna (Moisescu & Popescu 1980).

On the western border of the Transylvanian Basin (north of the Aries Valley) the lower Badenian Dej Formation outcrops at Borzesti, Livada and Padureni (Meszaros *et al.* 1992; Chira 1995, 1996). The formation was also investigated by Chira (1992) near Dej (Ciceu Corabia). In the region of Tureni and Copaceni, the equivalent Copaceni–Tureni Formation is characterized by the *Neopycnodonte navicularis* Biozone (Lubenescu *et al.* 1978; Chira 1996). The most frequent nannoplankton taxa are *Sphenolithus heteromorphus* Deflandre, *Coccolithus pelagicus* Wallich (Schiller), *Helicosphaera carteri* (Wallich) Kamptner, *Cyclococcolithus rotula* (Kamptner) Kamptner, *Discoaster variabilis* Martini & Bramlette, *Discoaster musicus*

Stradner and other species (Chira 1995, 1996).

According to Siesser (1993) one of the methods to determine relative palaeotemperatures is the investigation of the relative abundance of certain species of coccolithophores. Although individual coccolithophore species may be useful climatic indicators, more information can be gained from a study of the entire assemblage.

The warm water nannoplankton taxon *Sphenolithus heteromorphus* is the index species for NN5 zone, *Helicosphaera carteri* lives today in the tropical and sub-tropical waters of the Atlantic, at a temperature between 16°C and 26°C (McIntyre & Be 1967). *Cyclococcolithus rotula* prefers warm, tropical waters (Rahman & Roth 1990). Most species of *Discoaster* (e.g. *D. variabilis, D. exilis* and *D. musicus*) had an ecological preference for tropical and sub-tropical oceans, and are known from warm water masses throughout the Cenozoic (Rahman & Roth 1990). Aubry (1984) indicates that the *Discoaster*-producing organisms were essentially restricted to warm waters and one of their maximum diversity events (in the early Miocene) appears to correspond to warmer climatic conditions. *Discoaster variabilis* may have been a more temperate form, while *D. exilis* is either tolerant of, or exhibits a preference for, cooler waters.

Although *Coccolithus pelagicus* is a sub-polar species today, it is also very abundant in Lower Badenian sediments. This species evolved in the tropics during the early Cenozoic and migrated polewards during the mid-Cenozoic (Haq & Lohmann 1976). Some calcareous nannoplankton seem to be extremely rare or absent in tropical waters today, for example *Coccolithus pelagicus* amongst others. Other taxa, such as species of *Sphenolithus* and some *Helicosphaera* seem to avoid boreal waters (Martini 1971).

In the Badenian sediments of the Vienna Basin, Fuchs & Stradner (1977) showed that *Coccolithus pelagicus* characterizes sub-polar watermasses, while Rahman & Roth (1990) considered *C. pelagicus* as a long-ranging species which provides palaeoclimatic information for the latest Middle Miocene to the Pleistocene. *C. pelagicus* prefers cold nutrient-rich surface waters, with a temperature between 7°C and 14°C (McIntyre & Be 1967). It appears to be a good palaeoclimatic indicator (Haq *et al.* 1977). It seems that the species might have changed its ecological preference and may not have been a cold-water indicator during late Miocene times. Subsequent studies confirmed the earlier observations that *C. pelagicus* is indeed indicative of cold water even in the late Miocene (Haq *et al.*

1977; Rahman & Roth 1990). Because it is a resistant species, carbonate dissolution would increase the apparent frequency of *C. pelagicus* in sediments, giving a cold aspect to the assemblages (Rahman & Roth 1990).

Marunteanu (pers. comm.) presumes that the temperature of the water is important at the moment of first appearence of the calcareous nannoplankton as most of the species are adaptable.

On the western border of the Transylvanian Basin (south of Aries Valley) Filipescu & Girbacea (1994) have described two sequences of the Garbova de Sus Formation. These sequences show a significant difference between the sedimentation patterns and foraminiferal associations, and this is probably related to the first cooling event during the early Badenian. The first sequence, having a transgressive pattern, contains typical sub-tropical associations that are quite different from the relatively scarce associations from the upper, regressive sequence. Amongst the benthonic foraminifera there is an abundance of lagenids in the lower sequence. Many of the species are shared with the 'Upper Lagenid Zone' of the Badenian, thus allowing an easy correlation with the paratethyan faunas. The benthonic associations are accompanied by a large number of planktonic taxa belonging to the M5b–M6 Zone of the Berggren *et al.* (1995) scale.

The sub-tropical foraminifera are very frequently found close to the interval of maximum flooding in the lower sequence, but the frequency of these species markedly diminishes above the unconformity surface between the two sequences after the initiation of shallow carbonate sedimentation (mainly algal limestone). These sediments contain 'cibicidids', ephidiids and occasional large foraminifera.

The favourable life conditions for the larger foraminifera (*Planostegina costata* d'Orbigny, *Amphistegina hauerina* d'Orbigny, *Amphistegina mammilla* (Fichtel & Moll)) which occur in large numbers in the carbonate sands and sandstones, sometimes produced concentrations in 'lumachelle-like' beds. The ratio between richly and poorly ornamented *Planostegina* decreases in the upper part of the formation, as a result of a change in the environmental parameters (probably a slight cooling).

Taking into consideration the macrofauna, the *Neopycnodonte navicularis* Biozone is recorded in many Lower Badenian sections from Romania. In the Tureni–Copaceni area, Lubenescu *et al.* (1978) have recorded this zone in the northwestern part of the Transylvanian Basin. Chira (1992, 1996) recorded this biozone at

Borzesti, Livada, Padureni, Ciceu Corabia, while Filipescu (1996) has recorded it on the western border of the basin, south of Aries River. The biozone was also found at other localities in Romania by Moisescu & Popescu (1980). It was identified in the Mures Passage (Lapugiu de Sus, Faget Basin) as well as in the connection between the Transylvanian Basin and the Pannonian Basin. At this location is found one of the richest Lower Badenian molluscan faunas (about 1000 species). Some of the palaeoclimatic features of these assemblages have been described by Petrescu *et al.* (1990). The subtropical climate of the early Badenian was more like that of the early Miocene than the late Miocene. The average annual temperatures and rainfall in the area of the Gulf of Lapugiu were estimated to be in the range of 17–18°C and 2000 mm according to the palynological data. The nannoplankton assemblages here correspond to the NN5 zone, in which the more frequent forms are *Coccolithus pelagicus* (Wallich) Schiller, *C. eopelagicus* Bramlette & Riedel, *Cyclicargolithus floridanus* Roth & Hay, *Dictyococcites abisectus* Mueller, *Helicosphaera carteri* (Wallich) Kamptner, *Reticulofenestra pseudoumbilicus* Gartner, *Sphenolithus heteromorphus* Deflandre and *Discoaster variabilis* Martini & Bramlette.

The frequent occurrence of species from the families *Conidae, Cypraeidae, Strombidae, Muricidae, Fusidae, Veneridae* and *Carditidae* indicates a sub-tropical to tropical climate with water temperatures over 21°C Also, the presence of corals confirms these water temperatures, and suggests shallow waters of normal salinity at Lapugiu de Sus during the early Badenian. The whole fauna identified at Lapugiu de Sus is characteristic of littoral and sublittoral areas.

Neopycnodonte (Family *Pycnodontinae*) and *Acesta* (Family *Limidae*) are deep water genera which sometimes occur together in cold waters or at great depth. Both taxa were found at Ciceu Corabia (*Neopycnodonte navicularis* (Brocchi) and *Acesta miocenica* Sismonda (Chira 1992)) and Livada (*Neopycnodonte navicularis* (Brocchi) and *Acesta livada* Chira (Chira 1995, 1996)). Today *Neopycnodonte* lives at 12–14°C and at a depth of between 27 and 1500 m (Stenzel 1971). *Neopycnodonte* is considered an ecological and palaeoecological indicator for this temperature range and euryhaline environments of the circalittoral and bathyal levels (Freneix 1975).

Using significant groups of marine megafauna, it is possible to estimate the relative evolution of the temperature in the marine waters (Demarcq 1987). The Gastropoda fauna includes many families with warm water elements including *Apporhaidae, Strombidae, Cypraeidae, Pirulidae, Muricidae, Fasciolariidae, Olividae, Mitridae, Cancellariidae, Turridae* and *Conidae.* Demarcq (1987) also mentioned the presence of taxa which are polythermal, such as *Chlamys gigas,* or *Chlamys latissima, Chlamys scissa* and *Chlamys lilli.* The assemblage also includes *Flabellipecten leythajanus* (Partsch, in Hoernes), *Chlamys elegans* (Andrzejovsky), *Chlamys scissa* (Favre) and *Chlamys lilli* (Pusch). The *Chlamys latissima nodosiformis* Biozone is identified from northwestern Transylvania and from Lapugiu (Moisescu & Popescu 1980) and correlated with the upper part of the NN5 zone and the upper part of the M5 zone (= *Candorbulina universa–Globorotalia (T.) bykovae* Zone).

The microfaunal assemblage in the shallow gulf from Lapugiu de Sus contains coarse agglutinated and large calcareous foraminifera. The presence of *Planostegina, Amphistegina* and other species are important in this respect. The occurrence of *Candorbulina,* which belongs to the warm-water group of planktonic foraminifera, supports the hypothesis of a warm climate. The extinction of this extremely rich microfauna must have been related to an acute marine crisis generated by the onset of salt deposition at the end of the early Badenian (Wielician).

Brachipotherium brachipus (Lartet), one of the characteristic mammal species, was found in the Moravian (MN5) from the Hateg Depression, which functioned as a related area to the Transylvanian Basin (Codrea 1991). This indicates an extension of the same climatic conditions – warm, subtropical, high precipitation (an annual average precipitation of about 1800 mm) – to the base of the early Badenian, which allowed the development of several densely afforested areas, with marshy, even coal-generating characteristics.

Another rhinoceros, this time found in the south of the Transylvanian Depression, from Dobarca (Sibiu County), has been assigned to *Lartetotherium sansaniense?* (Lartet). The age of the deposits where it was found is also Moravian in age. This is younger than the Hateg rhinoceros (Codrea 1996b). Cerdeno & Nieto (1995) associate this rhinoceros with areas of open forests. This indicates, in comparison with of the base of the Badenian, a tendency of the climate to become continental, which led to a certain degradation and weathering of the forest.

This climatic change started in the upper part of the Dej Formation, where the sub-tropical foraminifera of M6 became scarce. The new marine microfauna from the Upper Badenian confirms the slight cooling of the environment.

The index species for the NN6 Biozone – *Discoaster exilis* – prefers cooler water (Bukry 1971). Popescu & Gheta (1984) indicated the presence of the NN6 zone in the 'Mires Beds' which are equivalent to the upper part of the 'Campie Beds'. NN6 corresponds to the late Badenian (Kosovian) (Marunteanu 1992). The presence of the NN6 zone is recorded in the 'Iclod Beds' (Meszaros 1992). There is a possibility of NN7 assemblages alongside the radiolaria mentioned by Ciupagea *et al.* (1970) and records of *Spiratella* (Meszaros 1992).

At Cheia, in the western part of the Transylvanian Depression, calcareous nannoplankton have been assigned to the Kosovian (NN6) (Ghergari *et al.* 1991; Popescu *et al.* 1995; Chira 1996).

At the base of the Serravalian (Kossovian) there is an important faunal change close to the base of the M7 zone of planktonic foraminifera or the boundary between the NN5 and NN6 nannoplankton zones. At this level, the Mediterranean-type fauna disappeared and a new marine assemblage, with endemic elements and boreal influences (Dumitrica *et al.* 1975) appeared.

At the top of the NN5 zone and M6 zone, corresponding to Wielician, Moisescu & Popescu (1980) identified the *Amussium denudatum* Biozone. This mollusc biozone corresponds, in Transylvania, to the Kossovian deposits; the upper part of 'Spirialis marls horizon'. From Garbova de Sus the same authors have described the *Chlamys wolfi–Chlamys scissa wulkae* Biozone. They included it in the Kossovian, but the species probably belong to the early Badenian.

The connection with the Indo-Pacific area in the Kossovian was preserved while the connections to the Mediterranean area were closed (Dumitrica *et al.* 1975). The species characteristic of the biozone are also described from Mesopotamia (Nicorici 1977).

At the boundary between the Kossovian and **Sarmatian** (base of the NN7 zone) the marine faunas disappeared and the endemic brackish faunas appeared and proliferated. In the late Bessarabian, close to the boundary with the Chersonian, the destruction of the marine environment excedeed the survival limit of the foraminifera and they disappeared entirely, at least in Transylvania.

Although this restriction of the connections with open marine areas makes global correlation difficult, the fossil association confirms that the early Sarmatian began with a short-lived warm period, followed again by a cooling episode.

Lower Sarmatian (Volhynian) sediments belong to the NN7 zone (*Discoaster kugleri* Biozone). As in other parts of the Paratethys *Braarudosphaera bigelowii* Gran & Braarud, *Calcidiscus leptoporus* Murray & Blackman and *Calcidiscus macintyrei* Bukry & Bramlette (Stradner & Fuchs 1979; Marunteanu 1992) are relatively common. It is likely that the Upper Volhynian deposits and, especially, the lower Bessarabian sediments contain the NN8–*Catinaster coalitus* Biozone. The interval around the Volhynian–Bessarabian boundary is, however, barren of nannofossils. Because of the adverse environmental conditions (a strong change in salinity and an important change of water chemistry) the normal development of calcareous nannoplankton in both the Dacian and Pannonian basins, only occurs at isolated stratigraphical levels and stratigraphical boundaries cannot be traced with certainty. The calcareous nannoplankton are often poorly preserved and reworked species are frequently encountered. The NN9–*Discoaster hamatus* Biozone, NN10–*Discoaster calcaris* Biozone and NN1–*Discoaster quinqueramus* Biozone are difficult to recognize in the Transylvanian Depression (Marunteanu 1992).

The nannoplankton assemblages from Aiton, Ceanu Mic and Turda belong to the Sarmatian (NN6/NN7, NN7, NN8, probably NN9) (Chira 1995, 1996). The lack of the characteristic marker fossils for the Sarmatian, given by Martini (1971) and Okada & Bukry (1980), precludes a correlation with the standard nannoplankton zonation. This has also been noted by Stradner & Fuchs (1979) in the Vienna Basin. Near Aiton, we found one of the richest Sarmatian (Volhynian–Lower Bessarabian) mollusc assemblages from the western Transylvanian Basin, which corresponds to the *Mohrensternia* (= *Rissoa*) Beds and *Ervilia* Beds (Chira 1995, 1996).

At Cojocna, in the northwestern part of the Transylvanian Depression, the nannoplankton are asigned to the Badenian–Volhynian (NN6, NN7) by Ghergari *et al.* (1991). On the basis of the calcareous nannoplankton (NN6, NN7) Meszaros *et al.* (1994) have assigned the Miocene deposits from the Lueta–Meresti area on the eastern border of the Transylvanian Depression to the late Badenian and Sarmatian.

Rado (1976) records the Lower Sarmatian deposits from the area of the Intra- and Extra-Carpathians as the equivalent of the *Mohrensternia* Beds (Volhynian). These are characterized by an assemblage of euryhaline species (*Cardium, Ervilia, Mohrensternia, Hydrobia, Pseudamnicola* and *Cerithium, Pirenella*) and also stenohaline genera from the Badenian (*Loripes, Tritonalia* and *Clavatula*) – and the

Ervilia Beds (Upper Volhynian).

Late Miocene

The rich bivalve populations suggest that the
Late Miocene marked the last significant warming interval.

At the beginning of the Pannonian the salinity
was probably close to 19‰. The aquatic realm
was progressively transformed into a series of
oligohaline lakes, with salinities down to 7‰.
This was lethal to most of the foraminifera, with
the exception of some agglutinated species.
Their place and importance was taken by
ostracods (*Amplocypris reticulata* (Hejjas), *Candona transilvanica* (Hejjas), *Loxoconcha rhombo-
valis* Pokorny and *Hemicytheria folliculosa*
(Reuss)). During the Pontian the surface area
of the lakes reduced considerably under the
influence of a dry climate (Filipescu 1996).

The Pannonian deposits of the Lopadea
Formation (Lubenescu & Lubenescu 1977)
contain a rich molluscan fauna with *Melanopsis,
Congeria* and *Limnocardium*. At Garbovita,
Filipescu (1996) recorded the presence of *Con-
geria banatica* Hoernes, *Paradacna lenzi*
(Hoernes), *Limnocardium* cf. *undatum* Reuss,
Undulotheca cf. *pancici* (Brusina), *Pseudocatillus*
sp., *Valenciennsia* sp. and a rich ostracod
assemblage belonging to zones a and b of the
Slavonian.

The faunas are equivalent to the *Congeria
banatica* Biozone of the entire Transylvanian
Depression (Vancea 1960; Lubenescu 1977;
Filipescu 1996).

From the same deposits, at Garbovita, the
presence of the NN10 Biozone is recorded
(Filipescu 1996).

In the southwestern Transylvanian Depression, Lubenescu (1981) described a rich molluscan fauna from the Vingard Formation, which
contains an assemblage that includes *Congeria
subglobosa/C. ungulacaprae*. The Upper Neogene deposits in the southwestern Transylvanian
Depression have been assigned to the Pannonian, equivalent to zones C, D, E of the Vienna
Basin. The Pontian species (*Congeria ungulacaprae* (Muenster), *Caladacna steindachneri* (Brusina)) have been interpreted as early Pontian in
age (Lubenescu 1981). The whole assemblage
indicates a latest Pannonian to Pontian age
(Popescu *et al.* 1995). The Pannonian formations
in the Transylvanian Depression were also
described by Marinescu (1985) who identified
the *Congeria banatica* and *Congeria subglobosa*
Biozones. The Pontian faunas were described
later by Lubenescu & Marinescu (1989).

A series of mammalian taxa, such as the fauna

found at Petrilaca (Codrea 1996*b*), indicates the
presence in the Transylvanian Basin of some
small-sized aceratheres at the level of the
Pannonian deposits (MN9). This might be
Alicornops simorrense (Lartet), a rhinoceros with
short and robust limbs, indicative of marshy
areas in a sub-tropical to temperate, though still
wet, climate. Moreover, this rhinoceros seems to
have existed also in the Zarand Depression, in
somehow older deposits (Volhynian) (MN7 +
8) at Minisu de Sus (Arad county) (Codrea
1992). The fauna and flora at Minis prove sub-
tropical climate, humid enough to permit the
development of certain areas with marshy
tendencies, in which such a rhinoceros could
have lived beside listriodons, deinotheres, water
chevrotains or turtles belonging to the *Trionyx*
genus.

From the same Pannonian deposits at Un-
gurei (Alba county) the existence of the *Acer-
atherium incisivum* (Kaup) species (Codrea
1996*b*) is known, which was adapted to the
same type of environment.

Conclusions

The fossil record corresponding to Miocene
times from the Transylvanian Depression,
demonstrates four significant warming episodes.
The warm intervals with sub-tropical palaeocli-
mate were recorded in the Eggenburgian, Lower
Badenian, Lower Sarmatian and Pannonian
Formation. These intervals alternated with three
cooling episodes recorded in the Ottnangian,
Upper Badenian to Sarmatian and probably
Pontian formations.

References

AUBRY, M.-P. 1984. Handbook of Cenozoic Calcar-
   eous Nannofossils, **1**, *Ortholithae (Discoasters).*
   Micropaleontology Press, 1–266.
BERGGREN, W. A., KENT, D. W., SWISHER, C. C. &
   AUBRY, M. P. 1995. A revised Cenozoic Geochro-
   nology and Chronostratigraphy. *In: Geochronology,
   Time Scales and Global Stratigraphic Correlations*
   SEPM Special Publication, **54**, 129–212.
BUKRY, D. 1971. *Discoaster* evolutionary trends.
   Micropaleontology, **17**, 43–52.
CERDENO, E. & NIETO, M. 1995. Changes in Western
   European *Rhinocerotidae* related to climatic varia-
   tions. *Palaeogeography, Palaeoclimatology, Palaeoe-
   cology*, **114**, 325–338.
CHIRA, C. 1992. Lower Badenian molluscs from the
   Dej tuff belonging to the Ciceu Pyroclastic Complex
   (Transylvanian Basin). *Studia Universitatis Babes-
   Bolyai, Seria Geologia*, **37/2**, 25–30.
———— 1994. Eggenburgian molluscs identified at
   Petrestii de Sus (Cluj county). *In*: NICORICI, E. (ed.)

The Miocene from the Transylvanian Basin. Carpatica, 71–74.

———— 1995. Calcareous nannoplankton and molluscs from a few Badenian–Sarmatian occurences of the Western Transylvanian Basin and Vienna Basin. *Romanian Journal of Stratigraphy. Xth Congress R.C.M.N.S. – Bucuresti*, **76**, Suppl. 7, *Abstracts*.

———— 1996. *Geologia si stratigrafia formatiunilor neogene si cuaternare din perimetrul Borzesti–Turda–Aiton–Comsesti pe baza studiului de moluste si nannoplancton.* PhD thesis, Universitatea Babes-Bolyai, Cluj-Napoca.

CIOFLICA, G. & POPESCU, G. 1971. Contributii la stratigrafia Miocenului mediu din nord–vestul Transilvaniei. *Studii si cercetari, Seria Geologie*, **16/2**, 531–539.

CIUPAGEA, D., PAUCA, M. & ICHIM, T. 1970. *Geologia Depresiunii Transilvaniei.* Ed. Academiei, Bucuresti.

CODREA, V. 1991. Some details concerning the discovery of *Brachypotherium brachypus* (Lartet) at Petros, Hunedoara district. *Studia Universitatis Babes-Bolyai, Seria Geologia*, **36/2**, 21–26.

———— 1992. New mammal remains from the Sarmatian deposits at Minisu de Sus (Taut, Arad county). *Studia Universitatis Babes-Bolyai, Seria Geologia*, **37/2**, 35–41.

———— 1996a. Données nouvelles concernant les Cétacés du Sarmatien de Cluj-Napoca. *Studies and Researches–Bistrita*, **1**, 91–97.

———— 1996b. Miocene rhinoceroses from Romania: an overview. *Acta zoologica cracoviae*, **39/1**, 83–88.

DEMARCQ, G. 1987. Paleothermic evolution during the Neogene in Mediterranea through the marine megafauna. *Annales Instituti Geologici Publici Hungarici*, **70**, 371–375.

DUMITRICA, P., GHETA, N. & POPESCU, G. 1975. Date noi cu privire la corelarea si biostratigrafia Miocenului mediu din aria carpatica. *Dari de Seama ale sedintelor, Institutul de Geologie si Geofizica*, **71/4**, 65–84.

FILIPESCU, S. 1996. *Stratigraphy of the Neogene from the western border of the Transylvanian Basin.* Studia Universitis Babes-Bolyai, Geologia **XLI/2**, 3–78, Cluj-Napoca.

———— & GIRBACEA, R. 1994. Stratigraphic Remarks on the Middle Miocene deposits from Garbova de Sus (Transylvanian Basin, Romania). *Studia Universitatis Babes–Bolyai, Seria Geologia*, **39/1–2**, 275–286.

———— & ———— 1997. Lower Badenian sea level drop on the western border of the Transylvanian Basin: foraminiferal palaeobathymetry and stratigraphy. *Geologica Carpatica*, **48/5**, 325–334.

FRENEIX, S. 1975. Au sujet du phylum *Neopycnodonte navicularis–Neopycnodonte cochlear. VI th Congress R.C.M.N.S.– Bratislava*, 443–449.

FUCHS, R. & STRADNER, H. 1977. Ueber Nannofossilien im Badenien (Mittemiozaen) der Zentralen Paratethys. *Beitraege zur Palaeontologie–Oesterreich*, **2**, 1–58.

GHERGARI, L., MESZAROS, N., HOSU, A., FILIPESCU, S. & CHIRA, C. 1991a. The gypsiferous formation at Cheia (Cluj county). *Studia Universitatis Babes–Bolyai, seria Geologia*, **36/1**, 13–28.

————, ————, MARZA, I., CHIRA, C., FILIPESCU, S. & IVAN, I. 1991b. Contributions to the petrographic and chronostratigraphic knowledge of the tuffs in the Cojocna area. *In*: MARZA (ed.) *The volcanic tuffs from the Transylvanian Basin*, 207–216.

———— & CHIRA, C. 1994. A new outcrop of Eggenburgian deposits in the Transylvanian Basin: Borzesti (Cluj county). *Studia Universitatis Babes–Bolyai, Seria Geologia*, **39/1–2**, 263–268.

GUERIN, C. 1980. Les rhinocéros (*Mammalia, Perissodactyla*) du Miocène términal au Pléistocène supérieur en Europe Occidentale. Comparaison avec les espèces actuelles. *Documents des Laboratoires de Geologie de la Faculte des Sciences de Lyon*, **79/1-3**, 1–1182.

HAQ, B. U. & LOHMANN, G. P. 1976. Early Cenozoic calcareous nannoplankton biogeography of the Atlantic Ocean. *Marine Micropaleontology*, **1**, 119–194.

———— PREMOLI-SILVA, I., LOHMANN, G. P. 1977. Calcareous plankton paleobiogeographic evidence for major climatic fluctuation in the early Cenozoic Atlantic Ocean. *Journal of Geophysical Research*, **82/27**, 3861–3876.

LUBENESCU, V. 1981. Studiul biostratigrafic al Neogenului superior din sud-vestul Transilvaniei. (Biostratigraphic study of the Upper Neogene in the South-West of Transylvania). *Anuarul Institutului de Geologie si Geofizica*, **58**, 123–202.

———— & LUBENESCU, D. 1977. Observatii biostratigrafice asupra Pannonianului de la Lopadea Veche (Depresiunea Transilvaniei). (Biostratigraphic observations on the Lopadea Veche Pannonian (Transylvania Depression). *Dari de Seama ale sedintelor, Institutul de Geologie si Geofizica*, **63/4**, 57–64.

———— & MARINESCU, F. 1989. L'equivalent du Pontien au Bassin de Transylvanie. *In: Chronostratigraphie und Neostratotypen*, **VIII**, Pontien, 264–267. Zagreb-Beograd.

———— PAVNOTESCU, V. & LUBENESCU, D. 1978. Badenianul la Copaceni–Tureni (NV Transilvaniei). 'Zona *Neopycnodonte navicularis*'. *Dari de Seama ale sedintelor, Institutul de Geologie si Geofizica*, **64/4**, 147–158.

MARINESCU, F. 1985. Der oestliche Teil des Pannonischen Beckens (Rumaenischer Sektor): Das Pannonien s. str. (Malvensien). *In: Chronostratigraphie und Neostratotypen*, **M6**, *Pannonien (Slavonien und Serbien)*, 144–154, Budapest.

MARTINI, E. 1971. Standard Tertiary and Quaternary calcareous nannoplankton. *In*: FARINACCI, A. (ed.). *Proceedings of the II. Planktonik Conference, Roma, 1970.* Tecnoscienza, 739–785.

MARUNTEANU, M. 1992. Distribution of the calcareous nannofossils in the Intra- and Extra- Carpathian areas in Romania. *In*: HAMRSMID, B. & YOUNG, G. R. (eds) *Proceedings of the Fourth INA Conference, Prague 1991*, Nannoplankton Research, II, *Knihovnicka ZPN*, **14 b/2**, 241–261.

———— 1993. Contents in calcareous nannoplankton of the marine lower and middle Miocene beds, east and north–east of Zalau. *Romanian Journal of Stratigraphy*, **75**, 99–103.

MCINTYRE, A. & BÉ, A. W. H. 1967. Modern *Coccolithophoridae* of the Atlantic Ocean. I. Placoliths and cyrtholiths. *Deep Sea Research*, **14**, 561–597.

MESZAROS, N. 1985. Les conditions ecologiques et l'importance stratigraphique du nannoplancton pour la stratigraphie du Tertiaire dans le Bassin de Transylvanie. *Evolution et adaptation*. II, Cluj-Napoca, 93–99.

—— 1992. Nannofossil Zones in the Paleogene and Miocene deposits of the Transylvanian Basin. *Proceedings of the Fourth INA Conference, Prague, 1991*, Nannoplankton Research, II, *Knihovnicka ZPN*, **14b/ 2**, 87–92.

——, GHERGARI, L. & CHIRA, C. 1992. The Badenian deposits at Borzesti. *Studia Universitatis Babes-Bolyai, Seria Geologia*, **37/2**, 71–78.

—— IANOLIU, C. & PION, N. 1976. Nannoplanctonul stratelor de Chechis si paralelizarea lor cu depozite similare ca virsta din Carpatii Orientali. *Analele Muzeului de Stiintele Naturii din Piatra Neamt, Seria Geologie, Geografie*, **3**, 213–218.

——, —— & PION, N. 1977. Nannoplanctonul din stratele de Hida si semnificatia lor stratigrafica. *Dari de Seama ale sedintelor, Institutul de Geologie si Geofizica*, **63/ 4**, 155–161.

——, LEBENZON, C., SURARU, N. & IANOLIU, C. 1975. Die mit des Nannoplanktons durchgefuerte Abgrenzung des Oligozaens in Tale des Almas (NW des Siebenbuergischen Beckens, Rumaenien). *Proceedings VIth RCMNS Congress, Bratislava*, 129–131.

—— MARZA, I., GHERGARI, L., CHIRA, C. & TOTH, L. 1994. Contributions to the stratigraphy of the Lueta–Meresti region (the Harghita county) and the age of the volcanic tuffs. *In*: NICORICI, E. (ed.) *The Miocene from the Transylvanian Basin*, 171–178.

MOISESCU, V. & POPESCU, G. 1980. Chattian–Badenian biochronology in Romania by means of molluscs. *Anuarul Institutului de Geologie si Geofizica*, **56**, 205–224.

NICORICI, E. 1977. Les *Pectinides* Badeniens de Roumanie. *Memoires, Institut de Geologie et de Geophysique*, **26**, 119–159.

—— & MESZAROS, N., 1994: Délimitation et subdivisions du Miocene en Europe et leur application sur certaines régions de Roumanie. *In*: NICORICI, E. (ed.) *The Miocene from the Transylvanian Basin*. Carpatica, 5–18.

——, PETRESCU, I. & MESZAROS, N. 1979. Contributii la cunoasterea miocenului inferior si mediu de la Coasta cea Mare (Cluj-Napoca). *Studii si cercetari de geologie, geofizica, geografie*, **24**, 103–137.

OKADA, H. & BUKRY, D. 1980. Supplementary modification and introduction of code numbers to the low-latitude *cocolith* biostratigraphic zonation (Bukry, 1973, 1975). *Marine Micropaleontology*, **4**, 321–325.

PETRESCU, I. 1990. *Perioadele glaciare ale Pamantului*. Ed. tehnica, Bucuresti.

—— MESZAROS, N., CHIRA, C. & FILIPESCU, S. 1990. Lower Badenian paleoclimate at Lapugiu de Sus (Hunedoara county), on account of palaeontological investigations. *Studia Universitatis Babes–Bolyai, Seria Geologia*, **35/2**, 13–22.

POPESCU, G., 1970. Planktonic foraminiferal zonation in the Dej Tuf Complex. *Revue roumaine de geologie, geophysique et geographie, serie de Geologie*, **14/2**, 189–203.

—— & GHETA, N. 1984. Comparative evolution of the marine Middle Miocene calcareous microfossils from the Carpathian and Pannonian areas. *Dari de Seama ale sedintelor, Institutul de Geologie si Geofizica*, **69/3**, 125–133.

—— & MARINESCU, F. 1978. Le Badenien de la Depression de Transylvanie et de la partie orientale de la Depression Intracarpatique. *In*: *Chronostratigraphie und Neostratotypen*, **M4**, *Badenien*. VEDA, Bratislava, VI, 86–89.

—— MARUNTEANU, M. & FILIPESCU, S. 1995. Neogene from Transylvania Depression. Guide to excursion A1 (X th R.C.M.N.S. Congress). *Romanian Journal of Stratigraphy*, **76/4**, Suppl. 3, 1–27.

RADO, G. 1976. Corelarea biostratigrafica a sarmatianului inferior din zonele intra si extracarpatice. *Analele Universitatii Bucuresti, Stiintele Naturii*, **25**, 97–106.

RAHMAN, A. & ROTH, P. H. 1990. Late Neogene paleoceanography and paleoclimatology of the Gulf of Aden region based on calcareous nannofossils. *Paleoceanography*, **5**, 1, 91–107.

ROEGL, F. 1996. Stratigraphic correlation of the Paratethys Oligocene and Miocene. *Mitteilungen der Gesellschaft der Geologie und Bergbaustudenten in Oesterreich*, **41**, 65–73.

SIESSER, W. G. 1993. Calcareous nannoplankton. *In*: LIPPS, J. H. (ed.). *Fossil prokariotes and protists*. Blackwell, Cambridge, 169–200.

STEININGER, F., BERNOR, R. L. & FAHLBUSH, V. 1990. European Neogene marine/continental chronologic correlations. *In*: LINDSAY *et al.* (eds) *European Neogene mammal chronology*. Plenum, New York, 15–46.

STENZEL, H. B. 1971. *Oysters*. *In*: MOORE, R. C. *et al.* (eds) *Treatise on Invertebrate Paleontology*, Part N, Mollusca 6, Bivalvia. GSA, Kansas University, Lawrence, Kansas, 953–1224.

STRADNER, H. & FUCHS, R. 1979. Ueber Nannoplanktonvorkommen im Sarmatien (Ober–Miozaen) der Zentralen Paratethys in Niederoesterreich und im Burgenland. *Beitraege zur Palaeontologie–Oesterreich*, **7**, 251–279.

SURARU, N. 1968. Contributii la cunoasterea macrofaunei argilelor de Chechis. *Studia Universitatis Babes-Bolyai, Seria Geologie–Geografie*, **2**, 47–58, Cluj.

VANCEA, A. 1960. *Neogenul din Bazinul Transilvaniei*. Ed. Academiei, Bucuresti.

Environmental changes in pre-evaporitic Late Miocene time in the Lorca Basin (SE Spain): diatom results

TH. JURKSCHAT [1,2], J. FENNER [2], R. FISCHER [1] & D. MICHALZIK [3]

[1] *Institut für Geologie und Paläontologie, Universität Hannover, Callinstr. 30, 30167 Hannover, Germany*

[2] *Bundesanstalt für Geowissenschaften und Rohstoffe, Stilleweg 2, 30655 Hannover, Germany*

(e-mail: t.jurkschat@t-online.de)

[3] *Institut für geologie, Ruhr-Universität Bochum, Universitätsstr. 150, 44780 Bochum, Germany*

Abstract: During the Late Miocene the Lorca Basin, which was an integrated part of the Betic Strait, underwent desiccation. The present investigation was completed to determine the response of the diatom flora to environmental changes before this event. Near the city of Lorca (Province of Murcia, SE Spain) the outcrop area around the Serrata ridge was studied and a stratigraphic section of 125 m has been measured and sampled (115 samples). The Serrata ridge is capped by a 25 m thick gypsum bed (Gypsum Member of the Serrata Formation). The profile below these gypsum layers comprises marl, sandstone, gypsum and partially laminated diatomite (Varied Member of the Serrata Formation). The frequency of radiolaria, siliceous sponge spicules, silicoflagellates, ebridians, diatoms and diatom species has been determined quantitatively. For the lower 87.5 m of the section, which contains sufficiently well preserved diatoms, an age of latest Miocene (6.6–5.3 Ma; Magnetic Epoch 6 and 5) can be assigned by the presence of the diatom species *Asterolampra acutiloba* and *Nitzschia reinholdii*, applying the stratigraphic ranges of the species used in the North and Equatorial Pacific and in the Atlantic. In total, 116 diatom species and species-groups have been identified. However, the number of species per sample varies only between 13 and 36 species (a minimum of 300 valves counted). This low diversity, together with a dominance of one species-group, *Thalassionema nitzschioides* (up to 80%), which tolerates considerable ecological changes (e.g. in salinity) indicate that environmental stress must have prevailed during the sedimentation of the Varied Member of the Serrata Formation. Abundance fluctuations of this species-group, as well as changes in the composition of the remaining diatom assemblage and decreases in abundances of the siliceous microfossil groups, reflect repeated periods of interruption of the still marine influenced deposits before the deposition of thick gypsum layers. Several repetitive changes are indicated, where each cycle is characterized by an increase in meroplanktonic species (e.g. *Paralia sulcata*, *Actinoptychus senarius*) and by a contemporary decrease in the holoplanktonic species towards the top, suggesting progressive environmental stress and a shallowing of the basin. Five of these cycles are clearly distinguished characterizing the initial stages of the evaporitic desiccation event in the Lorca Basin. Benthic diatoms are present only in low abundances (2–4%) and show no correlation with these fluctuations.

Late Miocene evaporatic sediments from continental outcrops (e.g. NW Africa, Crete, Sicily and N Italy, SE Spain) have been studied since the last century, and were interpreted to represent only a local phenomenon restricted to evaporitic periods in marginal basins of the Mediterranean. Since the cruises of the Deep Sea Drilling Project (DSDP Leg 13, Ryan *et al.* 1973; Leg 42A, Hsü *et al.* 1978) and extensive geophysical studies, these land deposits have been connected to the great desiccation event in the Mediterranean during Late Miocene time, the so-called Messinian Salinity Crisis (e.g. Schreiber *et al.* 1976; Montenat 1977; Rouchy 1982; Müller & Hsü 1987).

For this study of the reaction of siliceous-shelled microplankton to progressive desiccation of a Late Miocene basin, the Lorca Basin (Province of Murcia) in SE Spain has been chosen. This basin has already been studied by

From: HART, M. B. (ed.) *Climates: Past and Present.* Geological Society, London, Special Publications, **181**, 65–78, 1-86239-075-4/$15.00

different experts in the various fields of the geosciences; e.g. geochemistry (García-Veigas *et al.* 1994; Benali *et al.* 1995), palaeontology (Gaudant 1995) and sedimentology (Montenat *et al.* 1990*a*; Michalzik 1996), and multidisciplinary investigations have been carried out by Rouchy *et al.* (1998).

Within the framework of the German research project 'Evolution of Sequences and Biofacies in Neogene Deposits from SE Spain' (e.g. Dittert *et al.* 1993; Michalzik 1996), this investigation provides analyses of the siliceous microfossils (especially of diatoms) of the Serrata Formation. Because of the lack of palaeoecological interpretations, diatoms were selected, because this microfossil group sensitively reflects climatic and palaeoecological changes (e.g. Burckle 1978*a*; Hargraves & French 1983; Barron 1992). The Serrata Formation comprises a sequence of marls, diatomites, calcareous sandstones, dolomites and gypsum in the central part of the Lorca Basin deposited during Tortonian and Messinian time. The purpose of this investigation was to study these pre-evaporitic sediments so as to document the response of the diatom flora to climatic and ecological changes before the onset of the Messinian Salinity Crisis (Hsü *et al.* 1973).

Ehrenberg (1836, 1854) was the first to study Miocene diatoms from the Mediterranean region. At the beginning of the 20th century Forti (1908, 1913) studied the Miocene diatomites of Italy. In the last 20 years, several papers have been published that tried to correlate the Late Miocene pre-evaporitic sediments to the diatom zonation developed for the deep sea outside the Mediterranean; e.g. Burckle (1977, Sorbas Basin), Monjanel (1984, Guadalquivir Basin and Alicante region) and Müller & Schrader (1989, Fortuna Basin, see also Müller & Hsü 1987). Diatoms from the Lorca Basin were first mentioned by Azpeitia (1911). More recently, Gersonde (1980) analysed 16 samples from the Serrata Formation, and because of the dominance of the *Thalassionema nitzschioides* group he concluded that unfavourable palaeoecological conditions must be assumed for the time of deposition of these strata in the Lorca Basin.

Geological setting

The Lorca Basin, an intramontane basin within the Internal Zone of the Betic Cordillera, was part of a shallow seaway (Betic Strait) connecting the Atlantic Ocean and Mediterranean Sea during Late Miocene time (Santisteban & Taberna 1983).

The interconnected basins of the Betic Strait are in part fault-bounded and were formed by tectonic processes. Whether these basins were formed by transtensional collapse (Silva *et al.* 1993), extensional collapse (Paquet 1974; Mäkel 1985; Platt & Vissers 1989) or as strike-slip basins (e.g. the Lorca Basin along the Northern Betic and southern Alhama-de-Murcia shear zones) produced during compression between the African and Iberian plate (Montenat *et al.* 1987*a,b*, 1990*a,b*; Sanz de Galdeano 1990; Rodríguez Estrella *et al.* 1992; Sanz de Galdeano & Vera 1992) is still a matter of debate. The interpretation is complicated by the complex history of repeated extensional and compressional phases within the Betic Cordillera (Van der Beek & Cloetingh 1992). The restriction of these basins and the formation of evaporites at the end of Late Miocene time may be related to tectonic uplift of this region (Weijermars 1991) and/or a global sea-level fall as a result of the glaciation on Antarctica (e.g. Adams *et al.* 1977).

The basins of the Betic Strait are filled with a sequence of Neogene to Recent syn-orogenic and post-orogenetic sediments. Van der Beek & Cloetingh (1992) postulated that times of the initial phases of subsidence for the basins of the Betic Cordillera are not coeval. For the Lorca Basin they suggested subsidence since Tortonian time.

In the centre of the Lorca Basin these syntectonic sediments are more than 1500 m thick, and in marginal areas up to 400 m thickness has been documented (e.g. Wrobel 1993; Jurkschat 1995*a*). The sequence starts with marginal marine clastic sediments of Serravallian to Tortonian age that are followed by platform carbonates and hemipelagic marls. During Messinian times marly, diatomaceous and gypsiferous sediments were deposited. Evaporites are restricted to the centre of the basin. In the Lorca Basin, halite has been found only in drillholes, where it attains a thickness of more than 200 m (IGME 1982).

As a result of post-Miocene uplift the Upper Tortonian and Messinian pre-evaporitic sediments at Serrata now dip 25–30° NW, forming a resistant ridge with a massive gypsum layer along its crest. This ridge extends from the northern outskirts of the city of Lorca in a NNE direction (Fig. 1). Below the gypsum the ridge provides excellent exposures of the diatomite and marl alternations. The lithological units exposed in this ridge were termed the Serrata Formation by Geel (1976) and subdivided into three members: an < 100 m thick pre-evaporitic succession with intercalations of marl, diato-

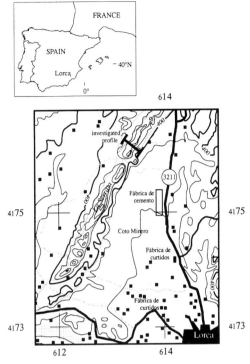

Fig. 1. Topographical map showing the location of the investigated stratigraphic section (modified after *Mapa Militar de España*, Lorca [953], coordinates after UTM).

mite, carbonates and gypsum: the Varied Member; an approximately 20–25 m thick evaporitic succession of two gypsum beds: the Gypsum Member; a post-evaporitic succession with intercalations of claystone, marl and gypsum: the Laminated Pelite Member.

Material

For this study a stratigraphic section through the Serrata Formation was sampled just north of the cement works (Fig. 1). At the base of the profile light, greyish yellow marl of the Hondo Formation (Geel 1976) is exposed. The first appearance of light grey diatomite was used to define the base of the Varied Member. The Varied Member of the Serrata Formation is characterized by alternating layers of marls and diatomites, with laminated diatomites occurring only in the lower part. Above the gypsum beds of the Gypsum Member, an alternation of laminated claystone and marly sediments (Laminated Pelite Member) follows. The sequence is covered by undifferentiated Quaternary sediments.

Biostratigraphy of Miocene deposits in the Mediterranean

The biostratigraphical age assignment for the Messinian deposits of the Mediterranean was made by using foraminifera from sediments below and above this sequence. After D'Onofrio *et al.* (1975) and Colalongo *et al.* (1979), who investigated deposits from type localities in Italy (Sicily), the base of the Messinian sequence is defined by the first appearance datum (FAD) of the Foraminifera species *Globorotalia conomiozea*. In sediments from the Northern Atlantic Berggren (1977) dated this event at 6.5 Ma. Nevertheless, the precise age of the base of the Messinian sequence in the Mediterranean region is not known because it cannot be proven unambiguously whether the first appearance of *G. conomiozea* in the Mediterranean can be considered an evolutionary process occurring at the same time as in the Atlantic or whether the immigration of this species into the Mediterranean region occurred later (Zachariasse 1979). Another biostratigraphical event used is a change in the coiling direction (from left to right) of the species *Neogloboquadrina acostaensis*, which occurs at the base of Magnetic Epoch 5 in the equatorial Pacific (Saito *et al.* 1975). This datum has been recognized by several workers in the Mediterranean area (Manuputty 1977; Colalongo *et al.* 1979). In the Lorca profile the change in the coiling direction of *N. acostaensis* takes place in the lower to middle part of the Varied Member of the Serrata Formation (Geel 1976; Manuputty 1977).

Schrader (1973, 1974) and Wornard (1973) were the first to use FAD's of planktonic diatoms from the northern Pacific and tropical Indian Ocean for age determinations. Burckle (1977) successfully compared stratigraphically useful diatom species from the diatomite deposits of southern Spain with those from the Equatorial Pacific. Gersonde (1980) assigned the assemblages from the Lorca Basin (Varied Member), with reservations, to lie within the time interval between upper Magnetic Epoch 6 and the top of Magnetic Epoch 5.

Methods

Sampling positions in the Varied Member were not selected at regular intervals, but material was collected whenever the lithology changed. At least two samples per lithological unit have been taken, and in beds thicker than 50 cm occasionally up to 10 samples. For all 115 samples, care was taken to sample unweathered

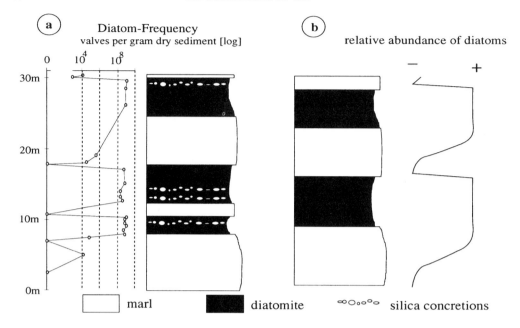

Fig. 2. (a) Observed diatom abundances in the lower part of the Varied Member. **(b)** Schematic drawing of supposed abundance fluctuation of diatoms in relation to the lithology.

sediment by digging at least 30 cm deep into the slope of the outcrop. All samples were investigated for siliceous microfossils and 55 of them, additionally, for diatom species abundances (see Fig. 6, below).

Quantitative preparation techniques for siliceous microfossils followed Battarbee (1973) and Fenner (1982), using Hyrax as a mounting medium. A Zeiss-Axioplan light microscope was used with an automatic camera to count a minimum of 300 diatom valves for each sample. Diatom resting spores, siliceous sponge spicules, phytoliths, and other siliceous microfossil groups such as silicoflagellates, radiolarians and ebridians, were recorded separately. Generally, an individual was counted if more than 50% of the valve or skeleton was preserved. In case of elongated diatoms (e.g. *Thalassiothrix*) every second end was counted. A scanning electron microscope (CAMSCAN CS2) was used to resolve difficult taxonomic questions.

Results

Abundance of siliceous microfossil groups

This study concentrates on the siliceous assemblages present in the Varied Member of the Serrata Formation. Samples taken from the sediments below the Varied Member (Hondo Formation) and the overlying deposits (Gypsum and Laminated Pelite Members) are barren of diatoms and other siliceous microfossils, or contain only extremely rare and poorly preserved specimens.

Among the biosiliceous microfossils marine diatoms are the dominant component. They make up more than 90% of the biosiliceous components, and in the marly part of the succession (from 85 m upsection) they decrease to 75%. Where diatom abundance decreases the relative abundance of other siliceous microfossils (mainly silicoflagellates and sponge spicules) increases. Occasionally, such a relationship can also be recognized in the normally rare microfossil groups such as ebridians, archeomonadaceans, actiniscids and radiolarians (only fragments encountered).

The preservation of the diatoms is moderate to poor. In observation with the light microscope and scanning electron microscope, many frustules show dissolution patterns such as broadened areolae and etched margins of the areolae. Nevertheless, species identification was never hampered. The fragmentation is low. Elongated diatom valves (up to 120 μm in length) of the species *Thalassionema nitzschioides* var. *nitzschioides* are commonly found. Chains of diatom frustules (e.g. of *Paralia sulcata*) are abundant, as well as

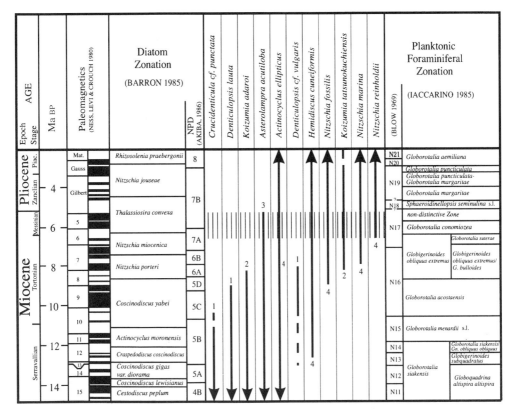

Fig. 3. Stratigraphically useful diatom species in the Lorca profile, their stratigraphic ranges as recorded from the Atlantic and Pacific Oceans using results from the literature (1, Yanagisawa & Akiba (1990); 2, Yanagisawa (1994); 3, Burckle (1978b), Koizumi (1985); 4, Barron (1985)), and their possible correlation with the biostratigraphical zonations for diatoms and Foraminifera developed by Barron (1985) and Iaccarino (1985), which are correlated with the palaeomagnetic reversal record after Ness et al. (1980).

complete frustules. The abundance fluctuations of diatoms are linked to lithological changes (Fig. 2) with high abundances in the diatomites (even the presence of nearby chert layers causes only a slight decrease) decreasing towards undetectable or low abundances (below 10^4 valves/g sediment) in the marls. The diatom assemblages from the Serrata profile contain a total of 116 species and species-groups. Of these, 69 species are described in the literature as marine and holoplanktonic, and four are marine and meroplanktonic. Thirty-nine species belong to marine and benthic habitats, and four species are of fresh-water origin. Marine planktonic specimens dominate the assemblages, with an average frequency of 95%. Only 2–4% are benthic species, and just 2–3% have been transported in from fresh-water environments. The abundance of fresh-water species increases towards the top of the profile. Throughout the section large numbers of diatom resting spores

are present. Silicoflagellates form the next most frequent siliceous microfossil group, reaching average numbers of 2–4% (maximum 7.9%) of the siliceous microfossils. Archaeomonadaceae-(chrysophycean-cysts), a group common in nearshore and neritic habitats, are rare, ranging from just present up to 2% (maximum 5%). Radiolarians have been found only in a few samples as fragments. Sponge spicules are present in certain intervals and provide evidence for the possibility of benthic life at these times. From 90 m upwards in the succession ebridians increase in numbers, indicating more oceanic conditions.

Biostratigraphic results of this study

The present study of the diatomites of the Varied Member of the Serrata Formation in the Lorca Basin provided 11 diatom species that are useful

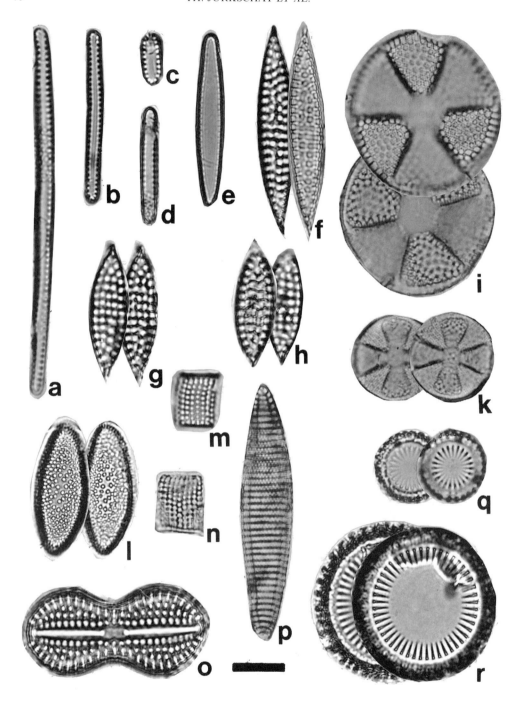

Fig. 4. Characteristic diatom species occurring in the Varied Member of the Serrata Formation, Lorca Basin. (**a**), (**b**), (**d**) *Thalassionema nitzschioides* var. *nitzschioides*; (**c**) *Thalassionema nitzschioides* var. *parva*; (**e**) *Thalassionema nitzschioides* var. *obtusa*; (**f**) *Koizumia adaroi*; (**g**), (**h**) *Koizumia tatsunokuchiensis*; (**i**) *Actinoptychus senarius*; (**k**) *Actinoptychus splendens*; (**l**) *Actinocyclus ellipticus*; (**m**), (**n**) *Aulacoseira* spp.; (**o**) *Diploneis bombus*; (**p**) *Nitzschia reinholdii*; (**q**), (**r**) *Paralia sulcata*. Scale bar 20 µm.

for biostratigraphy (Figs 3 and 4): *Actinocyclus ellipticus*, *Asterolampra acutiloba*, *Crucidenticula punctata*, *Denticulopsis lauta*, *D. vulgaris*, *Hemidiscus cuneiformis*, *Koizumia adaroi*, *K. tatsunokuchiensis*, *Nitzschia fossilis*, *N. marina* and *N. reinholdii*.

Nevertheless, dating the succession is still difficult because the first-order zonal marker species for the low and mid-latitudes of the Pacific and Atlantic Oceans, defined by Barron (1985), are missing (e.g. *Nitzschia jouseae* or *Thalassiosira oestrupii*).

Two well-established marker species that occur in the profile are *Nitzschia reinholdii* and *Asterolampra acutiloba*. *N. reinholdii* has its FAD in the upper *Nitzschia miocenica* Zone in Magnetic Epoch 6 (Barron 1985). As this species occurs from the base of the Serrata Formation upwards, the whole diatomaceous section cannot be older than Magnetic Epoch 6. In the Lorca profile *A. acutiloba* occurs upsection into the basal part of the last thick marl below the gypsum. As the diatom abundance in this part of the profile decreases and as *A. acutiloba* never occurs in larger numbers, the absence of this species in this part of the profile (with poor preservation) cannot be taken as the stratigraphical last occurrence. The last occurrence of *A. acutiloba* according to Burckle (1978*b*) and Koizumi (1985) coincides approximately with the boundary between Magnetic Epoch 5 and the Gilbert Magnetic Epoch. Thus, considering only the two species, *A. acutiloba* and *N. reinholdii*, the sediments of the Varied Member were deposited somewhere from upper Magnetic Epoch 6 to the top of Magnetic Epoch 5. In terms of the international diatom zonation the Varied Member lies within the interval from the uppermost *Nitzschia miocenica* Zone to the middle *Thalassiosira convexa* Zone.

The maximum stratigraphic time interval that is possible for the lower 89.5 m of the profile, which contains stratigraphic marker species, thus represents a time interval of 1.3 Ma (6.6–5.3 Ma).

There are, however, occurrences of species that, according to present knowledge, are known to have their last occurrence much earlier than the first occurrence of *N. reinholdii* (e.g. *Cruzidenticula punctata*, *Denticulopsis lauta* and *D. vulgaris*); two possibilities to explain their occurrence exist, as follows.

(1) These 'older' diatom valves are reworked. This interpretation would be in accord with other records of these species above their generally expected extinction level (e.g. Yanagisawa & Akiba 1990). The possibility that a few specimens of the genera *Cruzidenticula* and *Denticulopsis* have been reworked and redeposited into Upper Miocene sediments in the Lorca Basin is supported by finds of *Crucidenticula nicobarica*, *C. punctata* and *Denticulopsis hustedtii* as dominant species in the sediments of the Guadalquivir Basin (SW Spain) reported by Bustillo & López García (1997).

(2) *C. punctata*, *D. lauta* and *D. vulgaris* did not become extinct at that time, but continue to appear sporadically in younger sediments.

As specimens from all three species are just single finds, not much importance should be placed on their occurrence.

Abundance changes within the diatom assemblages

The diatom assemblages are dominated by *Thalassionema nitzschioides*. The percentage of this species ranges between 40 and 80% (Fig. 5). Being the dominant marine species, its abundance fluctuations control the diatom abundance per gram of sediment. In the marly sediments just below the Gypsum Member, a progressive decline in the abundance of diatoms is recognized. Inversely correlated with the abundance changes of *T. nitzschioides* are those of *Actinoptychus senarius* and of species of the genus *Rhizosolenia*. From 85 m upsection, oceanic species (e.g. *Koizumia* spp., *Azpeitia curvatulus*) show an increase in abundance.

Besides these abundance changes there are also systematic changes in diatom distribution correlating with the lithology (Fig. 2). At the base of each marl layer, diatom valves are absent or rare followed by a continuous increase in abundance. The maximum is reached just after the lithological change from marl to diatomite. Diatom abundance stays high throughout the diatomite layers. At the boundary with the next marl layer, the number of frustules decreases dramatically. The diatom flora disappears nearly, or completely, and the sequence begins again as described.

The frequency of each diatom species in the profile also reflects different intervals (Fig. 5). It is reasonable to divide the profile into several repetitive cycles for every species; however, for each species, these boundaries are different. This could also be explained by randomly taken samples. A repeated succession of diatom species in these cycles was not recognized except that after each drastic decrease in the frequency of meroplanktonic species, the number of individuals of *T. nitzschioides* increases.

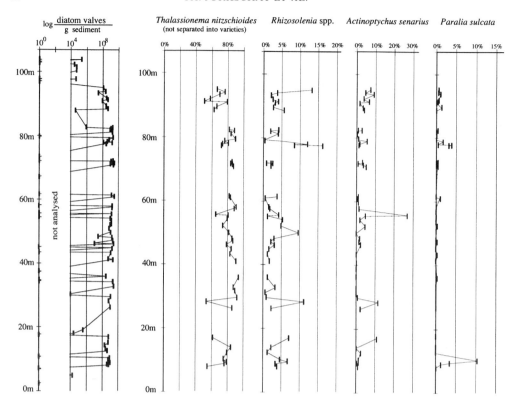

Fig. 5. Abundance fluctuation of diatom valves per gram sediment and of selected diatom species in the Varied Member up to the base of the Gypsum Member.

Discussion

Diatom biostratigraphy

Only secondary biostratigraphical diatom markers (e.g. *N. reinholdii* and *Asterolampra acutiloba*) are present (Fig. 3) and none of the first-order stratigraphical markers defined by Barron (1985) have been found. The age of the sediments in the investigated profile (0.0–89.5 m) cannot be narrowed down further than to lie within latest Tortonian to latest Messinian time (6.6–5.3 Ma). This confirms the age determination by Gersonde (1980), who placed the Varied Member of the Serrata Formation into Upper Magnetic Epoch 6 to the top of Magnetic Epoch 5. The sedimentation of the Varied Member probably represents a much shorter interval. Because of the palaeoenvironmental conditions in the Mediterranean during Messinian time only very few stratigraphically significant species are present, and no more precise an age determination than given in Fig. 3 is possible.

Palaeoecology

Fossil diatom assemblages are excellent indicators of palaeoenvironments and can be used to estimate palaeosalinity (e.g. Hustedt 1953; Barron 1992). Furthermore, they provide climatic proxies for palaeotemperature and palaeoproductivity.

Although planktonic diatoms are built to remain in suspension (Burckle 1978a) and because of their small size can easily be transported laterally by both bottom and surface currents (e.g. Burckle & Biscaye 1971), the diatom assemblages of the Varied Member can be assumed to be autochthonous. The observation of diatom-chains and unbroken, long, elongated frustules indicate that after settling to the sea floor, little or no resuspension or lateral transport took place. Additionally, sedimentary structures confirm deposition within a quiet environment. Fresh-water species and phytoliths were found in many samples. They have been brought into the basin by fluvial and/ or by aeolian transport. The potentially reworked specimens of *Crucidenticula* and *Denti-culopsis* are very rare.

Table 1. *Comparison of diatom abundances in surface sediments of recent upwelling areas and in diatomite layers of the Varied Member*

Region	Diatom frustules/g dry sediment (log)
Equatorial Pacific	7.2*
off Peru	7.0–7.5†
off SW-Africa	7.7–8.2‡
Antarctic Pacific	7.6–8.0*
Lorca basin (Serrata ridge); diatomites	8.0–9.0

* Lisitzin (1971). † Schuette & Schrader (1979). ‡ Richert (1975).

Productivity

The diatomite deposits of the Varied Member contain diatoms in abundances higher than in surface sediments of recent upwelling regions (Table 1). These high diatom abundances definitely reflect high productivity. The high numbers of diatoms per gram of dry sediment, especially in the diatomite deposits, can be caused by a lack of clastic input, low carbonate content or large numbers of small diatom valves and frustules. All these facts apply to the diatomites of the Varied Member. The high production of diatoms requires an input of nutrients, by fresh-water runoff from the continent, volcanic activity or upwelling. The first two of these sources may have contributed nutrients. Input from land is proven by the occurrence of fresh-water diatoms and phytoliths. Miocene volcanic activity is already known from SE Spain (De Clercq *et al.* 1975) and this may have been a source of the silica. During the lithological mapping of the profile and surrounding area, no volcanic deposits (e.g. ash layers) have been found. Nevertheless, numerous faults in this region (e.g. strike-slip faults; Hernandez *et al.* 1987) may have been pathways for hydrothermal solutions and provided silica. Whether upwelling of deeper nutrient-rich water may have played a role cannot be judged at present. This is because the palaeotopography is too poorly understood and it is not known whether deeper marine waters were able to reach the location of this basin within the Betic Strait.

The most dominant marine planktonic species in the diatom assemblages of the Varied Member, *T. nitzschioides*, has been described as a polyhalob and polytherme diatom species by Smayda (1958), who postulated a salinity range for this species from 4 to 45‰. As it is obviously a species that thrives in waters rich in nutrients, it is found in the upwelling zones of recent oceans (Richert 1975, off Africa; Schuette & Schrader 1979, off Peru) and appears to be common in the central and northern Pacific (Belyayeva 1970, 1971). Applying this knowledge to the Late Miocene diatomites of the Lorca Basin indicates that this was probably a nutrient-rich environment. The abundance fluctuations between *T. nitzschioides* and other marine holo- and mero-planktonic diatom species reflect repeated environmental changes from more normal marine to more stressful conditions. This is supported by the overall high abundance of resting cysts of the genus *Chaetoceros*. Species of this genus flourish particulary in nutrient-rich waters and react to unfavourable conditions by forming resting cysts.

More information on the palaeoecology has been obtained by interpreting the species diversity. The diversity of species within an ecological system reflects the environmental conditions and the stability of the environment. Actuopalaeontological studies identify the following possible reasons for fluctuations of diversity. The diversity of a biocoenosis is dependent on the number of occupiable ecological niches. The number of niches can be fixed primarily (articulation of the biotope), but there can be an increase in the course of ecological successions. Simpson (1964) described a relationship between an increase in diversity and the stability of the conditions of the surroundings. Sanders (1969) suggested that an increase in diversity depends on the duration of the ecological stability. Both workers described stress as a limiting factor. In the Varied Member the number of species per 300 counted valves is between 12 and 36. This low diversity appears to indicate a stressful environment. Another possibility to explain such a low-diversity diatom taphocoenosis is silica dissolution. This factor definitely plays a role throughout the Varied Member, as in its lower part chert layers occur. However, the repeated changes in the diatom assemblages are not bound to such changes in preservation or dissolution of biogenic silica (Fig. 6).

Also, the smaller-scale changes, which are correlated with lithological changes in the Varied Member (from diatomite to marl deposition), reflect repetitive changes in ecological conditions; e.g. in salinity and/or in the oxygen concentration of the sea water causing stagnant, anoxic conditions in the deeper parts of the basin. Diatoms, which live only in the uppermost layer of the water column (euphotic zone),

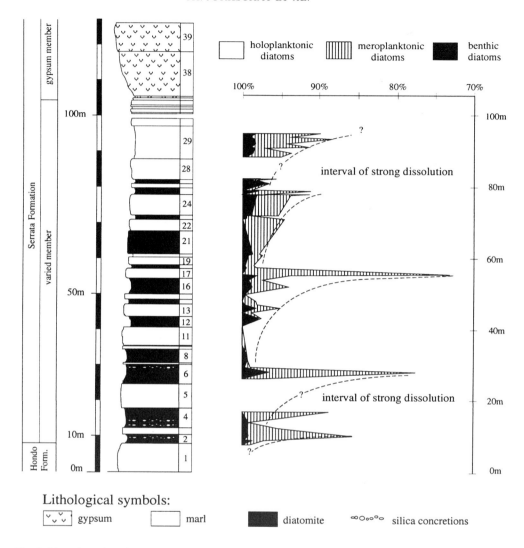

Fig. 6. Biogenic cycles reflected by abundance changes of ecologically significant diatom groups in the Varied Member.

are less affected by anoxic conditions if these affect only the deeper parts of the water column. Benthic organisms, as well as foraminifera, which live in the deeper levels of the water column, are severely affected and disappear when anoxic conditions appear (Jurkschat & Steffahn 1998).

The cyclical changes in abundance of the meroplanktonic diatom species (Fig. 6) are conspicuous, and overprint the systematic changes that correlate with the lithology. These cycles are inversely correlated with those of *T. nitzschioides* as well as with abundance fluctuations of diatom valves per gram of sediment.

They appear to be reflecting cyclical changes in environmental conditions. Diatom assemblages with high numbers of resting spores and meroplanktonic species represent more restricted conditions of the basin. The increase of holoplanktonic species characterizes an inflow of marine water re-establishing more normal marine conditions.

Environmental evolution

During the time of sedimentation of the lower part of the Varied Member frequent occurrences of anoxic conditions of the bottom water in the

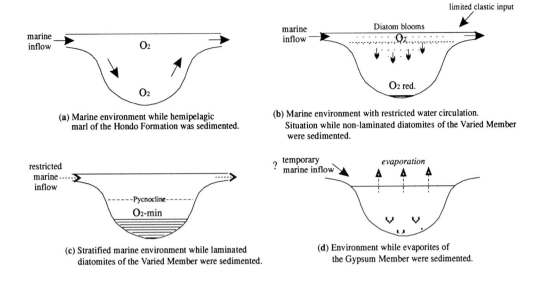

(a) Marine environment while hemipelagic
marl of the Hondo Formation was sedimented.

(b) Marine environment with restricted water circulation.
Situation while non-laminated diatomites of the Varied Member
were sedimented.

(c) Stratified marine environment while laminated
diatomites of the Varied Member were sedimented.

(d) Environment while evaporites of
the Gypsum Member were sedimented.

Fig. 7. Environmental changes in the Lorca Basin, a sedimentation model.

Lorca Basin are documented by the lamination of diatomite horizons as well as the lack of tests of benthic macro- and micro-organisms, absence of bioturbation and the very good preservation of fossil fish remains (Jurkschat 1995b). The presence of benthic lifeforms and/or carcass-feeding organisms would have caused a higher fragmentation of the fish skeletons found (Jurkschat 1995a).

Occurrence of anoxic conditions in the basin requires a barrier separating the basin from the rest of the Betic Strait, a stratified water column and, additionally, algal blooms (Fig. 7). Evaporation may have led to the formation of a halocline and a stratification of the water column, causing stagnation of the bottom water. The bacterial oxygenation of organic material induced an oxygen deficiency both in the bottom water and in the sediments (lamination).

In the middle and upper part of the Varied Member, laminated sediments are rare. This could be a result of the basin becoming too shallow and the wave base reached the basin floor resuspending the sediments, or there may have been better water replenishment and the barrier did not fully separate the depositional area.

Foraminiferal investigations have suggested a maximum basin depth of 180 m, because of the dominance of the genus *Nonion* (Steffahn 1995). This estimate of the palaeodepth corresponds to the rare occurrence of benthic diatoms, which must be derived from nearshore waters, and the

high abundance (up to 30%) of meroplanktonic diatom species, which live mainly in neritic regions. Most of these meroplanktonic species are able to form resting stages with resting spores (Hargraves & French 1983).

Conclusion

To conclude, the results of this study show that before the onset of the Messinian Salinity Crisis the Lorca Basin became repeatedly separated from the rest of the Betic Strait. After formation of a barrier, laminated sediments occurred during the deposition of the Varied Member whenever the deeper part of the basin was cut off from the circulation. Diatom assemblages show significant cyclical changes as a result of these periods of restriction and transgression of the Lorca Basin, indicating that there was not a progressive desiccation.

Special thanks are extended to the German Science Foundation for financial support (DFG-grant Mi 353/1-2). We are very grateful to the department B 3 of the Federal Institute for Geosciences and Natural Resources (BGR, Germany) for the possibility to use the diatom laboratory and the scanning electron microscope. We thank S. Stäger, B. Piesker, A. Bruns and K. Hoffmann (BGR, Germany) for technical support. We would like to thank J. Baldauf (Ocean Drilling Program, College Station, TX.), J. Barron (US Geological Survey, Menlo Park, CA), F. Akiba (Japex Research Centre, Chiba) and Y. Yanagisawa (Geological Survey Japan) for sending literature, and the last two for helpful comments during the study. Furthermore, we thank J. Baldauf for critically reading the manuscript, and J. Steffahn for many discussions and help in taking the samples. Helpful comments of anonymous reviewers are gratefully acknowledged.

References:

ADAMS, C. G., BENSON, R. H., KIDD, R. B., RYAN, W. B. F. & WRIGHT, R. C. 1977. The Messinian salinity crisis and evidence of the late Miocene eustatic changes in the world ocean. *Nature*, **269**, 383–386.

AKIBA, F. 1986. Middle Miocene to Quaternary diatom biostratigraphy in the Nankai Trough and Japan Trench and modified Lower Miocene through Quaternary diatom zones for middle-to-high latitudes of the North Pacific. *In*: KAGAMI, H., KARIG, D. E., BRAY, C. J. *et al.* (eds) *Initial Reports of the Deep Sea Drilling Project 87*. US Government Printing Office, Washington, DC, 393–481.

AZPEITIA, M. F. 1911. La Diatomología Española en los Comienzos del siglo XX. *Asociacion Española para el progreso de las Ciencas, Congreso de Zaragoza*, **4**, 1–320.

BARRON, J. A. 1985. Miocene to Holocene planktic diatoms. *In*: BOLLI, H. M., SAUNDERS, J. B. & PERCH-NIELSEN, K. (eds) *Plankton Stratigraphy*. Cambridge University Press, Cambridge, 763–809.

—— 1992. Diatoms. *In*: LIPPS, J. E. (ed.) *Fossil Prokaryotes and Protists*. Blackwell Science, Boston, MA, 155–167.

BATTARBEE, R. W. 1973. A new method for the estimation of absolute microfossil numbers, with reference especially to diatoms. *Limnology and Oceanography*, **18**, 647–652.

BELYAYEVA, T. V. 1970. Taxonomy and distribution patterns of planktonic diatoms in the equatorial Pacific. *Oceanology*, **10**, 101–107.

—— 1971. Quantitative distribution of planktonic diatoms in the western part of the tropical Pacific. *Oceanology*, **11**, 578–585.

BENALI, S., SCHREIBER, B. C., HELMAN, M. L. & PHILP, R. P. 1995. Characterization of organic matter from a restricted/evaporative sedimentary environment: Late Miocene of Lorca Basin, Southeastern Spain. *AAPG Bulletin*, **79**, 816–830.

BERGGREN, W. A. 1977. Late Neogene planktonic foraminiferal biostratigraphy of the Rio Grande Rise (South Atlantic). *Marine Micropaleontology*, **2**, 265–313.

BLOW, H. W. 1969. Late Middle Eocene to Recent planktonic foraminiferal biostratigraphy. *Proceeding of the 1st International Conference on Planktonic Microfossils*, Rome 1967, Vol. 1: 199–442.

BURCKLE, L. H. 1977. Diatom analysis of the pre-evaporitic facies in the neostratotype area of the Messinium. *Abstracts, Messinian Seminar*, Malaga, Vol. 3.

—— 1978a. Marine Diatoms. *In*: HAQ, B. U. & BOERSMA, A. (eds) *Introduction to Marine Micropaleontology*. Elsevier, New York, 245–266.

—— 1978b. Early Miocene to Pliocene diatom datum levels for the equatorial Pacific. *Geological Research and Development Centre, Special Publica-*

tions, **1**, 25-44.

—— & BISCAYE, P. 1971. Sediment transport by Antarctic Bottom Water through the eastern Rio Grande Rise. *Annual Meeting of the Geological Society of America*, **1971**, 518–519.

Bustillo, A. & López García, J. 1997. Age, distribution and composition of Miocene diatom bearing sediments in the Guadalquivir Basin, Spain. *Geobios*, **30**, 335–350.

COLALONGO, M. L., DI GRANDE, A. & D'ONOFRIO, S. 1979. A proposal for the Tortonian/Messinian boundary. *Annales Géologique des Pays Hélleniques*, **1**, 285–294.

DE CLERCQ, S. W. G., SMIT, J. & VEENSTRA, E. 1975. A marine tuffaceous sediment in the lower Miocene of the Velez Blanco–Lorca Region, Southeastern Spain. *Geologische Uitgaven Amsterdam Papers of Geology*, **1**, 105–114.

DITTERT, N., SCHRADER, S., SKOWRONEK, A., WROBEL, F. & MICHALZIK, D. 1993. Sedimentationszyklen im Messinium des Lorca-Beckens (SE-Spanien). *Zentralblatt für Geologie und Paläontologie, Teil I*, **1993**, 841–851.

D'ONOFRIO, S., GIANNELLI, L., IACCARINO, S. *et al.* 1975. Planktonic Foraminifera of the upper Miocene from some Italian sections and the problem of the lower boundary of the Messinian. *Bolletin della Società Paleontologica Italiana*, **14**, 177–196.

Ehrenberg, C. G. 1836. Über den aus mikroskopischen Kiesel-Organismen gebildeten Polirschiefer von Oran in Afrika. *Bericht der Verhandlung der königlich preußischen Akademie der Wissenschaften zu Berlin*, 59–61.

—— 1854. *Mikrogeologie. Das Erden und Felsen schaffende Wirken des unsichtbar kleinen selbstständigen Lebens auf der Erde*. Leopold Voss Verlag, Leipzig.

FENNER, J. 1982. *Diatoms in the Eocene and Oligocene Sediments off NW Africa, their stratigraphic and paleogeographic occurrences*. PhD thesis, University of Kiel.

FORTI, A. 1908. Primo elenco delle Diatomee fossili contenute nei calcari marnosi biancastri di Monte Gibbio (Sassuolo–Emilia). *Nuova Notarisia de rette di G.B. de Toni*, **XIX**, 3–8.

—— 1913. Contribuzioni diatomologiche XIII. Diagnoses diatomacearum quarundam fossilium italicarum. *Atti dell Real Istituto Veneto di Scienza, Lettere ed Arti*, **72**(2), 1535–1701.

GARCíA-VEIGAS, J., ORTí, F., ROSELL, L. & INGLES, M. 1994. Caracterización petrológica y geoquímica de la Unidad Salina messiniense de la cuenca de Lorca (sondeos S4 y S5). *Geogaceta*, **15**, 78–81.

GAUDANT, J. 1995. Nouvelles recherches sur l'ichtyofaune messinienne des environs de Lorca (Murcia, Espagne). *Revista Española de Paleontología*, **10**, 175–189.

GEEL, T. 1976. Messinian gypsiferous deposits of the Lorca Basin (Province of Murcia, SE Spain). *Memorie della Società Geologica Italiana*, **16**, 369–385.

GERSONDE, R. 1980. *Paläoökologische und biostratigraphische Auswertung von Diatomeenassoziationen aus dem Messinium des Caltanissetta-Beckens (Sizi-*

lien) und einiger Vergleichsprofile in SO-Spanien, NW-Algerien und auf Kreta. PhD thesis, University of Kiel.

HARGRAVES, P. E. & FRENCH, F. W. 1983. Diatom resting spores: significance and stratigies. *In*: FRYXELL, G. A. (ed.) *Survival Strategies of the Algae*, Cambridge University Press, Cambridge, 49–68.

HERNANDEZ, J., DE LAROUZIERE, F. D., BOLZE, J. & BORDET, P. 1987. Le magmatisme néogène bético-rifain et le couloir de déchrochement trans-Alboran. *Bulletin de la Societé géologique de France*, **8**, 257–267.

HSÜ, K. J., CITA, M. B. & RYAN, W. B. F. 1973. The origin of the mediterranean evaporites. *In*: RYAN, W. B. F., HSÜ, K. J., CITA, M. B. *et al.* (eds) *Initial Reports of the Deep Sea Drilling Project, 13*. US Government Printing Office Washington, DC, 1203–1231.

——, MONTADERT, L., BERNOUILLI, D. *et al.* 1987. History of the Mediterranean Salinity Crisis. *Initial Reports of the Deep Sea Drilling Project 42A*, US Government Printing Office, Washington, DC.

HUSTEDT, F. 1953. *Kieselalgen (Diatomeen)*. Kosmos Verlag, Stuttgart.

IACCARINO, S. 1985. Mediterranean Miocene and Pliocene planktic Foraminifera. *In* BOLLI, H. M., SAUNDERS, J. B. & PERCH-NIELSEN, K. (eds) *Plankton Stratigraphy*. Cambridge University Press, Cambridge, 283–314.

IGME 1982. *Ampliación de pizzaras bituminosas en la zona de Lorca (Murcia) (Fase II). Inscripciones n. Lorca 134 (Murcia) y Lorca biz 155 (Murcia)*. Instituto Geologico y Minero de España, Madrid.

JURKSCHAT, TH. 1995a. *Paläoökologische und biostratigraphische Untersuchungen an Diatomeen aus dem obersten Miozän des Lorca-Beckens (Provinz Murcia, SE-Spanien)*. Diploma thesis, University of Hannover.

—— 1995b. *Geologische Kartierung der nördlichen Umgebung der Stadt Lorca (Provinz Murcia, SE-Spanien)*. Diploma thesis, University of Hannover.

—— & STEFFAHN, J. 1998. Paleoecological investigation on diatoms and foraminifera from the Late Miocene Serrata section (Lorca Basin/SE Spain). *15th Sedimentological Congress in Alicante, Spain*, 456–457.

KOIZUMI, I. 1985. Diatom biochronology for the Late Cenozoic NW Pacific. *Journal of the Geological Society of Japan*, **91**(3), 195–211.

LISITZIN, A. P. 1971. Distribution of silicious micro-fossils in suspension and in bottom sediments. *In*: FUNELL, B. M. & RIEDEL, W. R. (eds) *The Micropaleontology of Oceans*. Cambridge University Press, Cambridge, 173–195.

MÄKEL, G. H. 1985. The geology of the Malaguide Complex and its bearing on the geodynamic evolution of the Betic–Rif orogen (southern Spain and northern Morocco). *Geologische Uitgaven Amsterdam Papers on Geology*, **22**.

MANUPUTTY, J. A. 1977. The late Miocene planktonic foraminiferal associations of the Lorca Basin (Province of Murcia, SE Spain). *Field Trip Guidebook, Messinian Seminar*, Vol. 3, 21–29.

MICHALZIK, D. 1996. Lithofacies, diagenetic spectra and sedimentary cycles of Messinian (Late Miocene) evaporites in SE Spain. *Sedimentary Geology*, **106**, 203–222.

MONJANEL, A.-L. 1984. *Les diatomées oligocène à holocènes de l'Atlantique nord et de la Méditerranée occidentale: biostratigraphie et paléocéanographie*. PhD thesis, Université de la Bretagne, Brest.

MONTENAT, C. 1977. *Les bassins néogènes de Levant d'Alicante et de Murcia (Cordillères bétiques orientales—Espagne). Stratigraphie, paléogéographie et évolution dynamique*. Documents des Laboratoires de Geologie de la Faculté des Sciences Lyon, **69**.

——, MASSE, P., COPPIER, G. & OTT D'ESTEVOU, P. 1990b. The sequence of deformation in the Betic Shear Zone (S.E. Spain).- *Annales Tectonicae*, **4**(2), 96–103.

——, OTT D'ESTEVOU, P., LAROUZIÈRE, F. D. & BEDU, P. 1987a. Originalité géodynamique des bassins néogènes du domaine bétique oriental (Espagne). *Notes et Mémoires TOTAL Compagnie Française des Pétroles*, **21**, 11–50.

—— & MASSE, P. 1987b. Tectonic–sedimentary characters of the Betic Neogene basins evolving in a crustal transcurrent shear zone (SE Spain). *Bulletin du Centre de Recherches Exploration–Production Elf-Aquitaine*, **11**(1) 1–22.

——, —— & DELORT, T. 1990a. Le bassin de Lorca. *Documents et Travaux, Institut Géologique Albert de Lapparent*, **12–13**, 261–280.

MÜLLER, D. W. & HSÜ, K. J. 1987. Event stratigraphy and paleoceanography in the Fortuna Basin (Southeast Spain): a scenario for the Messinian Salinity Crisis. *Paleoceanography*, **2**, 679–696.

—— & SCHRADER, H.-J. 1989. Diatoms of the Fortuna Basin, southeast Spain: evidence for the intra-Messinian inundation. *Paleoceanography*, **4**, 75–86

NESS, G., LEVI, S. & CROUCH, R. 1980. Marine magnetic anomaly timescale for the Cenozoic and late Cretaceous: a précis, critique, and synthesis. *Review of Geophysics and Space Physics*, **18**, 753–770.

PAQUET, J. 1974. Tectonique Éocène dans la Cordillères Bétiques: vers une nouvelle conception de la paléogéographie en Méditerranée occidentale. *Bulletin de la Société Géologique de France*, **16**, 58–73.

PLATT, J. P. & VISSERS, R. L. M. 1989. Extensional collapse of thickened continental lithosphere: a working hypothesis for the Alboran Sea and the Gibraltar arc. *Geology*, **17**, 540–543.

RICHERT, P. 1975. *Die räumliche Verteilung und zeitliche Entwicklung des Phytoplanktons, mit besonderer Berücksichtigung der Diatomeen, im NW-afrikanischen Auftriebswassergebiet*. PhD thesis, University of Kiel.

RODRÍGUEZ ESTRELLA, T., MANCHEÑO, M. A., GUILLÉN MONDÉJAR, F., LÓPEZ-AGUAYO, F., ARANA, R., FERNÁNDEZ TAPIA, M. T. & SERRANO, F. 1992. Tectonica y sedimentacion neogena en la cuenca de Lorca (Murcia). *III. Congreso Geologico de España y VIII. Congreso Latinoamerica de Geologia, Salamanca, Actas, Vol. 1*, 201–206.

ROUCHY, J. M. 1982. *La genèse des évaporites messiniennes de Méditerrannée*. Memoires de la

Muséum Nationale de l'Histoire Naturelle, **50**.

———, TABERNER, C., BLANC-VALLERON, M.-M. *et al.* 1998. Sedimentary and diagenetic markers of the restriction in a marine basin: the Lorca Basin (SE Spain) during the Messinian. *Sedimentary Geology*, **121**, 23–55.

RYAN, W. B. F., HSÜ, K. J., CITA, M. B. *et al.* 1973. *Initial Reports of the Deep Sea Drilling Project, 13*. US Government Printing Office Washington, DC.

SAITO, T., BURCKLE, L. H. & HAYS, J. D. 1975. Late Miocene to Pleistocene biostratigraphy of equatorial Pacific sediments.- *In*: SAITO, T. & BURCKLE, L. H. (eds) *Late Neogene Epoch Boundaries*. Micropaleontology, Special Publication **1**, New York, 226–244.

SANDERS, H. L. 1969. Benthic marine diversity and the Stability–Time hypothesis. *Brookhaven Symposia in Biology; Diversity and Stability in Ecological Systems*, **22**, 71–81.

SANTISTEBAN, C. & TABERNA, C. 1983. Shallow marine and continental conglomerates derived from coral reef complexes after desiccation of the deep marine basin: the Tortonian–Messinian deposits of the Fortuna Basin, SE Spain. *Journal of the Geological Society, London*, **140**, 401–411.

SANZ DE GALDEANO, C. 1990. Geological evolution of the Betic Cordilleras in the Western Mediterranean, Miocene to present. *Tectonophysics*, **172**, 107–119.

——— & VERA, J. A. 1992. Stratigraphic record and palaeogeographical context of the Neogene basins in the Betic Cordillera, Spain. *Basin Research*, **4**, 21–36.

SCHRADER, H.-J. 1973. Cenozoic diatom from the Northeast Pacific Leg 18. *In*: KULM, L. D., VON HUENE, R., DUNCAN, J. R. *et al.* (eds) *Initial Reports of the Deep Sea Drilling Project, 18*, US Government Printing Office, Washington, DC, 673–797.

——— 1974. Cenozoic marine planktonic diatom biostratigraphy of the tropical Indian Ocean. *In*: FISHER, R. L., BUNCE, E. T., CERNOCK, P. J. *et al.* (eds) *Initial Reports of the Deep Sea Drilling Project 24*, US Government Printing Office, Washington, DC, 887–967.

SCHREIBER, B. C., FRIEDMAN, G. M., DECIMA, A. & SCHREIBER, E. 1976. Depositional environments of Upper Miocene (Messinian) evaporite deposits of the Sicilian Basin. *Sedimentology*, **23**, 729–760.

SCHUETTE, G. & SCHRADER, H.-J. 1979. Diatom taphocoenoses in the coastal upwelling area off Western South America. *Nova Hedwigia, Beihefte*.

SILVA, P. G., GOY, J. L., SOMOZA, L., ZAZO, C. & BARDAJI, T. 1993. Landscape response to strike-slip faulting linked to collisional settings: Quaternary tectonics and basin formation in the eastern Betics, southeastern Spain. *Tectonophysics*, **224**, 289–304.

SIMPSON, G. G. 1964. Species density of North America recent mammals. *Systematical Zoology*, **13**, 57–73.

SMAYDA, T. I. 1958. Biogeographical studies of marine phytoplankton. *Oikos*, **9**(11), 158–191.

STEFFAHN, J. 1995. *Geologische Kartierung der nordöstlichen Umgebung der Stadt Lorca und Untersuchungen an Foraminiferen der obermiozänen Serrata-Formation im Lorca-Becken (Provinz Murcia/SE-Spanien)*. Diploma thesis, University of Hannover.

VAN DER BEEK, P. A. & CLOETINGH, S. 1992. Lithospheric flexure and the tectonic evolution of the Betic Cordilleras (SE Spain). *Tectonophysics*, **203**, 325–344.

WEIJERMARS, R. 1991. Geology and tectonics of the Betic Zone, SE Spain. *Earth-Science Review*, **31**, 153–236.

WORNARD, W. W. 1973. Correlation of the neostratotype of the Messinian stage Capodarso and Pasquaia Sections based on diatoms. *Nova Hedwigia, Beihefte*, **45**, 391–402.

WROBEL, F. 1993. *Das Neogen der nordwestlichen Umgebung von Totana (Provinz Murcia, SE Spanien)*. Diploma thesis, University of Hannover.

YANAGISAWA, Y. 1994. *KOIZUMIA YANAGISAWA* gen. nov., a new marine fossil araphid diatom genus. *Transactions and Proceedings of the Palaeontological Society of Japan, New Series*, **176**, 591-617.

———& AKIBA, F. 1990. Taxonomy and phylogeny of the three marine diatom genera, *Crucidenticula, Denticulopsis* and *Neodenticula*. *Bulletin of the Geological Survey of Japan*, **41**(5), 197–301.

ZACHARIASSE, W. J. 1979. The origin of *Globorotalia conomiozea* in the Mediterranean and the value of its entry level in biostratigraphic correlations. *Annales Géologique des Pays Helleniqués*, **3**, 1281–1292.

Late Miocene climatic cycles and their effect on sedimentation (west Hungary)

M. KORPÁS-HÓDI[1], E. NAGY[1], E. NAGY-BODOR[1], K. SZÉKVÖLGYI[2] &
L. Ó. KOVÁCS[3]

[1]Geological Institute of Hungary, PO Box 106, H-1442 Budapest, Hungary
[2]Enricom Kft, H-1132 Budapest, Váci út 50, Hungary
[3]Hungarian Geological Survey, PO Box 17, H-1440 Budapest, Hungary

Abstract: On the basis of sedimentological, well-log, palaeontological and pedological analyses of three Upper Miocene borehole sections from Hungary, a cyclic change of sediments from deltaic and alluvial depositional systems is presented, which is considered to have been driven by a cyclical change of the climate (temperature and precipitation). A delayed but positive correlation between the sedimentary cyclicity linked with fourth order sedimentary sequences, and a long-term cyclicity in temperature have been observed. Within this cyclicity, higher-order sedimentary changes are seen. They are interpreted as fifth order sedimentary cycles related, in turn, to short-term climate phases. Within a long-term cycle, four short-term climate phases are defined as a result of changes in temperature and precipitation: semi-arid, turning from warm-temperate to cold-temperate; humid, cold-temperate; warm-temperate with a warming trend, with average but decreasing precipitation; and warm-temperate, with average precipitation.

Sporadic information on the climate that prevailed during Late Miocene times in the area of Hungary has existed for some time in the relevant literature. Palaeobotanical (Pálfalvy 1952; Andreánszky 1955; Nagy & Pálfalvy 1961) and vertebrate-related palaeontological (Kretzoi 1969, 1983; Kretzoi et al. 1976) data supplied the first sporadic interpretations. Palynology was the first to allow an investigation of changes in climate over time, from a profile, by continuous sampling (Nagy 1958, 1970, 1992; Planderova 1962; Nagy & Planderova 1985). As a result of these investigations, three climate zones were distinguished (Nagy 1993). In recent years, local and calculated mean temperature curves for two Upper Miocene borehole profiles (boreholes Berhida-3 and Szombathely-II) were completed on the basis of changes in the structure of the taphocoenosis of sporomorphs (Nagy & Ó. Kovács 1996; Nagy-Bodor in Korpás-Hódi & Bohn-Havas 1998), and the temperature curve of the Szombathely borehole was presented as a long-term cyclic change in temperature during the Late Miocene.

The change in thicknesses of sand and lignite layers in Upper Miocene sections from Hungary also reveals cyclicity of a different order (Juhász et al. 1998) which can be correlated with the Milankovitch–Bacsák orbital cycles. The authors indicated that precession (19 ka) and obliquity (50 ka) caused climate effects that can be determined, for the most part, in the sediments but that the climate changes caused by longer periods of eccentricity (400 ka) are poorly detectable. Nevertheless, using seismic stratigraphy the third and fourth order depositional sequences can be regionally identified and related to climate changes (Pogácsás et al. 1990; Vakarcs et al. 1994). The sedimentological and sequence stratigraphical studies indicated above did not analyse the climate but interpreted the structure of the sedimentary cycles as an effect of the climate.

This paper aims to present the long-term change of Late Miocene climate in the Pannonian Basin through borehole profiles. Using palynological and pedological data the temperature and precipitation changes will be presented and the close correlation between climate and sedimentation cycles will be demonstrated.

Methods

The samples used in this investigation came from the exploratory wells of Nagylózs-1, Szombathely-II and Berhida-3 (Fig. 1) which

From: HART, M. B. (ed.) *Climates: Past and Present*. Geological Society, London, Special
Publications, **181**, 79–88, 1-86239-075-4/$15.00

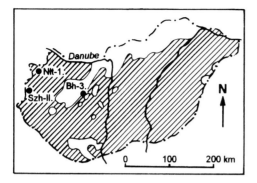

Fig. 1. Location of boreholes. Dashed line shows the national border of Hungary; hatching indicates the extent of deposits of the Late Miocene Pannonian brackish lake; Szh-II, Borehole Szombathely-II (basinal environment); Nlt-1, Borehole Nagylózs-1 (basinal environment); Bh-3, Borehole Berhida-3 (basin margin environment).

had previously been drilled for other purposes. In the deeper, lithologically uniform, sections of the boreholes the sampling interval was 10 m on average. In the upper sections sampling was carried out from each bed for grain size distribution determinations but only from the pelitic sediments for palynological analyses. The number of palynologically interpretable samples was reduced by the fact that sporomorphs were occasionally oxidized during burial or fossilization. A pedological analysis of only the upper 700 m thick section of the Nagylózs-1 borehole was carried out for evaluation of precipitation.

Calculation of temperature

In order to calculate the mean annual temperatures the simplified method of Nagy & Ó. Kovács (1996) was used. Sporomorphs were evaluated according to their climatic requirements and were assigned to three groups – tropical, sub-tropical and temperate climate. Cosmopolitan species were eliminated from the database. For each sample the average temperature values were calculated using a weighted average calculation based on the following formula:

$$T_{est} = \frac{A_{trop} \times T_{trop} + A_{subtr} \times T_{subtr} + A_{temp} \times T_{temp}}{A_{trop} + A_{subtr} + A_{temp}}$$

where T_{est} is the mean annual temperature calculated for the given sample, A_{trop}, A_{subtr}

and A_{temp} are the number of individuals of species of different climatic requirements and T_{trop}, T_{subtr} and T_{temp} are the mean temperatures of different climate belts.

A major source of error in the estimated mean temperature values may be the fact that the percentages of the sporomorph species in the taphocoenoses are dependent on their transportability and the pollen production capability of the area. In order to assess the level of inaccuracy of the calculations, two different methodologies are used to evaluate the local mean temperature values for a profile taken from borehole Berhida-3 (Fig. 2). The first method is based on the percentage distribution of temperate, sub-tropical and tropical species in the sporomorph taphocoenoses. The second method takes into consideration the ease of transport of species that have a greater pollen production capability. In the Late Miocene of Hungary, two groups require a temperate climate and, therefore, their percentage distribution in taphocoenoses considerably exceeds the role they play in climate indication. For this reason, the occurrences were only taken into consideration in one-fifth for Coniferae, and in one-third for *Sparganum* (Nagy & Ó.Kovács 1996). In Fig. 2 it can be seen that the two different ways of calculation may result in a difference of up to 2°C at particular sample points, but that they have no influence on the overall trend of the mean temperature curves gained by smoothing with a five-point moving average. Considering also that a breakdown of the temperate group is rather subjective, the first calculation method was used when the profiles taken from boreholes Nagylózs-1 and Szombathely-II were examined with regard to temperature.

The numerical mean temperature values are considered to be local since these reflect the effects of local microclimate. The temperature values are thought to be relative, since the calculated values express not the concrete annual mean temperature of the studied period but indicate only the character of change with time. The trend of long-term change of the annual mean temperature was determined by smoothing with a five-point moving average. The applicability of this mathematical method requires data with distance of the same size. In our example the deeper parts of the borehole profiles fulfil this precondition whereas the upper part does not. Therefore, in Figs 3, 5 & 6 manually corrected versions of the calculated curves are used. The individual calculated mean temperature values are also given in all figures.

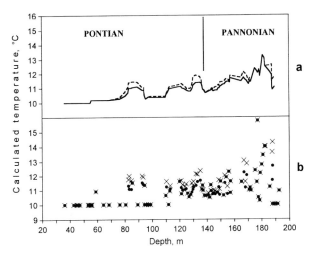

Fig. 2. Two different evaluations of mean temperature from the taphocoenoses of sporomorphs in the Berhida-3 borehole. In method 1, all sporomorphs are used with the same weighting while, in method 2, the weights for Coniferae and *Sparganum* are reduced. The temperature is calculated as the weigthed average of the optimum living temperature of the sporomorphs found. (**a**) Curves smoothed with a five-point moving average, for mean temperature calculated using method 1 (continuous curve) and 2 (dashed curve). (**b**) Calculated mean temperature values for individual samples using method 1 (•), and 2 (×).

Calculation of precipitation

According to Retallack (1994), the depth of $CaCO_3$ level that was leached out and then precipitated during soil development can be correlated with rainfall. That is, a greater amount of rainfall corresponds to a greater depth of calcitic horizon measured from the soil surface. The quantity of precipitation can be calculated according to the formula:

$$P = 139.6 - 6.388D - 0.01303D^2$$

where P is the annual precipitation (in mm) and D is the depth of pedogenic carbonate horizon from the top of the palaeosol horizon (in cm). The correlation coefficient (r) is 0.79; the dispersion is 141 mm.

The value of the calculated precipitation is reduced when the palaeosol profile is compacted or eroded and is increased when the level of atmospheric carbon dioxide is considerably different from that of the present. According to Crowley & North (1991), the atmospheric CO_2 concentration in the Neogene would have been very similar to that of today. Thus, based on the formula of Retallack, the calculation of precipitation can be carried out for the Late Miocene palaeosols. In our database the eroded palaeosol profiles are omitted, with only the complete palaeosol profiles being considered. The measure of palaeosol compaction proved to

be negligible and this factor was disregarded in our calculations. Consequently, the calculated values of precipitation can be regarded as only approximate values.

Magnetostratigraphical correlation

Magnetostratigraphical and biostratigraphical methods were used to correlate and determine the duration of the cycles. The magnetic polarity zones (Fig. 3) in the three borehole profiles show a very different pattern (Lantos et al. 1992; Lantos, pers. comm.). The boundaries for zone C5n can be identified in all of the three profiles with a high degree of certainty, although the correlation of the younger zones with the standard scale of Berggren et al. (1995) is rather unclear due to the lack of a marine fauna and insufficient radiometric data. The date of this upper section can be determined by the lack of the *Prosodacnomya* fauna. The occurrence of the genus *Prosodacnomya* is dated as 7.5 ± 0.5 Ma (Müller & Magyar 1992). In the boreholes Nagylózs-1, Szombathely-II and Berhida-3 this otherwise frequent genus is lacking in the delta–fluvial-plain sediments. In these sediments an older fauna is found. Consequently, the normal polarity section (520–310 m) of the Nagylózs-1 borehole, the section below 440 m in the Szombathely-II borehole and the section below 60 m in the Berhida-3 borehole can only be determined as older than the C4n polarity zone.

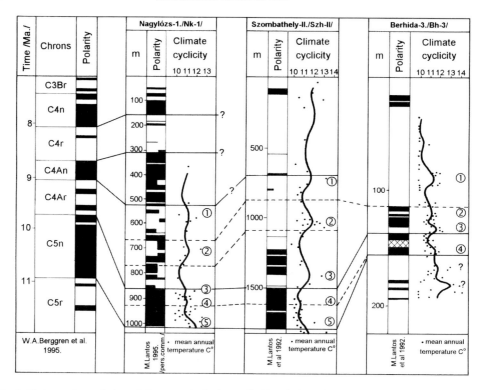

Fig. 3. Timing and correlation of climate cycles in the studied boreholes. The climate cycles are correlated on the basis of the magnetostratigraphy (modified after Lantos, pers. comm.; Lantos *et al.* 1992) and biostratigraphy (Sütő-Szentai, pers. comm.) identified by numbers.

Sedimentation

At the beginning of the Late Miocene the Pannonian Basin was a brackish-water lake and the basin was filled from the margins towards the centre by prograding deltas. The sedimentary sequences of the three boreholes preserve a record of this filling process. The Szombathely-II (1811–2 m) and Nagylózs-1 (1031–15 m) boreholes record a more basinal succession while the Berhida-3 (223–6 m) borehole penetrated more marginal sediments.

In the Szombathely-II borehole the sediments between 1811 m and 1003 m (lower section) record a lacustrine environment. As a result of delta progradation on the shoreline a thick prodelta succession was developed followed by a delta-front depositional system. Above 1003 m the sediments represent a delta–flood plain environment. The Upper Miocene sediments of the borehole rest on the Middle Miocene succession on top of an erosional unconformity. The stratigraphic position of the succession 2 m below the unconformity surface is debated as this displays transitional features. Above the

basal sandstone–conglomerate, about 800 m of (predominantly) mudstone is found with thin sandstone intercalations. At 1003 m the pelitic sedimentation is terminated by a thick sand layer with an erosional unconformity. Subsequently, the borehole succession consists of an alternation of sandstone, siltstone, claystone and lignite (Phillips *et al.* 1989). In the lower section (1811–1600 m) of the profile deep-water sublittoral environments are represented. The faunal succession records this decrease in the relative water depth and, between 1062 m and 1003 m, the sediments of the delta front were deposited in shallow, sublittoral environments. Above 1003 m the sediments of the delta–fluviatile plain were deposited in a littoral to supralittoral environment (Korpás-Hódi 1992).

The Nagylózs-1 borehole records a similar sedimentary sequence, although it is notable that the basal sand–conglomerate sequence is missing. Based on lithological and sedimentological evidence neither the unbroken sedimentation nor the phytoplankton zones indicate the presence of a hiatus (Sütő-Szentai, pers. comm.). Smaller erosional surfaces can be observed only at the bases of the thin turbidites, the channel deposits and on the surfaces of the palaeosol horizons. Based on the molluscan fauna, the section between 1024 m and 1017 m probably contains a break in sedimentation. In the borehole the characteristic mudstone lacustrine–prodelta sedimentation was replaced, at 785 m, by the delta front (sand and aleurite strata) and then, from 646 m, the sediments represent a variable delta–fluvial plain environment (sandstone, siltstone, claystone, variegated clay, lignite).

The succession in the Berhida-3 borehole (Kókay *et al.* 1991) differs markedly from that of the other boreholes. The sediments represent near-shore, littoral lacustrine–lagoonal and terrestrial environments. At the base of the Upper Miocene there is an alternating succession of clays and marls with root-imprints, aleuritic clays, gravels and calcareous marls (223–190 m). These are unconformably overlain by a claystone sequence. Above 168 m the sediment becomes coarser and the proportion of aleurite and sand increases. Above 31 m detrital sedimentation is replaced by lime-muds and marls. The Upper Miocene sediments are interpreted as representing a lagoonal environment with a water depth of several metres. This is followed by a relative increase in water depth and the development of a shallow, sub-littoral lacustrine environment (190–168 m). Above 168 m the decrease in relative water depth generated sediments that were deposited above wave base. From 31 m the sediments represent a supralittoral, freshwater, terrestrial environment.

Changes of temperature and precipitation

Relative and local mean temperature values were calculated on the basis of sporomorph analysis of the three boreholes, Nagylózs-1, Szombathely-II and Berhida-3. The curves of relative and local temperature changes drawn by averaging the temperature values of each sample show the repetition of shorter warm and longer cool periods (Fig. 3). These repeating phases are interpreted as climate cycles. The cyclic change of temperature based on numeric evaluation is also supported by the sporomorph index fossils. The occurrence of species of *Engelhardtia*, *Sapotacea*, *Reevisia*, *Zelkova*, *Sciadopitys* and *Rhus* indicates a warmer period, while cooling periods are associated with the predominance of species of *Quercus*, *Carpinus*, *Fagus*, *Alnus*, *Sparganium* and *Betula*.

The thickness (in metres) of the climate cycles varies both in different boreholes and within a single borehole. It follows from the sediment infilling process of the basin that the rate of sedimentation is slower in the deeper stratigraphic horizons (pelagic sedimentation), and faster in the upper section (littoral, onshore sedimentation). The differences between the boreholes are a consequence of the structurally defined local subsidence rates.

The duration of the postulated climate cycles can only be estimated. A magnetostratigraphic interpretation of the polarity zones in borehole Nagylózs-1 (modified after Lantos, pers. comm. 1995) was used to determine the periodicity of the cycles which was eventually estimated as *c.* 400 ka. In Fig. 3 it can be seen that the warming and cooling sections of the climatic cycles are in the same positions within the polarity and biostratigraphic zones. According to our interpretation, the illustrated cycles of temperature variations are in phase with one another and their periodicity corresponds to the orbital cycle of *c.* 400 ka.

Estimates of precipitation values are available from a roughly 300 m thick section of the Nagylózs-1 borehole (Fig. 4). In this borehole, above 450 m depth a period of 300–400 mm annual precipitation is followed by a drier phase with 200–250 mm annual precipitation and a 'rainy' phase of a maximum 1000 mm annual precipitation can be postulated. At a depth of 190 m the precipitation decreases to that observed at 450 m. The calculated values of precipitation shown in Fig. 4. Indicate the possibility of cyclic change of precipitation. Within a single cycle, low, medium, high, and then a decreasing mean rainfall can be suggested (Fig. 4).

Because of the infrequent sampling interval and the sporomorph oxidation we cannot prove the above statement by palynological investigations from the upper 500 m thick section of the Nagylózs-1 borehole. Nevertheless, the palynological studies performed in the three boreholes show that, as well as the temperature cycles, the distribution of species indicates dry and rainy climatic variations. The occurrence of drought-resistant species such as *Ephedra* and members of the Compositae is related to the warm, and from warm to cooling climate phases, while the cool and from cool to warming phases are associated with the abundance of species of *Alnus* and *Salix* that require a rainy climate. In the current state of research the relationship between rainfall and mean temperature variations can be examined indirectly, through the experience discussed above and the correlation of cyclic changes in the sediments and the temperature (Fig. 4).

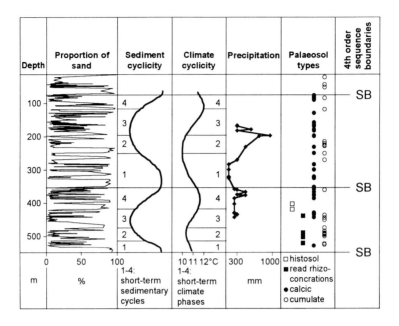

Fig. 4. Correlation of the cyclic changes in the sediment, estimated temperature, amount of precipitation and types of palaeosol in the alluvial plain depositional system (Nagylózs-1 borehole).

Cyclicity of sediments and the sedimentary environment

Sedimentary sequences were designated on the basis of well-logs, together with high-resolution sedimentological and facies analyses. Results from our investigations show that the sedimentary structures of fourth order sequences display a cyclic repetition of deltaic and alluvial environments. In Fig. 5, the cyclic changes in the sediment, the sedimentary environment and climate are illustrated using borehole Szombathely-II as an example, in the delta–fluvial plain environment. The percentage values of the sand fraction, smoothed with a five-point moving window, are used to indicate the changes in the sediment. The curve for the sedimentary cycle shows how the deposits become finer, then coarser, omitting any of the minor changes that are related to lateral movements of the river. Phytoplankton, ostracod (Sütö–Szentai & Szuromi-Korecz, pers. comm.) and molluscan (Korpás-Hódi 1992) investigations provide additional information on the sedimentary environment, including changes in salinity and relative water depth. Figure 5 also shows the calculated mean temperature values, a local climate curve gained by averaging the individual temperature values and, finally, the cyclic varia-

tion of the sedimentary environment (after Phillips *et al.* 1989). The boundaries of the sequences are marked by the major channel deposits. In the lower part of the sequence the deposits becomes finer whereas, in the upper part, they become coarser, with an intervening aggradation section. In the lower part, after an inundation by fluvial water, well-developed lateral accretion and fining-upwards bed-stackings are typical. In the upper part of the succession, vertical accretion and coarsening bed-stackings reflect the progradation of the river. Within these fourth order sedimentary sequences, four sedimentary cycles can be recognized. The beginning of each cycle is related to the progradation of a river.

The changes in the biota parallel the cyclicity of the sedimentation. The limnic and terrestrial species found near the boundaries of the sequences are followed by brackish ones in the middle of the sequences that indicate a cyclical variation in the position of the shoreline. In the first sequence (1003 to 611 m) shown in Fig. 5, the transgression of the brackish lake is visible over the limnic, fluvial facies that records the boundary; a shoreline–lower delta-plain facies was formed. In the upper part of the sequence, the progradation of fluvio-lacustrine facies can be identified. In the subsequent sequence (611 to 240 m), the changes in facies are the same but the

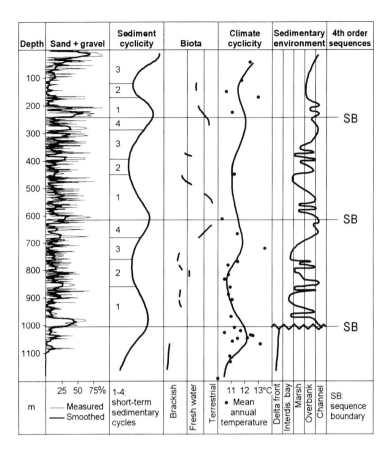

Fig. 5. Cyclical changes in sediment, sedimentary environment and climate in the delta–fluvial-plain sedimentary environment (Szombathely-II borehole).

importance of limnic and terrestrial sediments has increased as compared to those deposited in brackish environments. In the uppermost sequence (above 240 m), only species indicating a terrestrial–limnic environment were found.

Borehole Nagylózs-1 (Fig. 6) can be used to show the changes in environment from brackish, basin floor to fluvio-lacustrine. Sediments in the lower part of the borehole were deposited on a basin floor followed by prodelta, delta-slope and delta-front environments (1031 to 646 m). A delta-plain environment (646 to 550 m) was then formed and this was followed by a fluvio-lacustrine environment (550–15 m). The cyclical changes in sedimentation are shown through the percentage variations of the sand content, and the well-log data (Fig. 6). Here the cycle boundary also coincides with the boundary of a fourth order sedimentary sequence. The boundary is marked by the coarsest sediment

which terminates the coarsening bed-stackings and, above which, typically fining-upward bed-stacking sequences are found. The cyclic pattern is seen more clearly in the shallow-water environments of the delta-slope–front–plain (two cycles in the range of 875 to 550 m) than in the deeper-water sediments (1031–875 m). In shallow-water deltaic and alluvial environments (above 875 m) within the fourth order sequences, higher-order cycles marked by the progradation of a river can be observed (Fig. 4). Palaeontological investigations have detected the cyclical variation of the environment in the shallow-water onshore delta facies but not in the deep-water section. In the interval between 760 m and 500 m, the cyclical changes in the sedimentary environment are shown by the change in the structure of taphocoenoses of molluscs and phytoplankton (Sütö-Szentai, pers. comm.).

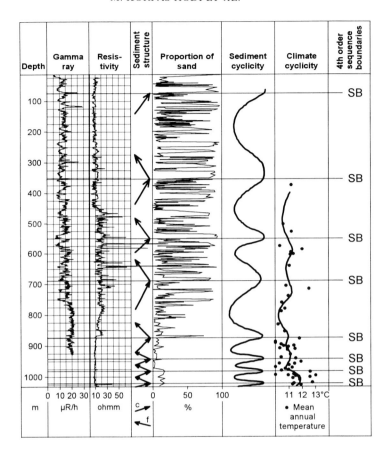

Fig. 6. Cyclical changes in sediment and climate in the different depositional systems (Nagylózs-1 borehole). c, coarsening upwards; f, fining upwards.

Correlation of sedimentation and climate cycles

In Figs 5 and 6 the relationship between the cyclical changes in sedimentation and mean temperature can be seen in the two borehole logs. It is clearly shown that the change in sedimentation follows, with a time delay, the change in temperature. This phase shift is probably due to the combined influence of changes in precipitation and temperature.

The data on rainfall distribution were obtained from pedological tests using the Nagylózs-1 borehole. From the same section of the borehole, however, no data on temperature are available. If one extends the relationship between sediment and temperature seen elsewhere to this part of the borehole, a pattern of temperature change can be determined. In Fig.

4, the cyclical variation of fluvial-plain sedimentation is shown, together with the changes in precipitation and estimated temperature. The precipitation and temperature conditions identify four short-term climate phases (Fig. 4). In the Nagylózs-1 borehole (alluvial-plain environment) four fluvial sediment cycles are identified. Each cycle is interpreted as being the result of an individual short-term climate phase.

The sedimentary cycles are as follows:

- sections 550–513 m and 351–251 m represent a semi-arid, cooling warm-temperate climate, characterized by sand layers with a fining-upward grain size, decreasing thicknesses, poorly developed overbank sediments and a calcitic soil. Higher in the succession, sediments of a prograding river are the result of an increase in precipitation. The sandbody is characterized by

coarse-grained sand at the base and lateral accretion at the top.
- sections 513–474 m and 250–196 m represent a humid, cold-temperate climate and are characterized by thicker overbank deposits, fewer channel deposits and cumulate soil horizons in mudstone and siltstone.
- sections 474–418 m and 196–118 m represent a warming climate with an average, but decreasing, precipitation and are characterized by well-developed overbank deposits with mudstone and siltstone layers, crevasse splay sediments and calcitic soil horizons, between thick sandbodies marking the boundaries of the cycles.
- sections 418–351 m and 118–70 m represent a warm-temperate climate with average precipitation and are characterized by a thick alluvial sandbody preceded by extended vertical accretion linked with calcitic soil horizons.

In the deltaic–fluvial-plain environment the same four climate-controlled, short-term sedimentary cycles can be identified (Fig. 5). The cycle boundaries are marked by the sandbody. The progradation of the river is the result of climate-controlled water-level change and variations in the amount of transported clastic material.

Conclusions

Using a numerical evaluation of palynological and pedological investigations a long-term climatic cyclicity of about 400 ka has been determined in the Upper Miocene sediments of the Pannonian Basin. The interpretation of temperature and precipitation changes has allowed the identification of four climate phases within the cycle.

The correlation of climate cycles with the fourth order sedimentary sequences developed in the delta–fluviatile sedimentary environment has shown the control of sedimentation by the fourth order climate cycle. The temperature maximum was followed by a maximum coarsening of the sediment supply. Sedimentary cycles distinguished within the fourth order sedimentary sequence are in correlation with the climate phases.

Within a single climate cycle, the change in relative water level of the lake was estimated from the biota as several metres. This change in delta–fluvial flood-plain sediments meant a significant change in habitat and this was followed by corresponding changes in the biota.

Using the magnetostratigraphic data, the long-term cycles have a period of about 400 ka and are related to variations in eccentricity. Within each 400 ka cycle, the four climate phases are interpreted as results of Milankovitch-type eccentricity cycles of the about 100 ka.

We express our thanks to P. Scharek for supplying us with the core samples and the results of analyses on these samples, and for supporting our continued work on the palaeoclimate. Thanks are also due to M. Süto-Szentai for permitting us to use her still unpublished phytoplankton research results, to M. Lantos for providing us with unpublished magnetostratigraphic data and to L. Bucsi Szabó for the digitized well-logs. A. Mindszenty helped in the study of the palaeosols. All the above-mentioned colleagues were extremely helpful in subsequent discussions. Comments of the referees were most helpful and we also acknowledge of the help of the Editor in finalizing the manuscript

This research was sponsored by the Paleoenvironmental Project of the Geological Institute of Hungary and by the OTKA (National Scientific Research Fundation, Project 026276).

References

ANDREÁNSZKY, G. 1955. Klimate der ungarischen Tertiärfloren. *Annals of the Hungarian Geological Institute*, **44**, 209–231.
BERGGREN, W. A., KENT, D. V., SWISHER III, C. C. & AUBRY, M. P. 1995. A revised Cenozoic geochronology and chronostratigraphy. *In*: BERGGREN, W. A., KENT, D. V. & HARDEBOL, F. (eds) *Geochronology, timescale and global stratigraphic correlations. A unified temporal framework for an historical geology*. SEPM, Society for sedimentary Geology, Special Publication, **54**, 129–211.
CROWLEY, T. M. & NORTH, G. R. 1991. *Paleoclimatology*. Oxford University Press, New York.
JUHÁSZ, E., A. KOVÁCS, L., MÜLLER, P. TÓTH-MAKK, Á. PHILLIPS, L. & LANTOS, M. 1997. Climatically driven sedimentary cycles in the Late Miocene sediments of the Pannonian Basin, Hungary. *Tectonophysics*, **282**, 257–276.
KÓKAY, J., HÁMOR, T., LANTOS, M. & MÜLLER, P. 1991. A Berhida-3. sz. fúrás paleomágneses és földtani vizsgálata. *Annual report of the Hungarian Geological Institute of 1976*, 45–63.
KORPÁS-HÓDI, M. 1992. The Pannonian (*s.l.*) Molluscs of borehole section Szombathely-II. *Annual Report of the Hungarian Geological Institute of 1990*, 505–525.
——— & BOHN HAVAS, M. 1998. Miocene evolution of sedimentary environments and paleogeographic relations of Pannonian (Carpathian) Basin. *Geografia Fisica e Dinamica Quaternaria*, **21**, 49–54.
KRETZOI, M. 1969. Sketch of the Late Cenozoic (Pliocene Quarternary) terrestrial stratigraphy in Hungary. *Geographical Review*, **17**, 197–204.
——— 1983. Kontinenstörténet és biosztratigráfia a

felso harmadkor és a negyedidoszak folyamán a Kárpát-medencében és korrelációi. *Geographical Review*, **31**, 230–240.

———— KROLOPP, E., LORINCZ, H. & PÁLFALVY, I. 1976. Flora, Fauna und stratigrafische Lage der unterpannonischen Prehominiden-Fundstelle von Rudabánya (NO-Ungarn). *Annual Report of the Hungarian Geological Institute of 1974*, 365–394.

LANTOS, M., HÁMOR, T. & POGÁCSÁS, Gy. 1992. Magneto- and seismostratigraphic correlations of Pannonian s.l. (Late Miocene and Pliocene) deposits in Hungary. *Paleontologia i Evolució*, **24–25**, 35–36.

MÜLLER, P. & MAGYAR, I. 1992. Stratigraphic significance of the Late Miocene lacustrine cardiid Prosodacnomya (Kötcse section, Pannonian basin, Hungary). *Bulletin of the Hungarian Geological Society*, **122**, 1–38.

NAGY, E. 1958. Palynologische Untersuchung der am Fusse des Mátra-Gebirges gelagerten Oberpannonischen braunkohle. *Annals of the Hungarian Geological Institute*, **47**, 45–353.

———— 1970. Hungary's Neogene climate on the basis of palynological researches. *Giornale di Geologia*, **35**(1), 91–104.

———— 1992. A comprehensive study of Neogene sporomophs in Hungary. *Geologica Hungarica, Series Palaeontologica*, **53**, 1–379.

———— 1993. Paleofloristic and paleoclimatic changes in the Hungarian Neogene. *Proceeding of Symposium on Paleofloristic and Paleoclimatic changes during Cretaceous and Tertiary, Slovalian Academy*, 31–132.

———— & Ó. KOVÁCS, L., 1996. Climate curve construction from palynological data of the Pannonian section in borehole Berhida-3. *Acta Geologica Hungarica*, **40**, 101–108.

———— & PÁLFALVY, I. 1961. Plantes du Pannonian supéieur dans les environs de Rudabánya. *Annual Report of the Hungarian Geological Institute of 1957–1958*, 417–426.

———— & PLANDEROVA, E. 1985. Palynologische Auswertung der Floren des Pannonien. *In*: PAPP, A., JÁMBOR, Á. & STEININGER, F. F. (eds) *Chronostratigraphie und Neostratotypen, Miozän der Zentralen Paratethys VII. M.6 Pannonien (Slavonien, Serbien)*. Hungarian Academy, Budapest, 586–615.

PÁLFALVY, I. 1952. Plantes fossiles du Pliocene inferiur des environs de Rózsaszentmárton. *Annual Report of the Hungarian Geological Institute of 1949*, 63–65.

PHILLIPS, R. L., JÁMBOR, Á. & RÉVÉSZ, I. 1989. Depositional environments and facies in continuous core from the Szombathely-II. well (0–2150 m), Kisalföld basin, Western Hungary. US Geological Survey, Open-file Report **92–250**, 1–14.

PLANDEROVA, É. 1962. Bemerkungen zur Entwicklung der Flora und zu den Klimartischen Veränderungen im Neogen der SW Slowakei.- *Geologicke Pracepravy*, **63**, 147–156.

POGÁCSÁS, Gy., JÁMBOR, Á., MATTICK, R. E., ELSTON, D. P., HÁMOR, T., LAKATOS, L., SIMON, E., VAKARCS, G., VÁRKONYI, L. & VÁRNAI, P. 1990. Chronostratigraphic relations of Neogene formations of the Great Hungarian Plain based on interpretation of seismic and paleomagnetic data. *International Geology Review*, **32**, 449–457.

RETALLACK, G. J. 1994. The Environmental factor approach to the interpretation of paleosols. *In*: AMUNDSON, R., HARDEN, J. & SINGER, M. (eds) *Factors of Soil Formation: A Fiftieth Anniversary Retrospective*. Soil Science Society of America, Special Publication, **33**, 31–64.

VAKARCS, G., VAIL, P. R., TARI, G., POGÁCSÁS, Gy., MATTICK, R. E. & SZABÓ, A. 1994. Third-order Middle-Miocene–Pliocene depositional sequences in the prograding delta complex of the Pannonian Basin: an overview. *Tectonophysics*, **240**, 81–106.

Climatic oscillations versus environmental changes in the interpretation of Tertiary plant assemblages

ZLATKO KVAČEK

Charles University, Faculty of Science, Albertov 6, 128 43 Prague 2, Czech Republic
(e-mail: kvacek@prfdec.natur.cuni.cz)

Abstract: Zonal plant communities are the best climatic indicators because they thrive under specific climatic conditions. However, palaeoclimatic estimates based on mere comparisons of local fossil plant assemblages, even if assessed by multivariate methods, may be misleading, if the taphonomical bias is neglected. The fossil record expresses only limited parts of the vegetation of a given period. The example of the Lower Miocene section at Bílina, North Bohemian Basin, shows how quickly the composition of the plant megafossil record changes as a result of different environmental factors and sedimentary settings. These changes can imitate climatic oscillation because swamp, fluvial and upland facies represented within this site have different vegetation–palaeoclimatic aspects. On the other hand, if one compares two assemblages originating from the same sedimentary and environmental setting (preferably mesophytic conditions), the resulting palaeoclimatic changes can be estimated more safely. Two local assemblages from the Oligocene volcanic area of the České středohoří Mts in North Bohemia are of mesophytic character. Both are embedded in diatomites deposited in crater–maar lakes. The assemblage of Suletice (20–29 Ma), with prevailing thermophilous elements, indicates more warm-temperate conditions, whereas the assemblage of Bechlejovice (27 Ma), composed mostly of deciduous broad-leaved trees and shrubs, corresponds to a temperate climate. Thus, together, these two floral assemblages suggest that a climatic oscillation occurred within Late Oligocene time in Central Europe.

In its composition and physiognomy zonal vegetation reflects accurately regional climatic conditions on land. Hence, fossil records of this type would seem to be the best source of information for palaeoclimatic estimates in non-marine regions. However, problems arise when attempts are made to reconstruct ancient zonal vegetation. Taphonomic processes mix up various portions of plant cover, the more so when the deposition sites lie far from the source areas of the fossils. Fragments of both intrazonal (azonal) and zonal associations close to the depositional sites (lake, alluvial plain, mire) may reach places of burial or fossilization in different ways and with more or less incomplete representation of various organs (Ferguson 1996). Two approaches have been recently improved by statistical techniques to obtain better palaeoclimatic estimates for evaluation of the Tertiary and Late Cretaceous plant record.

The improved method of the Nearest Living Relative (NLR) is based on multivariate analysis of mean autecology (coexistence range) derived from the living counterparts (species or genus) of a given local flora (Mosbrugger & Utescher

1993, 1997; Mosbrugger 1995). The accuracy of the proxies obtained depends to a large extent on a reliable understanding of the taxonomy and evolution of the floristic elements. Not only exact identifications of the fossils are required. There is growing evidence that many Late Tertiary plants, notably in Europe, have drastically changed their autecology during Cenozoic time or have become extinct without known descendants (Belz & Mosbrugger 1994; Wolfe 1995). Striking examples are the following common elements of Tertiary floras.

Tetraclinis Mast. (Cupressaceae) is today a monotypic relict, restricted to the western Mediterranean, but during Tertiary time was represented by two species and one variety in humid broad-leaved evergreen as well as mixed mesophytic forests in Europe and North America (Kvaček 1989; Kvaček *et al.* 2000).

Glyptostrobus Endl. (Taxodiaceae) is today monotypic, restricted to the evergreen broad-leaved forest in Yunnan and in other parts of the SE China (it is not hardy enough to be cultivated in Central Europe), but during Tertiary time was widespread in the Northern Hemisphere, and

From: HART, M. B. (ed.) *Climates: Past and Present.* Geological Society, London, Special
Publications, **181**, 89–94, 1-86239-075-4/$15.00

was represented by several species associated with mainly warm-temperate or even temperate vegetation (e.g. in Late Pliocene time in Central Europe; Bůžek *et al.* 1985).

Craigia W. W. Smith & Evans (Tiliaceae) today has two species restricted to the evergreen broad-leaved forest of SW China, but during Tertiary time was represented by at least two species throughout the Northern Hemisphere, mainly in association with deciduous broad-leaved woody elements, one species surviving into Early Pliocene time in Europe (Bůžek *et al.* 1989; Kvaček *et al.* 1991, 2000).

A number of other examples can be presented to demonstrate the drawbacks of the NLR method. In the case of Late Cretaceous and Early Tertiary assemblages, in which most of the elements belong to extinct genera, this method fails to produce acceptable palaeoclimatic estimates.

To avoid dependence on autecology of Recent taxa, an improved leaf physiognomy method, the Climate–Leaf Multivariate Program (CLAMP), has been devised (Wolfe 1993). It evaluates sets of leaf morphological characters of woody dicots (e.g. character of the margin, leaf base, leaf apex, leaf size and many others) and calibrates their relationship to environmental variables by correspondence (Wolfe 1993) and, more recently, by canonical correspondence analysis (Wolfe 1995; Kovach & Spicer 1996). Although the causes of correlation between leaf physiognomy and climate are not fully understood, the improved CLAMP method provides surprisingly convincing arguments for palaeoclimatic interpretations of the Tertiary and even Late Cretaceous floras in the Northern Hemisphere, as well as Africa (Jacobs & Deino 1996). It offers a more reliable tool for plant palaeoclimatology than the currently used general physiognomic principles and comparisons with forest vegetation that have been used so far (Wolfe 1979).

The CLAMP method needs accurate identifications at the species level, which may become an obstacle when the fossils are not studied by cuticular analysis (for instance, when lumping or splitting is used, particularly of laurophyllous elements). Taphonomic processes may also distort the physiognomic signal of a given assemblage (Ferguson 1985; Mosbrugger 1989; Spicer 1989).

Both the NLR and CLAMP methods, although producing objective and repeatable results, in fact neglect vegetational differentiation of the plant cover into zonal and intrazonal communities. These may differ in palaeoclimatic signal, even when they are coeval. In this way,

biased basic information will be transferred but not eliminated by multivariate statistics. The two following examples from the Tertiary deposits of Central Europe, although based on mere conventional comparisons of local assemblages, illustrate the need for such vegetational studies.

Environmental bias in palaeoclimatic interpretation of the Early Miocene flora of Bílina

The vast Lower Miocene section of the open mine Bílina in the North Bohemian Basin has yielded thousands of plant megafossils (Bůžek *et al.* 1992, 1993). At the moment, three levels of sedimentary environment can be recognized there: the lignite seam and its roof; fluvial–alluvial sandy clays; lake clay deposits.

Table 1. *Bílina swamp forest*

Species	Foliage
Conifers	
Glyptostrobus europaeus	evergreen
Quasisequoia couttsiae	evergreen
Taxodium dubium	deciduous
Angiosperm trees	
Acer tricuspidatum	deciduous dentate
Alnus julianiformis	deciduous dentate
Alnus ?menzelii	deciduous dentate
Cercidiphyllum crenatum	deciduous dentate
Craigia bronnii–Dombeyopsis lobata	deciduous entire
Fraxinus bilinica	deciduous dentate
Laurophyllum saxonicum	evergreen entire
Nyssa haidingeri	deciduous entire
Ulmus pyramidalis	deciduous dentate
Quercus rhenana	evergreen entire
Angiosperm shrubs or lianas	
Berchemia multinervis	deciduous entire
Calamus daemonorhops	evergreen entire
Decodon	deciduous entire
Diospyros brachysepala	deciduous entire
Myrica integerrima	evergreen entire
Paliurus tiliaefolius	deciduous dentate
cf. *Rhus*	deciduous entire
Rubus merianii	deciduous dentate
Sabal major	evergreen entire
Salix varians	deciduous dentate
Smilax	deciduous entire

Conifers 3, evergreen 2 (66.6%); angiosperms 21, evergreen 5 (23.8%); dicots 18, entire-margined 9 (50.0%).

Table 2. *Bílina riparian forest*

Species	Foliage
Conifers	
Glyptostrobus europaeus	evergreen
Quasisequoia couttsiae	evergreen
Taxodium dubium	deciduous
Angiosperm trees	
Acacia beneschii	deciduous entire
Acer dasycarpoides	deciduous dentate
Acer integerrimum	deciduous dentate
Acer tricuspidatum	deciduous dentate
Alnus gaudinii	deciduous dentate
Alnus julianiformis	deciduous dentate
Alnus ?menzelii	deciduous dentate
Betula sp.	deciduous dentate
Carpinus grandis	deciduous dentate
Carya serrifolia	deciduous dentate
Cercidiphyllum crenatum	deciduous dentate
Craigia bronnii–Dombeyopsis	
lobata	deciduous entire
Daphnogene polymorpha	evergreen entire
'Ficus' truncata	deciduous entire
Fraxinus bilinica	deciduous dentate
'Juglans' acuminata	deciduous entire
Koelreuteria reticulata–	
Sapindus falcifolius	deciduous entire
Liquidambar europaea	deciduous dentate
cf. *Magnolia kristinae*	evergreen entire
Nyssa haidingeri	deciduous entire
Podocarpium podocarpum	deciduous entire
Populus populina	deciduous dentate
Populus zaddachii	deciduous dentate
Tilia brabenecii	deciduous dentate
Ulmus pyramidalis	deciduous dentate
Ulmus sp.	deciduous dentate
Zelkova zelkovifolia	deciduous dentate
Angiosperm shrubs or lianas	
Berberis berberidifolia	deciduous dentate
Berchemia multinervis	deciduous entire
Crataegus sp.	deciduous dentate
Decodon	deciduous entire
Diversiphyllum aesculapi	deciduous dentate
Mahonia bilinica	evergreen dentate
Paliurus tiliaefolius	deciduous dentate
Parrotia pristina	deciduous dentate
Phyllites kvacekii	deciduous dentate
Pungiphyllum cruciatum	?evergreen dentate
Rosa europaea	*deciduous dentate*
Rubus merianii	deciduous dentate
Sabal major	evergreen entire
Salix haidingeri	deciduous dentate
Smilax weberi	deciduous entire
Vitis stricta	deciduous dentate
Wisteria cf. *fallax*	deciduous entire

Conifers 3, evergreen 2 (66.6%); angiosperms 43, evergreen 5 (11.3%); dicots 41, entire-margined 12 (29.2%).

Table 3. *Bílina and Litvínov upland–shore forest*

Species	Foliage
Conifers	
Glyptostrobus europaeus	evergreen
Pinus engelhardtii	evergreen
Pinus ornata	evergreen
Quasisequoia couttsiae	evergreen
Taxodium dubium	deciduous
Angiosperm trees	
Acer tricuspidatum	deciduous dentate
Alnus gaudinii	deciduous dentate
Alnus julianiformis	deciduous dentate
Alnus ?menzelii	deciduous dentate
Cercidiphyllum crenatum	deciduous dentate
Craigia bronnii–Dombeyopsis	
lobata	deciduous entire
Daphnogene polymorpha	evergreen entire
Engelhardia orsbergensis–	
macroptera	decidous dentate
Fraxinus bilinica	decidous dentate
Gordonia sp.	evergreen dentate
Laurophyllum nechranicense	evergreen entire
Laurophyllum pseudoprinceps	evergreen entire
Laurophyllum pseudovillense	evergreen entire
Laurophyllum saxonicum	evergreen entire
Nyssa haidingeri	deciduous entire
Platanus neptuni	evergreen dentate
Quercus rhenana	evergreen entire
Symplocos sp.	evergreen dentate?
Tilia brabenecii	deciduous dentate
Trigonobalanopsis rhamnoides	evergreen entire
Ulmus pyramidalis	deciduous dentate
Zelkova zelkovifolia	deciduous dentate
Angiosperm shrubs or lianas	
Comptonia difformis	deciduous dentate
Mahonia bilinica	evergreen dentate
Myrica lignitum	evergreen dentate/entire
Paliurus tiliaefolius	deciduous dentate
Sabal major	evergreen entire
Vaccinioides lusatica	evergreen entire

Conifers 5, evergreen 4 (80%); angiosperms 28, evergreen 14 (50.0%); dicots 27, entire-margined 11 (40.7%).

These three superimposed facies encompass three vegetation types: swamp forest (Table 1), riparian–levee forest (Table 2) and upland–lake-shore forest (Table 3), which can be traced elsewhere in the basin (Bůžek *et al.* 1987; Boulter *et al.* 1993). When examined together they imitate climatic oscillation with a cooling phase shown by the riparian forest level. The swamp forest is composed of 23.8% evergreen woody angiosperms and 50% entire-margined dicotyledonous species, suggesting conditions of nearly subtropical to warm–temperate climate. The upland–shore forest is similar, with 46.4% evergreen angiosperms and 40.7% entire-margined dicotyledonous species. In contrast, the riparian forest has only 11.3% evergreen angiosperms and 29.2% entire-margined dicotyledonous species, suggesting a rather cooler, temperate climate. Although the floristic composition of each facies is characteristic of local environment, taken together the three vegetation types represent a single floristic complex (the Bílina–Brandis complex *sensu* Mai (1996), dated as MN 3 zone). The coexistence of palms with deciduous elements, such as *Parrotia*, and the general physiognomy suggest a warm-temperate humid (Cfa) climate with a mean annual temperature of about 13°C. The riparian assemblage contains 'cooler' elements. If such an assemblage is taken as a 'flora', even interpretation of its age may become erroneous. Thus the coeval site Čermníky (Bůžek 1971) in the western part of the North Bohemian Basin, which also yielded mostly riparian assemblages, was dated to mid-Miocene time on the basis of statistical comparisons of floral lists for Europe (Günther & Gregor 1992).

Climatic change within the Late Oligocene time in Central Europe

The plant megafossil record in volcanic localities of the České středohoří Mts, North Bohemia, shows a variable floristic composition that reflects palaeoclimatic change. Except for the oldest site Kuclín, which is of Late Eocene age and yields subtropical plant remains, all the sites are of Oligocene age and include variable proportions of modern Arcto-Tertiary elements. Two of these sites, although not far from each other and both preserved in crater–maar lake diatomite deposits, differ pronouncedly in general aspect and in the share of deciduous broad-leaved elements.

Suletice (Kvaček & Walther 1996) is dated between 20 and 29 Ma (Bellon *et al.* 1999) and has a warm–temperate signature (Table 4). Thermophilous elements prevail at this site (*Sloanea, Platanus neptuni, Laurophyllum, Engelhardia*). The majority of plants at this locality are mesophytic, whereas plants tolerating moist soils are far less diversified. Swamp elements are lacking. The assemblage corresponds to the Mixed Mesophytic Forest (*sensu* Wolfe 1979). Evergreen woody angiosperms total 31.6%, and entire-margined dicotyledonous species 37.9%.

Table 4. *Suletice mesophytic forest*

Species	Foliage
Conifers	
Calocedrus suleticensis	evergreen
cf. *Cephalotaxus* sp.	evergreen
Tetraclinis salicornioides	evergreen
Angiosperm trees	
Acer dasycarpoides	deciduous dentate
Acer pseudomonspessulanum	deciduous dentate
Acer palaeosaccharinum	deciduous dentate
Acer tricuspidatum	deciduous dentate
Ailanthus sp.	deciduous dentate
cf. *Betula* sp.	deciduous dentate
Carpinus grandis	deciduous dentate
Carya serrifolia	deciduous dentate
Celtis sp.	deciduous dentate
Craigia bronnii–Dombeyopsis sp.	deciduous entire
cf. *Cyclocarya cyclocarpa*	deciduous dentate
Daphnogene cinnamomifolia	evergreen entire
Dicotylophyllum maii	?evergreen entire
'*Elaeodendron*'	evergreen dentate
Engelhardia orsbergensis–macroptera	deciduous dentate
Hydrangea microcalyx	deciduous dentate
Laurophyllum cf. *acutimontanum*	evergreen entire
Laurophyllum cf. *medimontanum*	evergreen entire
Laurophyllum cf. *pseudoprinceps*	evergreen entire
Leguminosae gen. et sp. 1	deciduous entire
Leguminosae gen. et sp. 2	deciduous entire
Leguminosae gen. et sp. 3	deciduous entire
Liriodendron haueri	deciduous dentate
Magnolia sp.	evergreen entire
cf. *Matudaea* sp.	evergreen entire
Mimosites haeringianus	deciduous entire
Ostrya atlantidis	deciduous dentate
'*Palaeolobium*'	deciduous entire
Platanus neptuni	evergreen dentate
Populus zaddachii	deciduous dentate
Sloanea artocarpites	evergreen dentate
Zelkova zelkovifolia	decidous dentate
Angiosperm shrubs or lianas	
Ampelopsis hibschii	deciduous dentate
Cornus studeri	deciduous entire
Mahonia sp.	evergreen dentate
Pungiphyllum cruciatum	?evergreen dentate
Rosa lignitum	deciduous dentate
Smilax sp.	deciduous entire

Conifers 3, evergreen 3 (100%); angiosperms 38, evergreen 11 (31.6%); dicots 37, entire-margined 14 (37.9%).

The second locality, Bechlejovice (Bůžek *et al.* 1990; Knobloch 1994), is older than 27 Ma (Bellon *et al.* 1999) and has yielded a flora without conifers, composed of mostly deciduous broad-leaved angiosperms (evergreen species

24.3%, entire-margined dicotyledonous species 19.4%), suggesting temperate climatic conditions (Table 5). Both composition and physiognomy correspond to the ecotone between the Mixed Broad-leaved Deciduous Forest and Mixed Northern Hardwood Forest (*sensu* Wolfe 1979). Thus a climatic oscillation from warm-temperate to temperate conditions, i.e. a shift in the mean annual temperature from 13°C to about 10°C, would appear to have occurred during the deposition of these volcanic strata in Late Oligocene time.

Table 5. *Bechlejovice mesophytic forest*

Species	Foliage
Angiosperm trees	
Acer dasycarpoides	deciduous dentate
Acer palaeosaccharinum	deciduous dentate
Acer pseudomonspessulanum	deciduous dentate
Acer subcampestre	deciduous dentate
Acer tricuspidatum	deciduous dentate
Alnus gaudinii	deciduous dentate
Betula buzekii	deciduous dentate
Carpinus grandis	deciduous dentate
Carya serrifolia	deciduous dentate
Cercidiphyllum crenatum	deciduous dentate
Craigia bronnii–Dombeyopsis lobata	deciduous entire
cf. *Cyclocarya cyclocarpa*	deciduous dentate
Daphnogene sp.	evergreen entire
'Elaeodendron'	evergreen dentate
Engelhardia	deciduous dentate
Laurophyllum acutimontanum	evergreen entire
Laurophyllum sp.	evergreen entire
Mimosites haeringianus	deciduous entire
Nyssa ?	evergreen entire
Ostrya atlantidis	deciduous dentate
Platanus neptuni	evergreen dentate
Populus zaddachii	deciduous dentate
Pterocarya paradisiaca	deciduous dentate
Tilia gigantea	deciduous dentate
Ulmus fischeri	deciduous dentate
Zelkova zelkovifolia	deciduous dentate
Angiosperm shrubs or lianas	
Ampelopsis hibschii	deciduous dentate
Cornus studeri	deciduous entire
Crataegus cf. *pirskenbergensis*	deciduous dentate
Mahonia	evergreen dentate
Paliurus tiliaefolius	deciduous dentate
?Pyracantha	evergreen dentate
Pungiphyllum cruciatum	?evergreen dentate
Rosa lignitum	deciduous dentate
cf. *Rhus herthae*	deciduous dentate
Smilax grandifolia	deciduous entire
Vitis stricta	deciduous dentate

Conifers 0%; angiosperms 37, evergreen 9 (24.3%); dicots 36, entire-margined 7 (19.4%).

Conclusions

(1) Palaeoclimatic aspects of Tertiary plant assemblages depend not only upon regional climatical conditions, but are also influenced by other environmental factors (ground water table, substrate). Hence differences between zonal and intrazonal vegetation may imitate climatic changes, if the corresponding plant assemblages are compared mechanically.
(2) Sedimentary settings and resulting vegetational analysis of the sites should be incorporated into both the Nearest Living Relative and CLAMP palaeoclimatic approaches as correcting factors.
(3) It is desirable to combine various means of palaeoclimatic assessment based on fossil plant assemblages with the aid of these newly developed powerful statistical methods in connection with deciphering taphonomical processes.

The study was supported by Charles University Grants 266/97/B-GEO/PřF and 13/98:113100006 and funding by Northbohemian Mines (Severočeské doly, a.s.) Chomutov. Steady interest and comments during the field excursions in the North Bohemian Tertiary region offered by D. K. Ferguson and R. A. Spicer are gratefully acknowledged.

References

BELLON, H., BŮŽEK, C., GAUDANT, J., KVAČEK, Z. & WALTHER, H. 1999. The České středohoří magmatic complex in northern Bohemia: ^{40}K–^{40}Ar ages for volcanism and biostratigraphy of the Tertiary freshwater formations. *Newsletters on Stratigraphy* **36**, 77–103.

BELZ, G. & MOSBRUGGER, V. 1994. Systematisch-paläoökologische und paläoklimatische Analyse von Blattfloren im Mio/Pliozän der niederrheinischen Bucht (NW-Deutschland). *Palaeontographica*, **B, 233**, 19–156.

BOULTER, M. C., HUBBARD, R. N. L. B. & KVAČEK, Z. 1993. A comparison of intuitive and objective interpretations of Miocene plant assemblages from north Bohemia. *Palaeogeography, Palaeoclimatology, Palaeoecology*, **101**, 81–96.

BŮŽEK, C. 1971. Tertiary flora from the northern part of the Pětipsy area (North-Bohemian Basin). *Rozpravy Ústředního Ústavu Geologického*, **36**, 1–118.

———, DVOŘÁK, Z., KVAČEK, Z. & PROKŠ, M. 1992. Tertiary vegetation and depositional environment of the 'Bílina delta' in the North Bohemian brown-coal basin. *Časopis pro Mineralogii a Geologii*, **37**, 117–134.

———, FEJFAR, O., KONZALOVÁ, M. & KVAČEK, Z. 1990. Floristic changes around Stehlin's grande coupure in Central Europe. *In*: KNOBLOCH, E. & KVAČEK, Z. (eds) *Palaeofloristic and Palaeoclimatic Changes in the Cretaceous and Tertiary*. Geological Survey, Prague, 167–181.

————, HOLÝ, F. & KVAČEK, Z. 1985. Late Pliocene palaeoenvironment and correlation of the Vildštejn floristic complex within Central Europe. *Rozpravy Československé akademie věd, řada matematických a přírodních věd*, **95**, 1–72.

————, ———— & ———— 1987. Evolution of main vegetation types in the lower Miocene of NW Bohemia. *In*: POKORNÝ, V. (ed.) *Contribution of Czechoslovak Palaeontology to Evolutionary Science 1945–1985*. Univerzita Karlova, Praha, 150–161.

————, KVAČEK, Z., DVOŘÁK, Z. & PROKŠ, M. 1993. Tertiary vegetation of the North Bohemian brown-coal basin in view of new data. *Zpravodaj SHD, Most*, **II/93**, 38–51 [in Czech].

————, KVAČEK, Z. & MANCHESTER, S. R. 1989. Sapindaceous affinities of the *Pteleaecarpum* fruits from the Tertiary of Eurasia and North America. *Botanical Gazette, Chicago*, **150**, 477–489.

FERGUSON, D. K. 1985. The origin of leaf-assemblages—new light on an old problem. *Review of Palaeobotany and Palynology, Amsterdam*, **46**, 117–188.

———— 1996. Actuopalaeobotany—a taphonomic peep-show?—Summary of workskop discussion. *Neues Jahrbuch für Geologie und Paläontologie, Abhandlungen*, **202**, 149–158.

GÜNTHER, TH. & GREGOR, H.-J. 1992. Computeranalyse neogener Frucht- und Samenfloren Europas. Band 3. Übereinstimmungen von Florenlisten und ihre stratigraphisch-geographischen Beziehungen. *Documenta Naturae*, **50** 1–244.

JACOBS, B., F. & DEINO, A., L. 1996. Test of climate–leaf physiognomy regression models, their application to two Miocene floras from Kenya, and ^{40}Ar/^{39}Ar dating of the Late Miocene Kapturo site. *Palaeogeography, Palaeoclimatology, Palaeoecology*, **123**, 259–271.

KNOBLOCH, E. 1994. Einige neue Erkentnisse zur oligozänen Flora von Bechlejovice bei Dèčin. *Vestník Ceského Geologického Ustavu*, **69**, 63–67.

KOVACH, W. L. & SPICER, R. A. 1996. Canonical correspondence analysis of leaf physiognomy: a contribution to the development of a new palaeoclimatological tool. *Palaeoclimates*, **2**, 125–138.

KVAČEK, Z. 1989. Fosilní *Tetraclinis* Mast. (Cupressaceae). *Časopis Národního Muzea, řada přírodovědná*, **155**, 45–53.

———— & WALTHER, H. 1996. The Oligocene volcanic flora of Suletice–Berand near Ústí nad Labem, North Bohemia—a review. *Acta Musei Nationalis Pragae*, B, **50**, 25–54.

————, BŮŽEK, C. & MANCHESTER, S. R. 1991. Fossil fruits of *Pteleaecarpum* Weyland—tiliaceous, not sapindaceous. *Botanical Gazette, Chicago*, **152**, 522–523.

————, MANCHESTER, S. R. & SCHORN, H. E. 2000. Cones, seeds and foliage of *Tetraclinis salicornioides* (Cupressaceae) from the Oligocene and Miocene of western North America: a geographic extension of the European Tertiary species. *International Journal of Plant Science*, **161**, 331–344.

MAI, H. D. 1996. *Tertiäre Vegetationsgeschichte Europas*. Fischer, Jena.

MOSBRUGGER, V. 1989. Zur Gliederung und Benennung von Taphozönosen. *Courier Forschungs-Institut Senckenberg*, **109**, 17–28.

———— 1995. New methods and approaches in Tertiary palaeoenvironmental research. *Abhandlungen des Staatlichen Museums für Mineralogie und Geologie zu Dresden*, **41**, 41–52.

———— & UTESCHER, T. 1993. Quantitative terrestrial palaeoclimate reconstructions in the Tertiary—possibilities and limitations of a new approach. *Terra Nova, Abstract Supplement*, **5**, 727.

———— & ———— 1997. The coexistence approach—a method for quantitative reconstructions of Tertiary terrestrial palaeoclimate data using plant fossils. *Palaeogeography, Palaeoclimatology, Palaeoecology*, **134**, 61–86.

SPICER, R. A. 1989. The formation and interpretation of plant megafossil assemblages. *Advances in Botanical Research*, **16**, 95–191.

WOLFE, J. A. 1979. Temperature parameters of the humid and mesic forests of eastern Asia and relation to forests of other regions of the Northern Hemisphere and Australasia. *US Geological Survey, Professional Papers*, **1106**, 1–37.

———— 1993. A method of obtaining climatic parameters from leaf assemblages. *US Geological Survey Bulletin*, **2040**, 1–71.

———— 1995. Paleoclimatic estimates from Tertiary leaf assemblages. *Annual Reviews in Earth and Planetary Science*, **23**, 119–142.

An integrated micropalaeontological approach applied to Late Pleistocene–Holocene palaeoclimatic and palaeoenvironmental changes (Gaeta Bay, Tyrrhenian Sea)

F. O. AMORE, G. CIAMPO, V. DI DONATO, P. ESPOSITO, E. RUSSO ERMOLLI & D. STAITI

Dipartimento di Scienze della Terra, Università di Napoli Federico II, Largo San Marcellino, 10, 80138 Napoli, Italy (e-mail: ciampo@unina.it)

Abstract: An integrated micropalaeontological approach has been used to characterize the last climatic cycle in the Tyrrhenian Sea. Three gravity cores, drilled in the Gaeta Bay continental shelf, have been analysed by means of quantitative methods applied to different taxonomic groups. In particular, calcareous nannofossils, planktonic Foraminifera, ostracods and palynomorphs have been studied to detect palaeoclimatic trends. The Last Glacial period has been identified in the lower part of two cores (C8 and C9), and an older cold period is represented in the lower part of the more proximal core (C5). The upper part of all the cores clearly records the Holocene period, whereas the Late Glacial period is not always easily detectable. Pollen seems to precede the Holocene climatic optimum. Climatic curves have been reconstructed for all the taxonomic groups and high-resolution palaeobathymetric reconstruction has been inferred on the basis of the ostracod assemblages.

Planktonic Foraminifera and pollen are largely used in palaeoclimatological studies. Only recently have data on calcareous nannofossils become available, but the environmental significance of many species is still controversial. Attempts to reconstruct high-resolution curves of the sea level based on the response of biotic communities have never been made. The analysis of calcareous nannofossils, Foraminifera, Ostracoda and pollen assemblages, carried out on three gravity cores (G93-C5, -C8, and -C9) recovered in the Gaeta Bay continental shelf (Tyrrhenian Sea) (Fig. 1), represents a contribution to an understanding of how glacial–interglacial cycles may influence microcommunities.

Materials and methods

Senatore (1996) interpreted high-resolution seismic profiles identifying two prograding successions separated by an erosional surface. Another erosional surface separates the upper prograding succession from an overlying Late Glacial–Holocene wedge. These two surfaces have been referred (Senatore 1996) to two transgressive events, the first occurring at the 6–5 IS boundary, and the second at the 1–2 IS boundary. On the basis of the reflection geometries the prograding successions have been included in two fourth-order depositional sequences.

The more proximal core (G93-C5) was recovered at 111 m bsl (below sea level) (41°05′41″N, 13°35′56″E). It is 5 m long and consists of two units separated by a discontinuity surface (Amore *et al.* 1996). The lower unit consists of silts and silty sands, and the upper unit consists of clays and sandy silts. The C8 core, 3.62 m long, was taken at 126 m bsl (41°04′31″N, 13°32′18″E). It consists of three sedimentological units. Its sediments are moderately coarser than those collected in the C5 core. The basal unit consists of silts and silty sands. The intermediate unit consists of bioclastic silty sands. This unit gradually passes upward to the clayey–sandy silt upper unit. The more distal Core (C9), 212 m bsl (41°02′4″N, 13°29′24″E), is 4.7 m long and consists of a single clayey–silty

From: HART, M. B. (ed.) *Climates: Past and Present.* Geological Society, London, Special Publications, **181**, 95–111, 1-86239-075-4/$15.00

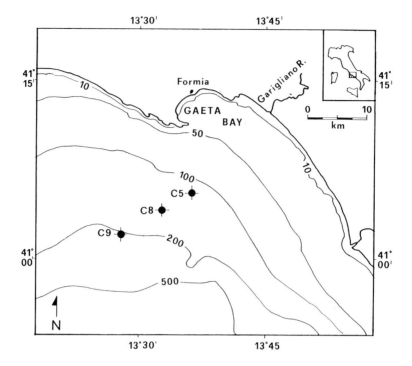

Fig. 1. Location map.

clay unit. Because of the geometry of the sedimentary bodies detected from the seismic profiles the lower interval of the C5 core is older than those recorded in the C8 and C9 cores.

Quantitative methods have been used for each taxonomic group on samples spaced 0.10 m apart. On the basis of this investigation three climatically different intervals have been recognized: glacial intervals in the lower part of the cores; warmer, but climatically variable intervals (Late Glacial) in the intermediate part; warm interval (Holocene) in the upper part.

Climatic curves have been obtained for calcareous nannofossils, Foraminifera and pollen by using the formula $(a-b) \times 100/(a+b)$ where a represents the warm indicators and b the cold indicators (Blanc-Vernet *et al.* 1983; Violanti *et al.* 1991; Sanvoisin *et al.* 1993).

Calcareous nannofossils (F. O. Amore & P. Esposito)

The calcareous nannofossil assemblages recorded in the Gaeta Bay samples are similar to those recorded in the Mediterranean Sea (Violanti *et al.* 1987, 1991; Sanvoisin *et al.* 1993; Flores *et al.* 1997). The assemblages are gen-

erally diverse and well preserved. Poorer preservation and an increase in reworked specimens have been observed in association with increased terrigenous inputs. A vicariance has been observed between calcareous nannofossils and high numbers of diatoms, probably linked to a runoff increase.

The assemblages mainly consist of *Coccolithus pelagicus* (Wallich) Schiller, *Emiliania huxleyi* (Lohmann) Hay & Mohler, *Syracosphaera pulchra* Lohmann, *Helicosphaera carteri* (Wallich) Kamptner, *Gephyrocapsa oceanica* Kamptner, *G. caribbeanica* Boudreaux & Hay, *Calcidiscus leptoporus* (Murray & Blackman) Loeblich & Tappan, *Umbilicosphaera sibogae* (Weber–van Bosse) and *Rhabdosphaera clavigera* Murray & Blackman.

Helicosphaera wallichii (Lohman) Boudreaux & Hay, *Helicosphaera hyalina* Gaarder, *Rhabdosphaera xiphos*, *Umbellosphaera irregularis* Paasche, *Schyphosphaera* spp., *Pontosphaera* spp., *Braarudosphaera bigelowii* (Gran & Braarud) Deflandre and *Ceratolithus* spp. (mainly *C. cristatus* Kamptner) are scattered and rare.

R. clavigera, U. sibogae, C. leptoporus, Syracosphaera spp. and *Ceratolithus* spp. have been considered warm water species in agreement with the available data on the climatic meaning

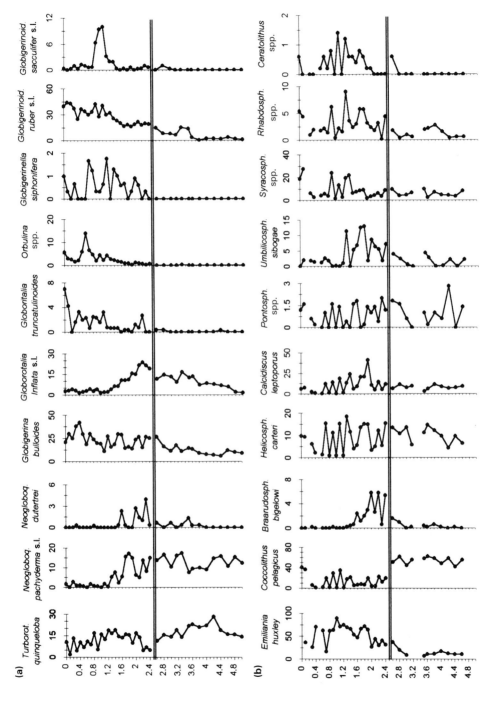

Fig. 2. C5 core: selected planktonic Foraminifera (**a**) and calcareous nannofossil (**b**) percentages plotted against depth (m).

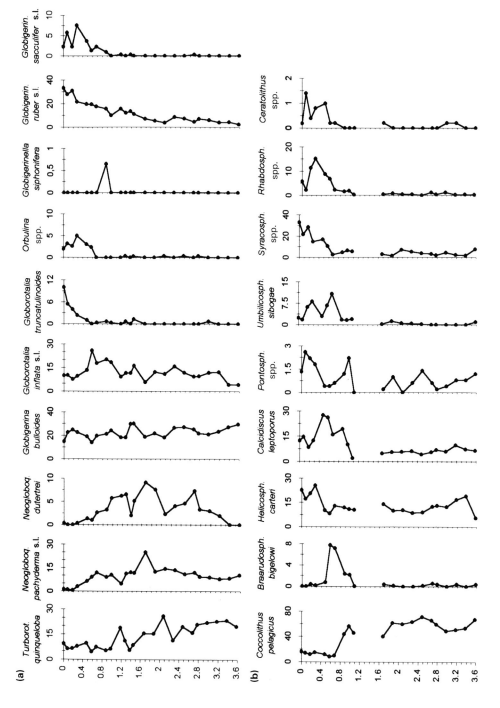

Fig. 3. C8 core: selected planktonic Foraminifera (**a**) and Calcareous nannofossil (**b**) percentages plotted against depth (m).

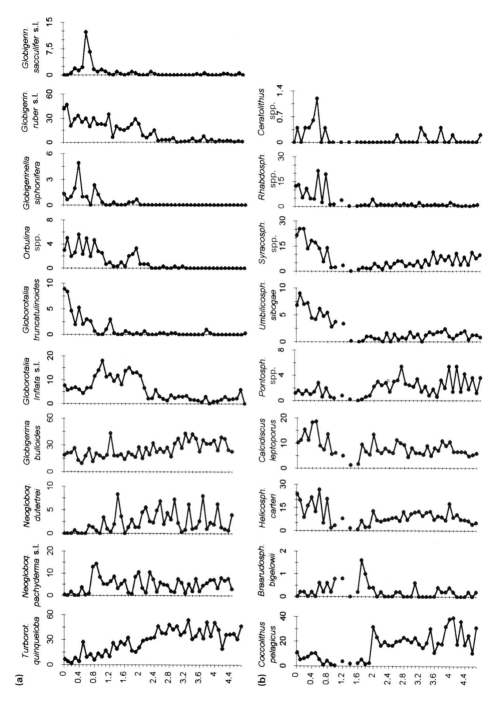

Fig. 4. C9 core: selected planktonic Foraminifera (**a**) and calcareous nannofossil (**b**) percentages plotted against depth (m).

of coccolithophores (e.g. McIntyre & Bé 1967; Okada & McIntyre 1979; Knappertsbusch 1993) and thus used as warm indicators (*a*) in the 'palaeoclimatic curves', *C. pelagicus* has been considered a cold-water indicator (McIntyre & Bé 1967; Bartolini 1970; Müller 1979), but it is also controlled by nutrient enrichment of sea water (Roth 1994; Young 1994; Cachao & Moita 1995). It represents the cold indicator (*b*) for the reconstruction of the 'palaeoclimatic curves'. The quantitative analysis consists of counting 500 coccoliths per slide.

High percentages of *C. pelagicus* and low percentages of warm-water species characterize the lower part of the three cores (Figs 2b, 3b and 4b). In the C5 core, at 2.9 m bsf (below sea floor), an increased abundance of *E. huxleyi* has been observed. The zonal boundary between *E. huxleyi* Acme and *E. huxleyi* Zones (Gartner 1977) has been identified at 2.7 m bsf. To detect this event, *E. huxleyi* has been counted separately to a minimum total of 300 specimens. Some selected samples have been examined by scanning electron microscope. The midpoint between the values below and above 20% has been considered the 'increase of *E. huxleyi*' event, in agreement with Rio *et al.* (1990).

The Late Glacial period is characterized by a decrease of *C. pelagicus* and by a first increase of warm-water taxa together with a further *E. huxleyi* increase.

B. bigelowii is generally absent within the Glacial and Holocene intervals of the cores or it is represented by very low values. However, peaks of abundance occur in all the cores during the Late Glacial period; the highest values have been found in the C5 and C8 nearshore cores. The ecological meaning of this species is still controversial. In the Mediterranean Sea this species seems to be related to the turbidity of nearshore waters and/or to low-salinity values in surface waters as a result of an intensified regional rainfall and a consequent runoff increase (Müller 1979). Nevertheless, the occurrence of *B. bigelowii* in the Black Sea and in the Persian Gulf indicates that the species may be also influenced by other ecological factors (Roth 1994).

The assemblages of the upper part of the cores are characterized by a dominance of warm-water species, and *C. pelagicus* shows a decrease. High values of *U. sibogae* and *Rhabdosphaera* spp. and peaks of *C. leptoporus* and *H. carteri* have also been observed. Flores *et al.* (1997), in the Mediterranean Sea, recorded the highest values of these taxa during interglacial periods, and a relationship with nutrient enrichment has also been suggested for *C. leptoporus* and *H. carteri*

by Violanti *et al.* (1991) and Pujos (1992).

Syracosphaera spp. have been recorded with high percentages in the Holocene interval of the cores. These become, together with *C. leptoporus*, the best represented species of the assemblages in the uppermost Holocene interval (except *E. huxleyi*). Peaks of *Ceratolithus* spp. are recorded in all the cores within the Holocene sequence.

Planktonic Foraminifera (V. Di Donato & D. Staiti)

Planktonic foraminiferal assemblages have been analysed by washing samples on a 106 μm sieve, and counting at least 300 specimens in each sample. Taking into account the small mean size of the specimens constituting our assemblages in comparison with the Atlantic ones, we preferred to utilize a 106 μm sieve rather than the 'standard' 125 μm sieve. This also allowed us to not underestimate the percentages of small species such as *Turborotalita quinqueloba* (Natland). Ecological features of living species have been used to obtain some information about the palaeoceanographical and palaeoclimatological significance of fossil assemblages.

Globigerina bulloides d'Orbigny is a typical North Atlantic species (Ottens 1991). It also occurs in tropical warm waters with high nutrient levels, such as upwelling areas (e.g. Thiede 1983; Hembleben *et al.* 1989) and close to river mouths (Van Leeuwen 1989). Although it is generally more abundant during cold intervals in the three cores, its distribution is not easily interpretable because of the occurrence of peaks in abundance within the Holocene sequence.

Globigerinoides ruber (d'Orbigny) is a surface-dwelling species, which prefers warm oligotrophic waters (Hembleben *et al.* 1989; Pujol & Vergnaud Grazzini 1995). In the studied cores it is almost exclusively represented by *G. ruber* var. *alba*, including *Globigerinoides elongatus* (d'Orbigny). Scattered specimens of *G. ruber* var. *rosea* have been found only in the Holocene interval.

Globorotalia inflata (d'Orbigny) is a grazing species requiring a rather cool and deep mixed layer (Pujol & Vergnaud Grazzini 1995). It occurs with both typical and *Globorotalia oscitans* Todd morphotypes. The latter are more abundant within colder intervals. In agreement with Muerdter & Kennett (1984) and Jorissen *et al.* (1993), this species is missing in the lower interval of the C9 core, showing an increase during the Late Glacial period. This trend has been also recorded by Pujol & Vergnaud

Grazzini (1989) in the Alboran Sea, and by Capotondi (1995) in the Corsica Basin.

Pujol & Vergnaud Grazzini (1995) suggested that the primary control factor on the distribution of *Globorotalia truncatulinoides* (d'Orbigny) in the Mediterranean Sea is the winter convection and vertical mixing. In our assemblages this species is mainly represented by left-coiled specimens. It is absent during the Glacial period, like *Globigerinella siphonifera* (d'Orbigny), and it increases in the upper part of the cores. The abundances of *G. truncatulinoides* and *G. ruber* are strictly correlated. Thus, in fossil assemblages, their distribution seems to be chiefly controlled by climatical changes.

Globorotalia scitula (Brady) thrives in deep and cool waters (Bé 1969). Rare specimens are found in the colder intervals of the cores.

Neogloboquadrina dutertrei (d'Orbigny) is a salinity dependent species (Bé & Tolderlund 1971). However, its abundance is closely linked to the development of a deep chlorophyll maximum (Rohling & Gieskes 1989). It shows a cyclical occurrence in the C9 core, becoming rare in the Holocene interval (Fig. 4a).

Neogloboquadrina pachyderma (Ehrenberg) also includes *N. incompta* (Cifelli) morphotypes. Right-coiled specimens are always more abundant than left ones. Although this species is typical of polar and subpolar regions, its abundance in the cores seems to be influenced more by productivity than by temperatures.

In the reconstruction of the 'palaeoclimatic curves' the warm assemblages (*a*) include *Globigerinoides* spp., *Globoturborotalita* spp., *Globigerina calida* and *Orbulina* spp. whereas the cold assemblages (*b*) include *Globigerinita uvula*, *Globorotalia scitula*, *Neogloboquadrina pachyderma* and *Turborotalita quinqueloba*.

The clearest data have been obtained from the C9 core (Fig. 4a), whereas those acquired from the C8 core (Fig. 3a), characterized by a conspicuous sandy fraction, are not always easily interpretable.

The lower parts of the cores are characterized by the dominance of the cold-water species *T. quinqueloba* and *G. bulloides*. Percentages of the genus *Globigerinoides*, which is considered a good indicator of sea surface temperatures (SSTs), are very low. *G. inflata* is represented mainly by *G. oscitans* morphotypes.

At the end of this cold interval a decrease of *T. quinqueloba*, together with a first increase in the *G. ruber* percentages, marks the end of the Late Glacial period, and indicates a first rise in SSTs. The simultaneous increase of *G. inflata*, whose abundance should be related, in nearshore cores, to the presence of a shallow chlorophyll maximum (Pujol & Vergnaud Grazzini 1995), suggests strong seasonal changes in the composition of surface-water assemblages.

In the upper parts of the cores a further increase of the warm-water species has been recorded. Nevertheless, increased percentages of *N. pachyderma* have been detected in association with high peaks of *C. leptoporus* and *H. carteri*. The highest percentages of *G. truncatulinoides* and high values of *G. ruber* characterize the samples from the surface of the sea bottom, and the neogloboquadrinids and *T. quinqueloba* are very rare.

Ostracods (G. Ciampo)

Depth, type of substratum, salinity, vegetation at the bottom, food supply and temperature mainly determine the composition of benthic and nectobenthic marine ostracod assemblages. Specific quantitative analysis can distinguish, to a certain extent, the influence of the different environmental factors on ostracod associations.

Many species are limited to infralittoral environments (upper and/or lower) as indicated by *Cytheretta adriatica* Ruggieri, *Cytherois uffenordei* Ruggieri, *Leptocythere levis* (G. W. Müller), *Leptocythere ramosa* (Rome), *Loxoconcha turbida* G. W. Müller and others. These species are usually dominant throughout the cold lower units of the cores.

In contrast, species of the genera *Parakrithe*, *Polycope*, *Henrhyowella*, *Cytheropteron* and others, are markers of circalittoral and/or bathyal environments. They dominate ostracod associations in the warmer intervals of the cores.

A sharp reverse of dominance of the circalittoral species in respect to the infralittoral ones occurs in the Mediterranean Sea between 60 and 80 m depth. This event has been recognized in the C5 and C8 cores.

To reconstruct the palaeobathymetry along the three studied cores, the following data have been considered: the present bathymetric distribution of the determined ostracod species; the bathymetric constraint of the marker species and their numbers in different environments (Bonaduce *et al.* 1975; Breman 1976; Whatley & Coles 1991); the connections among depth, type of substratum, presence and quality of vegetable cover (Lachenal & Bodergat 1990).

The bathymetric models from the three cores are disturbed by reworking or by barren or very poor samples. The reworking events (Fig. 5b) have been recognized by the ratio between the numbers of right valves and left valves (usually different in weight) and the total number of valves of selected species. In the three cores

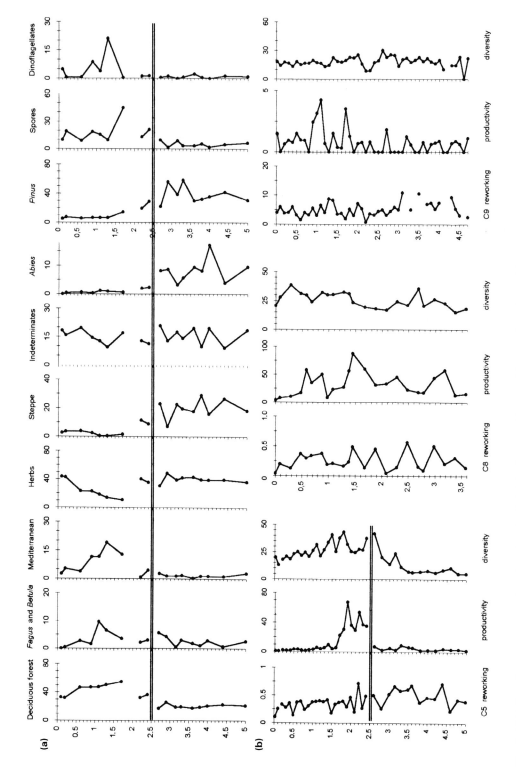

Fig. 5. (a) Pollen class percentages plotted against depth (m) in C5 core. **(b)** Reworking, productivity and diversity of the C5, C8 and C9 cores based on ostracod assemblages.

reworking is more common within the Glacial phases, just before the sea-level rise, and at the base of the Holocene interval, probably because of the increase in terrigenous inputs. Reworking, which seems to be linked to high sedimentation rates, is particularly widespread in the C8 core, and is common in the C5 core. In the Glacial interval of the C9 core many samples are barren, or ostracod assemblages are very poor, probably because of anoxic conditions on the bottom.

The bathymetric curves, however, show good agreement with the palaeotemperature trends obtained from Foraminifera, calcareous nanno-fossils and pollen (Fig. 7, see below). The inferred depths correlate with the Glacial Maximum of the IS 2 range between -126 m and -115 m with respect to the present sea level. The base of the Holocene interval is characterized by a sharp rise in sea level.

The productivity (number of specimens/g) is greater within the Late Glacial period. Species diversity patterns in ostracod assemblages have been calculated by the reversed formula of Simpson, $N(N-1)/\Sigma n(n-1)$, where N is the number of specimens in a sample and n_1, n_2, etc., are the number of specimens of each species in that sample. Diversity shows maximum values in cool, temperate–cool intervals, where the calculated depths are 80–100 m. Productivity and diversity are also influenced by reworking and terrigenous inputs in agreement with Breman (1978). Predation by gastropods is rather common in the C5 and C8 cores at the inferred depths of 40–60 m. Favourite prey species seem to be *Pontocythere turbida* (G. W. Müller) and *Cytheridea neapolitana* Kollmann.

Bythocythere turgida Sars, *Flexuocythere* sp., *Krithe praetexta* (Sars) and *Macrocypris adriatica* Breman have their last occurrence at different levels of the studied time interval.

Palynology (E. Russo Ermolli)

The analysed samples are rather poor in pollen grains; many levels were barren, especially in the C5 and C8 cores. Standard chemical treatments were used to concentrate the organic material. Because of the generally low concentration values it was not always possible to reach a pollen total of 300 grains, although a minimum of 150 grains were counted in the poorest samples.

Pollen analysis results are presented in frequency curves (Figs 5a, 6a and b) where the identified taxa are grouped in 10 classes following their presumed ecological and climatic requirements (Suc 1984).

Class 1 includes all the elements of the deciduous forest dominated by deciduous *Quercus*, accompanied by *Carpinus*, *Alnus* and, to a minor extent, *Ulmus*, *Tilia* and *Corylus*. This association was probably present, as today, on the slopes surrounding the Garigliano river plain up to about 500 m (Ozenda 1975). This vegetational association lives in warm and humid atmospheric conditions. *Alnus* was probably more abundant on wetter soils like those found in the plain or in association with minor flow lines. Class 2 includes *Fagus* and *Betula*, elements of the altitudinal forest belt. This vegetational association today occupies the highest forested belt in the Apennine chain (Ozenda 1975); it lives in a wet climate and can tolerate low temperatures. Class 3 includes the Mediterranean xerophytes, here mainly represented by *Quercus* type *ilex*. They now extend along the sea coast but can sometimes be mixed with the deciduous forest in the lowest levels of the supra-mediterranean belt (Ozenda 1975). Class 4 is represented by all the herbaceous elements (mainly Poaceae, Asteraceae and Chenopodiaceae), whose abundance in some levels is indicative of an open landscape in dry and cold climatic conditions. The same indication is given by the steppe elements of Class 5 dominated by *Artemisia*, a typical glacial climate indicator at our latitudes (Suc & Zagwijn 1983; Van der Hammen *et al.* 1971). In Class 6 are included not only the indeterminate grains, but also those taxa having no particular ecological meaning, such as Rosaceae, *Vitis* and water plants. *Pinus* and the indeterminate Pinaceae are the only elements of Class 7. As a result of the generic level of determination it is difficult to give *Pinus* a precise place in the vegetation associations. It was, perhaps mixed in the deciduous forest or occupied some stands in open spaces. Class 8 is represented by *Abies*, an altitudinal forest element now very rare in the Apennine chain. It has been excluded from Class 2, although it has the same ecological meaning, because of the problematic interpretation of bisaccate pollen in marine sediments. In fact, the abundance of *Abies* and *Pinus* pollen grains in marine records can mislead the interpretation of pollen spectra, because their frequencies can be disrupted by long-distance transport, typical for bisaccate pollen grains. It is not by chance that the highest percentages of both *Pinus* and *Abies* fall within cold periods where the absence of other trees in the landscape makes the bisaccate pollen relatively more abundant. Class 9 is represented by monolete and trilete spores. They give no particular climatic indication but their abundance may be linked to an increase of fluvial influx. Class 10 is represented by dinocysts,

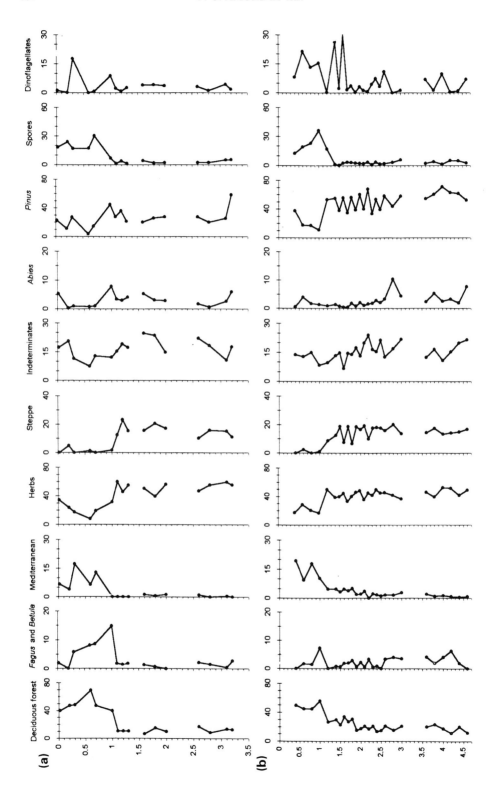

Fig. 6. Pollen class percentages plotted against depth (m) in the C8 (**a**) and C9 (**b**) cores.

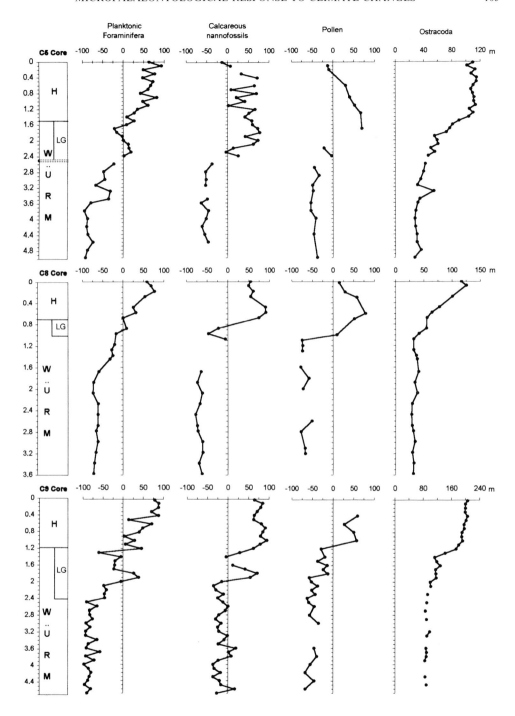

Fig. 7. Climatic curves based on planktonic Foraminifera, calcareous nannofossils and pollen assemblages. Bathymetric reconstruction based on ostracod assemblages.

whose increase could indicate an environmental change linked to a sea-level rise, a temperature rise or high nutrient availability. Class 1–6 percentages are calculated on a pollen sum in which *Pinus*, *Abies*, spores and dinocysts are excluded.

The warm climate indicators used for the reconstruction of the 'palaeoclimatic curves' (Fig. 7) are the deciduous forest and the Mediterranean elements (*a*), whereas the herbaceous and steppe elements have been used as cold indicators (*b*). The role of deciduous *Quercus* as a good terrestrial climate indicator has already been shown in a Tyrrhenian record (Rossignol-Strick & Planchais 1989).

The recorded changes in the vegetation associations throughout the cores have been correlated with the climatic changes that occurred during the last climatic cycle. In particular the upper parts of the cores represent the Holocene period whereas the lower parts are indicative of a glacial period that has been correlated with the Last Glacial period, where no sedimentation hiatus was recorded. In the basal parts of the cores herbs and steppe elements dominate the pollen spectra; the deciduous forest elements reach a maximum of about 20%. *Pinus* is the main arboreal taxon. In the upper parts of the cores a different situation can be outlined. The rapid rise of deciduous forest elements (up to 70%) precedes the increase of Mediterranean taxa and seems to be contemporaneous with the decrease of herbs and steppe elements. *Pinus* and *Abies* definitely decrease in frequency. This change in the vegetation pattern can be correlated with the warming trend that characterizes the Holocene period.

The first climatic improvement is recorded in the Mediterranean region between 15 ka (Rossignol-Strick & Planchais 1989; Rossignol-Strick 1995) and about 13 ka BP (Cheddadi *et al.* 1991; Rossignol-Strick *et al.* 1992; Zonneveld 1996), and the beginning of the Holocene period dates to 10 ka BP. The rapid vegetational change recorded in the eastern Mediterranean region (Cheddadi *et al.* 1991; Rossignol-Strick *et al.* 1992) and in the Adriatic Sea (Zonneveld 1996) in response to the Late Glacial–Holocene climate change shows, on the whole, the same pattern as recorded in Gaeta Bay. However, the observed differences in the floral composition of the pollen spectra can be explained by the different source area of pollen rains.

Discussion

On the basis of the interpretation of seismic profiles (Senatore 1996), the sediments underlying the Late Glacial–Holocene covering may be correlated with IS 2 in the C9 core, with IS 2–3 in C8 core, and with IS 4–5 in the oldest C5 core.

In the lower unit of the C5 core the base of the *E. huxleyi* Acme Zone has been recorded at 2.8 m; Gartner (1977) dated this event at 70 ka BP. This event is equivalent to the reversal dominance of *Gephyrocapsa caribbeanica–E. huxleyi* (Rio *et al.* 1990), which is a diachronous event, and is recorded in transitional waters at *c.* 73 ka BP (Thierstein *et al.* 1977). Flores *et al.* (1997) identified, in a cold interval just after IS 5, a reversal in abundance of *G. mullerae–E. huxleyi* from the western Mediterranean; this event is equivalent to the reversal dominance of *G. caribbeanica–E. huxleyi* of Thierstein *et al.* (1977), and is dated at 70–73 ka BP. On this basis, the lower unit of the C5 core may be correlated with the lower part of IS 4 near the 4–5 Stage boundary. Both planktonic Foraminifera and calcareous nannofossils show the dominance of cold-water species. Herbs and steppe elements dominate the pollen spectra.

Towards the top of the unit, just below the unconformity, a slight increase of warm-water species occurs. The infralittoral species *Cytherois uffenordei* and *Leptocythere ramosa* dominate ostracod assemblages. The calculated depth in the lower part of the unit is between −85 m and −80 m compared with present sea level. After the sea level rises, and just below the unconformity, it is −65 m with respect to the present. Reworking is conspicuous, and the diversity and productivity at the bottom are low.

The glacial interval of the C8 and C9 cores may be correlated with IS 2. Higher percentages of cold-water species such as *C. pelagicus* and *T. quinqueloba*, together with the general low values of warm-water species among both calcareous nannofossils and planktonic Foraminifera, suggest a colder climatic event than that recorded in the C5 core. In the cores the anomalously high percentages of *C. pelagicus*, a species that seems to be favoured by eutrophic conditions (Cachao & Moita 1995), might be also related to the high nutrient availability. The estimated palaeo-depths range from −90 to −100 m in the C8 core and from −115 to −126 m in the C9 core with respect to the present sea level. These different values have probably felt the effects of the strong reworking recorded in the C8 core and of the poor (or barren) ostracod associations in the C9 core. In the C9 core cyclical peaks of *N. dutertrei* are recorded. The abundance of this species has been related to low salinity of the surface waters (Vergnaud Grazzini *et al.* 1977; Thunnell & Lohmann 1979) or to the development of a 'deep chlorophyll maximum' (Rohling & Gieskes 1989). Fresh-water inputs from the continent and the consequent sea-water stratification may be assumed because of the scattered presence of fresh-water (*Candona*) or hypohaline (*Ilyocypris gibba*) ostracods, and the presence of spores usually linked to fluvial influxes. This water stratification and a poor oxygenation at the bottom may be indicated, in the more distal C9 core, by the very low productivity of ostracods, and the high amount of iron pyrite associated with *N. dutertrei* peaks.

The presence of hyposaline ostracods and spores falls within a restricted interval in both cores. In this interval the minimum depth has been calculated at about −100 m in the C8 core and −126 m in the C9 core. This interval might be correlated with the Last Glacial Maximum. At the same core depth, the maximum values of *C. pelagicus* and minima of warm-water species among the calcareous nannofossils occur. The ostracod species *Flexuocythere* sp. disappears within this interval.

The first gradual warming, correlated with the onset of the Late Glacial period, can be observed in the C8 core at −1 m and in the C9 core at −2 m. In the C5 core the Late Glacial period marks a transgressive event (−2.4 m) and its base is missing. The beginning of the Late Glacial period is characterized, among calcareous nannofossils, by a dramatic reduction in *C. pelagicus* and by an increase in warm-water taxa together with a further increase in *E. huxleyi*. At the same time, planktonic Foraminifera show a decrease in *T. quinqueloba* together with a first increase in *G. ruber*.

Only Foraminifera give some clues about the Older Dryas Event in the C9 core (2.4–2.05 m). Effectively, on the basis of ostracods, a sea-level rise of about 10 m has been calculated in the same interval.

The Late Glacial period shows variations in temperature and, consequently, in the sea level. Foraminiferal associations are characterized by a remarkable increase in *G. ruber* and *G. inflata*, and by the first appearance in the cores of *Orbulina universa* d'Orbigny, whereas *T. quinqueloba* and *N. pachyderma* decrease. Calcareous nannofossils are often absent during the Late Glacial period, probably because of diatom blooms.

The short Younger Dryas Event has clearly been recognized only in the C9 core. Pollen record the event by a decrease in the deciduous forest elements. A fall of about 10 m in sea level has been calculated. Planktonic Foraminifera and calcareous nannofossils show a slight decrease in warm-water species and a parallel increase in cold-water species.

Throughout the Late Glacial period trophic phenomena have been recorded in the water column. The grazing species *N. pachyderma*, and especially *G. inflata*, increase, and relative peaks of the philoeutrophic *C. pelagicus, C. leptoporus* and *H. carteri* occur. *N. dutertrei* records its last high peaks. Diatoms and dinocysts become common, especially in the C9 core. Spores have a maximum in abundance (C5 and C8 cores) and ostracod productivity increases sharply. The high nutrient content of the sea water in all the cores might be explained by the rise in temperature and by a runoff increase throughout the humid Late Glacial period. This is also suggested by the maximum in abundance of *Braarudosphaera bigelowii*, a calcareous nannofossil species probably linked to low salinity resulting from an increase of regional rainfall and/or eutrophic environments. Within the Late Glacial period the ostracod species *Bythocythere turgida* and *Krithe praetexta* appear to become extinct.

The Pleistocene–Holocene boundary has been detected in all the cores; some uncertainty exists with regard to the C8 core because of reworking and high terrigenous inputs. Among planktonic Foraminifera and calcareous nannofossils a noteworthy change in the composition of assemblages takes place by a general increase in warm-water species.

G. truncatulinoides and *G. siphonifera*, which are absent throughout the glacial intervals, and high values of *U. sibogae, R. clavigera* and *Syracosphaera* spp. suggest the boundary. At the same time, deciduous forest elements increase rapidly whereas herbs and steppe elements show, in contrast, a decrease. A sharp and conspicuous rise in the sea level is suggested by the replacement of infralittoral ostracod assemblages by circalittoral ones (C5 and C8 cores), and of circalittoral by outer circalittoral–epibathyal associations (C9 core). In the C8 core part of the Holocene sequence seems to be missing.

The Climatic Optimum is an elusive and disputable event. In all the cores, within the Holocene period, high peaks of *Globigerinoides* spp., associated with the presence of *G. sacculifer*, peaks of *R. clavigera* and the tropical genus *Ceratolithus* occur, indicating a further warming. The relative abundance of *G. sacculifer* has been recognized, with some uncertainty regarding the age, throughout the Mediterranean Sea around 5 ka BP (Pujol & Vergnaud Grazzini 1989; Capotondi 1995; Capotondi *et al.* 1996).

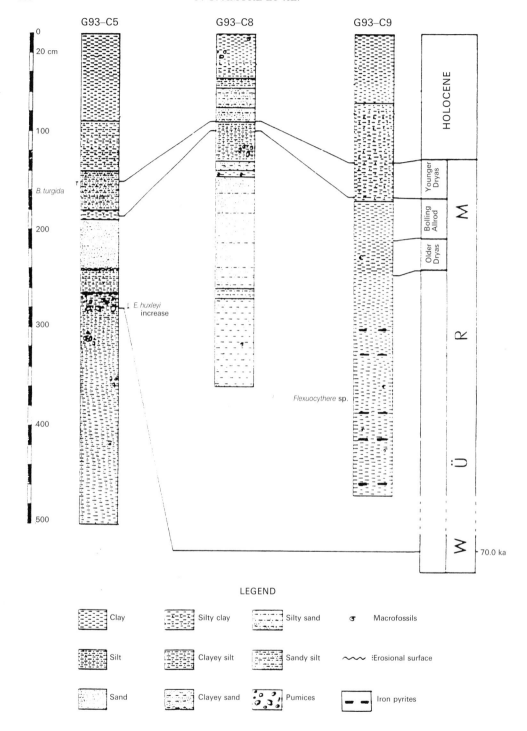

Fig. 8. Lithology and correlations of G93–C5, –C8 and –C9 cores.

G. sacculifer and *G. ruber* may be considered as competitive species. Salinity seems to be the primary factor controlling the domination of one species over the other (Bé & Tolderlund 1971). In fact, *G. sacculifer* lives today in the Balearic Basin and off the North African coast, whereas in the Levantine Basin, characterized by higher salinity values, this species is absent and *G. ruber* dominates the assemblages (Pujol & Vergnaud Grazzini 1995). In our opinion, the relative abundance of *G. sacculifer* recognized in the cores should not be directly linked to the rise in temperature but to an enhanced Atlantic influx, giving rise to a slight decrease of salinity in the surface water. In the same interval, ostracod assemblages show the highest presence of deep-water species, although without any effect on the calculated sea level.

Near these events *Macrocypris adriatica*, an ostracod species that, in agreement with Breman (1976), becomes extinct during the Atlantic stage, disappears. In all the cores the highest values of deciduous forest elements always predate this further warming.

Conclusions

Clearly demonstrated in the present paper is the usefulness of integrated micropalaeontological studies in the achievement of interpretations of palaeoclimatological events and of the consequential influence on environmental factors and on sea level.

The three cores from Gaeta Bay, encompass the last 70 ka, but sediments between about 60 and 20 ka have not been recorded. The more proximal C5 core shows, below the Late Glacial–Holocene cover, glacial sediments that may correlate with the 4–5 IS boundary, as a result of the recognition of the *Emiliania huxleyi* Acme Zone. This event is dated to 70-73 ka BP (Gartner 1977; Flores *et al.* 1997).

The palaeoclimatic curves (Fig. 7) show a satisfying agreement both among themselves and with the bathymetric curves inferred from the ostracod assemblages. The outcome of the research is particularly satisfying because it represents the first attempt at the construction of a climatic curve on the basis of calcareous nannofossils.

In the more distal C8 and C9 cores, glacial sediments have been related to the Last Glacial period. This is a colder event compared with that recorded by the C5 core, as indicated by planktonic Foraminifera and calcareous nannofossils. Sea level might be −126 m (C9 core) relative to the present, as calculated on the basis of the ostracod associations. The Late Glacial

period has been recorded in all the cores. The Younger Dryas Event has been observed only in the C9 core.

The Late Pleistocene–Holocene boundary is characterized by a remarkable increase of warm-water species among Foraminifera and calcareous nannofossils, and warm elements among pollen. The sea level rises rapidly. The Climatic Optimum might correspond to the contemporaneous peaks of *Globigerinoides* spp., including *G. sacculifer*, and of *Ceratolithus* spp. and *R. clavigera* (Fig. 8).

Anoxic events at the bottom occur, especially in the C9 core, through IS 2, as the very low productivity of ostracod associations and the abundant iron pyrite testify. The phenomenon might be triggered by fresh-water inputs from the continent, as peaks of the salinity-sensitive *N. dutertrei*, and the occurrence of spores and of hyposaline ostracod taxa indicate. The Late Glacial period and the earliest Holocene period provide a very productive intervals. Spores, dinocysts, diatoms and eutrophic species among Foraminifera and calcareous nannofossils become abundant, and ostracod productivity sharply increases. Throughout the Holocene period ostracod productivity and diversity again decrease. Reworking is widespread, especially within Glacial and Late Glacial intervals.

The authors thank T. S. Pescatore, who kindly supplied the cores.

References

AMORE, F. O., CIAMPO, G., DI DONATO, V. *et al.* 1996. A multidisciplinary study of Late Quaternary sediments in the Gaeta Bay continental shelf (Southern Italy). *Il Quaternario*, **9**, 521–532.

BARTOLINI, C. 1970. Coccoliths from sediments of the western Mediterranean. *Micropaleontology*, **16**, 129–154.

BE, A. W. H. 1969. Distribution of selected groups of marine invertebrates in waters south of 35° latitude. Antarctic Map Folio Service, **11**, 9–12.

—— & TOLDERLUND, D. S. 1971. Distribution and ecology of living planktonic foraminifera in surface waters of the Atlantic and Indian Oceans. *In*: FUNNELL, B. M. & RIEDEL, W. R. (eds) *The Micropaleontology of Oceans*. Cambridge University Press, Cambridge, 105-149.

BLANC-VERNET, L., FROGET, C. & SGARRELLA, F. 1983. Paléoclimatologie d'une carotte de la Mer Tyrrhénienne. *Géologie Méditerranéenne*, **10**, 93–104.

BONADUCE, G., CIAMPO, G. & MASOLI, M. 1975. Distribution of Ostracoda in the Adriatic Sea. *Pubblicazioni Stazione Zoologica Napoli*, **40**, 1–154.

BREMAN, E. 1976. *Distribution of ostracodes in the bottom sediments of the Adriatic Sea*. Vrije Universiteit Amsterdam, Offset Krips Repr. Meppel, 1–165.

—— 1978. Species diversity of ostracode shells in bottom sediments of the Adriatic Sea. *Palaeogeography, Palaeoclimatology, Palaeoecology*, **25**, 277–313.

CACHÃO, M. & MOITA, M. T. 1995. Coccolithus pelagicus a sort of productivity proxy? *6th International Nannoplankton Conference, Copenhagen Abstracts*, 33–34.

CAPOTONDI, L. 1995. Biostratigraphic changes and paleoceanographic record in the Corsica basin (Tyrrhenian Sea) during the last 18000 years. *Paleopelagos*, **5**, 215–226.

——, MORIGI, C., CESPUGLIO, G. & BORSETTI, A. M. 1996. Biostratigraphic and paleoceanographic records during the last deglaciation in the southern Adriatic Sea. *In*: EVANS, S. P., FRISIA, S., BORSATO, A., CITA, M. B., LANZINGER, M., RAVAZZI, C. & SALA, B. (a cura di). *Modificazioni climatiche ed ambientali tra il Tardiglaciale e l'Olocene antico in Italia. Abstracts Convegno AIQUA -MTSN, Trento, 7–9 Febbraio 1996*, 146–147.

CHEDDADI, R., ROSSIGNOL-STRICK, M. & FONTUGNE, M. 1991. Eastern Mediterranean palaeoclimates from 26 to 5 ka BP documented by pollen and isotopic analysis of a core in the anoxic Bannock Basin. *Marine Geology*, **100**, 53–66.

FLORES, J. A., SIERRO, F. J., FRANCÉS, G., VASQUEZ, A. & ZAMARREÑO, I. 1997. The last 100,000 years in the western Mediterranean: sea surface water and frontal dynamics as revealed by coccolithophores. *Marine Micropaleontology*, **29**, 351–366.

GARTNER, S. 1977. Calcareous nannofossil biostratigraphy and revised zonation of the Pleistocene. *Marine Micropaleontology*, **2**, 1–25.

HEMLEBEN, C. H., SPINDLER, M. & ANDERSON, O. R. 1989. *Modern Planktonic Foraminifera*. Springer, New York.

JORISSEN, F. J., ASIOLI, A., BORSETTI, A. M. *et al.* 1993. Late Quaternary central Mediterranean Biochronology. *Marine Micropaleontology,* **21**, 169–189.

LACHENAL, A. M. & BODERGAT, A. M. 1990. Les ostracodes et les variations paléoeustatiques du Golfe de Gabès (Méditerranée) depuis 30.000 ans. *Bulletin de la Société Géologique de France*, **1**, 113–122.

KNAPPERTSBUSCH, M. 1993. Geographic distribution of living and Holocene coccolithophores in the Mediterranean Sea. *Marine Micropaleontology*, **21**, 219–247.

McINTYRE, A. & BÉ, A. W. H. 1967. Modern coccolithophorids of the Atlantic, I. Placoliths and cyrtoliths. *Deep-Sea Research*, **14**, 561–597.

MUERDTER, D. R. & KENNETT, J. P. 1984. Late Quaternary planktonic foraminiferal biostratigraphy, Strait of Sicily, Mediterranean Sea. *Marine Micropaleontology*, **8**, 339–359.

MÜLLER, C. 1979. *Les nannofossiles calcaires*. La Mer Pelagienne, 210–220.

OKADA, H. & McINTYRE, A. 1979. Seasonal distribution of modern coccolithophores in the Western North Atlantic Ocean. *Marine Biology*, **54**, 319–328.

OTTENS, J. J. 1991. Planktic Foraminifera as North Atlantic water indicators. *Oceanologica Acta*, **14**, 123–140.

OZENDA, P. 1975. Sur les étage de végétation dans les mantagnes du Bassin Méditerranéen. *Documents de cartographie écologique*, **16**, 1–32.

PUJOS, A. 1992. Calcareous nannofossils of Plio-Pleistocene sediments from the northwestern margin of tropical Africa. *In*: SUMMERHAYES, C. P., PRELL, W. L. & EMEIS, K. C. (eds) *Upwelling Systems: Evolution since the Early Miocene*. Geological Society, London, Special Publications, **64**, 343–359.

PUJOL, C. & VERGNAUD GRAZZINI, C. 1989. Paleoceanography of the last deglaciation in the Alboran Sea (Western Mediterranean). Stable isotopes and planctonic foraminiferal records. *Marine Micropaleontology*, **15**, 153–179.

—— & —— 1995. Distribution patterns of live planktic foraminifers as related to regional hydrography and productive systems of the Mediterranean Sea. *Marine Micropaleontology*, **25**, 187–217.

RIO, D., RAFFI, I. & VILLA, G. 1990. Pliocene–Pleistocene calcareous nanofossil distribution patterns in the western Mediterranean. *Proceedings of the Ocean Drilling Program*, **107**. Ocean Drilling Program, College Station, TX, 514–531.

ROHLING, E. J. & GIESKES, W. W. C. 1989. Late Quaternary changes in Mediterranean intermediate water density and formation rate. *Paleoceanography*, **4**, 531–545.

ROSSIGNOL-STRICK, M. 1995. Sea–land correlation of pollen records in the eastern Mediterranean for the glacial–interglacial transition: biostratigraphy versus radiometric time-scale. *Quaternary Science Reviews*, **14**, 893–915.

—— & PLANCHAIS, N. 1989. Climate patterns revealed by pollen and oxygen isotope records of a Tyrrhenian sea core. *Nature*, **342**, 413–416.

——, ——, PATERNE, M. & DUZER, D. 1992. Vegetation dynamics and climate during the deglaciation in the south Adriatic basin from a marine record. *Quaternary Science Reviews*, **11**, 415–423.

ROTH, P. H. 1994. Distribution of coccoliths in oceanic sediments. *In*: WINTER, A. & SIESSER, W. (eds) *Coccolithophores*, Cambridge University Press, Cambridge, 199–218.

SANVOISIN, R., D'ONOFRIO, S., LUCCHI, R., VIOLANTI, D. & CASTRADORI, D. 1993. 1 Ma Paleoclimatic record from the Eastern Mediterranean–Marflux Project: first results of micropaleontological and sedimentological investigation of a long piston core from the Calabrian Ridge. *Il Quaternario*, **6**, 169–188.

SENATORE, M. R. 1996. Cicli eustatici di IV e V ordine in depositi del Pleistocene medio superiore-Olocene del Golfo di Gaeta (Mar Tirreno). *Atti della Riunione del Gruppo di sedimentologia del CNR. Catania 10–14 Ottobre 1996*, 242–244.

SUC, J. P. 1984. Origin and evolution of the Mediterranean vegetation and climate in Europe. *Nature*, **307**, 429–432.

—— & ZAGWIJN, W. H. 1983. Plio-Pleistocene correlations between the northwestern Mediterranean region and northwestern Europe according to recent biostratigraphic and paleoclimatic data. *Boreas*, **12**, 153–166.

THIEDE, J. 1983. Skeletal plankton and nekton in

upwelling water masses off northwestern South America and northwest Africa. *In*: SUESS, E. & THIEDE, J. (eds) *Coastal Upwelling, A*. Plenum, New York, 183–207.

THIERSTEIN, H. R., GEITZENAUER, K. R., MOLFINO, B. & SHACKLETON, N. J. 1977. Global synchroneity of late Quaternary coccolith datum levels: validation by oxygen isotopes. *Geology*, **5**, 400–404.

THUNNELL, R. C. & LOHMANN, G. P. 1979. Planktonic Foraminifera fauna associated with eastern Mediterranean Quaternary stagnations. *Nature*, **281**, 211–213.

VAN DER HAMMEN, T., WIJMSTRA, T. A. & ZAGWIJN, W. H. 1971. The floral record of the Late Cenozoic of Europe. *In*: TUREKIAN, K. (ed.) *The Late Cenozoic Glacial Ages*. Yale University Press, New Haven, CT, 291–494.

VAN LEEUWEN, R. J. V. 1989. *Sea-floor Distribution and Late Quaternary Faunal Patterns of Planktonic and Benthic Foraminifera in the Angola Basin*. Utrecht Micropaleontological Bulletin, **38**.

VERGNAUD GRAZZINI, C., RYAN, W. B. F. & CITA, M. B. 1977. Stable isotopic fractionation, climate change and episodic stagnation in the eastern Mediterranean during the Late Quaternary. *Marine Micropaleontology*, **2**, 353–370.

VIOLANTI, D., GRECCHI, G. & CASTRADORI, D. 1991. Paleoenvironmental interpretation of core Ban88-11Gc (Eastern Mediterranean, Pleistocene–Holocene) on the grounds of foraminifera, Thecosomata and calcareous nannofossils. *Il Quaternario*, **4**, 13–39.

——, PARISI, E. & ERBA, E. 1987. Fluttuazioni climatiche durante il Quaternario nel Mar Tirreno, Mediterraneo occidentale (carota PC-19 BAN 80). *Rivista Italiana Paleontologia e Stratigrafia*, **92**, 515–570.

WHATLEY, R. C. & COLES, G. P. 1991. Global change and biostratigraphy of North Atlantic Cainozoic deep water Ostracoda. *Journal of Micropalaeontology*, **9**, 119–132.

YOUNG, J. R. 1994. Functions of coccoliths. *In*: WINTER, A. & SIESSER, W. G. (eds) *Coccolithophores*. Cambridge University Press, Cambridge, 63–82.

ZONNEVELD, K. A. F. 1996. Palaeoclimatic reconstruction of the last deglaciation (18–8 ka BP) in the Adriatic Sea region; a land–sea correlation based on palynological evidence. *Palaeogeography, Palaeoclimatology, Palaeoecology*, **122**, 89–106.

Cooling evidence from Pleistocene shelf assemblages in SE Sicily

I. DI GERONIMO, R. DI GERONIMO, R. LA PERNA, A. ROSSO &
R. SANFILIPPO

Catania University, Dip. Scienze Geologiche di Oceanologia e Paleoecologia, Corso Italia, 55, I–95129 Catania, Italy

Abstract: A short Lower Pleistocene section, cropping out in SE Sicily, is studied. The sequence mainly consists of richly fossiliferous sandy layers with *Arctica islandica*, deposited in mid-shelf environments. Molluscs, bryozoans, serpuloideans and calcareous algae are investigated, to identify species with palaeoclimatological implications. Apart from some well-known North Atlantic molluscs, other species are identified as palaeoclimatological tools. Cooling evidence is provided by such palaeoclimatological indicators and by the increasing shell size in some species.

Since the last century, some North Atlantic species have been recorded from the Quaternary sediments of the Mediterranean region and are currently known as Boreal Guests (BGs). Deposits bearing BGs crop out largely in Southern Italy. Such species, used early in Quaternary biostratigraphy (Mars 1963; Ruggieri *et al.* 1976), have been more recently used as palaeoclimatological tools (Di Geronimo, 1974, Malatesta & Zarlenga 1986; Raffi 1986; Taviani *et al.* 1991).

Molluscs provide most BGs, such as the best-known one, *Arctica islandica*. Other benthic invertebrates have been only sporadically investigated with palaeoclimatological and palaeobiogeographical aims. Besides molluscs, three additional benthic groups (bryozoans, spirorbids and calcareous algae), never previously studied in Quaternary Mediterranean palaeoclimatology, are dealt with in the present work. Data from these groups are used to construct distribution curves of 'cold' species through the section, from which a palaeoclimatological trend may be inferred.

Description of the section

The studied section crops out in SE Sicily, *c.* 7 km SW of Augusta, Syracuse (Fig. 1). It is located along the NE edge of the Hyblean Plateau, a Mesozoic–Tertiary carbonate sequence (Pedley 1981; Grasso *et al.* 1982). Two Pleistocene sedimentary cycles are recognized in the area (Carbone 1985): the older consists of yellowish calcarenites and biogenic sands, grading up into bluish clays, and the younger consists of biocalcarenites ('panchina'). The older cycle was deposited during the later part of Early Pleistocene time, the younger during mid-Pleistocene time. On the Ionian side of SE Sicily, Pleistocene sediments were deposited in small palaeobasins controlled by NW–SE fault systems (Grasso 1993). An outer shelf (80–100 m) environment is suggested by micro- and macrofaunal assemblages for the bluish clays (Carbone 1985). Sandy beds, bearing *Arctica islandica*, occur within the clays (two of them were reported by Carbone (1985)).

The present study deals with one of the sandy *Arctica* beds, cropping out along the west side of the Marcellino river (Fig. 1). The studied section, 1.35 m thick, is well exposed along a recently excavated drain (Fig. 2). The sequence (Fig. 3) can be subdivided into three parts. The basal part (0.48 m thick) consists of fine sands (75–80%) with minor muddy and gravelly fractions. The macrofaunal content is relatively low, with two *Aequipecten opercularis* beds at the bottom, and an *Ensis ensis* layer in the middle. The mid-part of the section (0.64 m thick), is made up by fine sands (60–80%) with a low mud content. It is richly fossiliferous with obvious *Arctica islandica* layers. In the upper part (0.23 m thick) a remarkable muddy fraction (up to 60%) occurs together with a rare coarse-grained fossiliferous content.

The macrofaunal composition points to mid-shelf (30–80 m) sedimentation, with high bio-

From: HART, M. B. (ed.) *Climates: Past and Present.* Geological Society, London, Special Publications, **181**, 113–120, 1-86239-075-4/$15.00

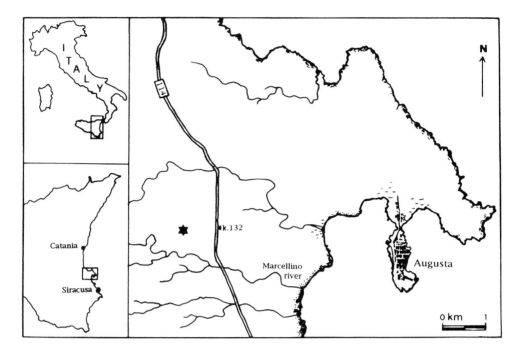

Fig. 1. Location of the Marcellino section.

genic production. The middle part of the section was deposited under alternating low and high hydrodynamic phases, the former leading to the development of *Arctica* communities on muddy–sandy bottoms, the latter causing scouring of the substrate and coarse-grained biogenic sedimentation. During this time, the *Arctica* valves became heavily colonized by bryozoans, serpulids and spirorbids.

Taphonomic evidence mainly points to *in situ* deposition and burial, as testified by some infaunal bivalves in life position or with joined valves (*Arctica islandica, Ensis ensis, Dosinia lincta, Acanthocardia mucronata*), and by the lack of signs of abrasion. Although most fossils are broken (i.e. the thick *Arctica* valves and the brittle bryozoan colonies), this is largely due to diagenesis, as fragments often are still in very close proximity.

Material and methods

A continuous bulk sampling of the section (Fig. 3) was planned by means of a preliminary field survey. The section was sampled along a column 0.5 m wide and 0.2 m deep. Fourteen bulk-samples were obtained, according to evident textural and macrofaunal changes. Samples 1–4 came from the basal part, 5–12 from the mid-part, and 13 and 14 from the top part. Samples were used for palaeontological and sedimentological analyses; the former devoted to species of palaeoclimatological interest. Four benthic taxa were examined: molluscs, bryozoans, serpuloideans and calcareous algae.

Three groups of species were distinguished based on their palaeoclimatological interpretation: Boreal Guests (BG), Residual Boreal Guests (RBG) and species of Atlantic affinity (Aa). As generally agreed, BGs are North Atlantic shelf species, whose latitudinal range shifted southwards into the Mediterranean during cold Pleistocene phases. Some BGs became permanent Mediterranean components, not withdrawing during interglacial phases (RBG). Their present-day Mediterranean distribution is limited to the coldest waters (Alboran Sea, Western Basin, Adriatic Sea). Other species show taxonomic affinities to Recent North Atlantic taxa and are known only from Pleistocene Mediterranean shelf deposits (species of 'Atlantic affinity'). For each taxonomic group, log-graphs of BG, RBG and Aa species occurrence and distribution through the section are given (Fig. 4). Finally, other species are at present distributed both in Atlantic and Mediterranean waters, but they are notably abundant in Pleistocene sediments with Boreal

Fig. 2. Section outcrop (**a**) and close-up of an *Arctica* bed (**b**).

Guests. The rare Recent records of these species, or their limited distribution (compared with the Pleistocene distribution) suggest a marked cold-water preference. They are therefore identified as 'Cold-water' species (Cw).

Results

Eight BG molluscs were detected, namely, the bivalves *Arctica islandica* (Linnaeus), *Dosinia lincta* (Pultney), *Macoma obliqua* (Sowerby), *Spisula elliptica* (Brown), *Pseudamussium septemradiatum* (Müller), *Camptonectes tigerinus* (Müller) and the gastropods *Acmaea testudinalis* (Müller) and *Neptunea contraria* (Linnaeus). *A.*

islandica occurs from sample 4 to the top, being abundant in the mid-part. *D. lincta* occurs from base to top but is more abundant in the lower to mid-part. The other species are less abundant and most co-occur in the mid-part of the section (*M. obliqua* from samples 8 to 11, *S. elliptica* from 6 to 13, *P. septemradiatum* from 6 to 8, *C. tigerinus* from 7 to 14, *A. testudinalis* from 3 to 10, *N. contraria* from 5 to 10). The present-day distribution of these species ranges from Arctic latitudes south to the British Isles (see review by Malatesta & Zarlenga (1986)), except for *M. obliqua*, which is extinct (Bellomo & Raffi 1991). All of them are well known from Pleistocene shelf deposits, except for *D. lincta*, which has

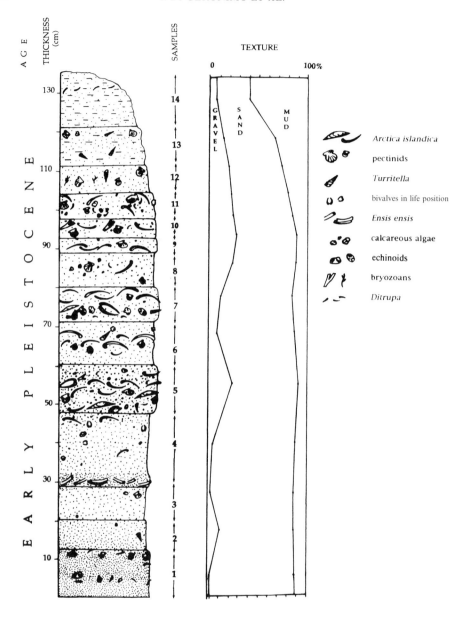

Fig. 3. Stratigraphic and sedimentological log of the studied section.

been only rarely regarded as a BG (Di Geronimo 1979–1980). It has often been reported as a larger and thick-shelled variety or subspecies of *D. lupinus* (Linnaeus) (see Sganga (1963), plate 7, fig. 4a and b). Although the Atlantic distribution of *D. lincta* is poorly known (it has been often synonymized with *D. lupinus*) its full specific status should be stressed. The larval shell of *D. lincta* points to a long planktonic

development (planktotrophic), whereas that of *D. lupinus* indicates a shorter larval stage (probably lecitotrophic).

Two BG species, *Spirorbis spirorbis* (Linnaeus) and *Spirorbis corallinae* de Silva & Knight-Jones, were detected among the serpuloideans. They are widespread throughout the section from samples 2 to 12, being especially frequent in the mid-part. They are Boreal–Arctic

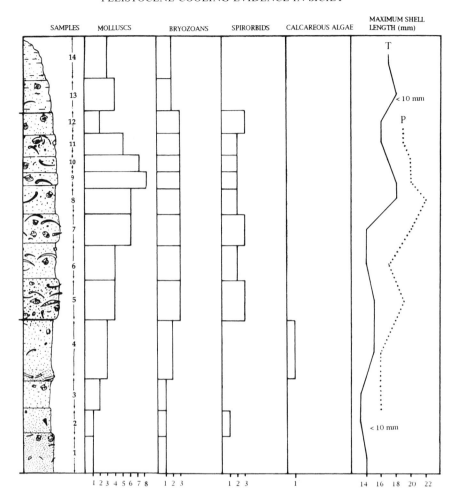

Fig. 4. Palaeoclimatical indicators (BG, RBG and Aa species) through the section (bars indicate the number of palaeoclimatical indicators per sample) and maximum shell length of *Timoclea ovata* (T) and *Plagiocardium papillosum* (P).

taxa, the former ranging from the Arctic Sea south to Nova Scotia, the latter known from the southern British Isles and the Barents Sea (Knight-Jones *et al.* 1991; Rzhavsky 1992). Both spirorbids have a shallow shelf distribution and have been only recently reported as BGs (Sanfilippo 1998).

Bryozoans are represented by a few specimens of *Haplota clavata* (Hincks) from the mid-part of the section (samples 6 and 7). This inconspicuous species is at present known (Ryland & Hayward 1977) as living on boreal North Atlantic shelves (St Lawrence Gulf to Cape Cod and North Sea). *H. clavata* could also be regarded as a Pleistocene Mediterranean BG.

A sole BG was detected among the calcareous algae. *Phymatolithon laevigatum* (Foslie) Foslie, occurs only in sample 4. It is a Boreal–Artic species, distributed both in the eastern and western North Atlantic shelves (Adey 1964; Parke & Dixon 1964; Irvine & Chamberlain 1994).

The bryozoan *Palmiskenea skenei* (Ellis & Solander) and the bivalve *Venerupis rhomboides* (Pennant) represent the RBG group. The former was found in samples 4–14, and is common in samples 8 and 9. Its RBG interpretation was identified by Rosso & Di Geronimo (1998). *V. rhomboides* occurs in sample 5. Its present-day Mediterranean occurrence as an RBG was discussed by Di Geronimo (1975) and Panetta & Dell'Angelo (1977).

The Aa group comprises the bivalve *Crenella rosariae* La Perna, the spirorbid *Circeis* sp. and the bryozoans *Amphiblestrum* sp. 1 and *Phylactella* sp. 1. Both bryozoans seem to be new species, known only from the Marcellino section; the former in samples 8–12, the latter throughout but more common in the basal and top parts. *Amphiblestrum* sp. 1 has no Recent Mediterranean counterpart, but shows strong affinities with two Recent North Atlantic species: *A. solidum* (Packard) (*sensu* Hayward & Ryland 1998) and *A. flemingi* (Busk). The latter is also known as a fossil from the Mediterranean area, where reliable records are only from cold Pleistocene phases (Rosso, pers. comm. 1999). *Phylactella* sp. 1 is similar to the Northeastern shelf species *P. labrosa* (Busk). Another, probably distinct, *Phylactella* is known from the Pleistocene and Recent Italian region (Rosso 1996: as *P. labrosa*). *Crenella rosariae* has been recently described from the present section (La Perna 1998), where it occurs in samples 10 and 11. It is very similar to the Boreal–Arctic *C. decussata* (Montagu) and is the first *Crenella* so far known from the Mediterranean region. *Circeis* sp., found in samples 2–12, still needs to be compared with the North Atlantic *C. armoricana* Saint-Joseph, *C. spirillum* (Linnaeus) and *C. guiganovae* Rzhavsky. *Circeis* is unknown from the Recent Mediterranean (Knight-Jones *et al.* 1991) and should, therefore, be regarded as a BG or Aa species.

The Cw group comprises the serpulid *Hydroides norvegicus* Gunnerus and the bivalve *Lepton squamosum* (Montagu), both Boreal–Mediterranean species. *H. norvegicus* is abundant throughout the section, especially in the mid-part. This species, although common in the Recent Mediterranean (Zibrowius 1972), is more abundant in Pleistocene shelf sediments. *L. squamosum* is common through the section. This species is rather common in Pleistocene sediments, whereas at present it is known only from the Western Mediterranean (Bucquoy *et al.* 1887–1898; Glibert & Van de Poel 1971).

The algal species *Lithothamnion bornetii* (Foslie) Foslie, from samples 2 and 5, is an Atlantic species known from Boreal waters (Parke & Dixon 1964; Irvine & Chamberlain 1994), except for a few records from southern latitudes. Although the distribution of this species needs revision, this first Mediterranean record is important. Its palaeoclimatological significance could be compared with that of Aa and/or Cw species.

Some species (e.g. the bivalves *Timoclea ovata*

(Pennant) and *Plagiocardium papillosum* (Poli), the serpuloidean *Pomatoceros triqueter* (Linnaeus) and the echinoid *Echinocyamus pusillus* (Müller)) attain a markedly large size, 2–3 times larger than the Recent Mediterranean specimens. The two bivalves show an increasing trend in shell length from the base to the mid–upper part of the section (Fig. 4).

Conclusions

Three molluscan BG floods during Early Pleistocene time are known (Raffi 1986). They partially correspond to the chronostratigraphic divisions of Early Pleistocene time (Santernian, Emilian, Sicilian). The arrival of the BGs follows an increasing cooling trend and the third coldest flood, during which *S. elliptica* first appeared in the Mediterranean, occurred at *c.* 0.8 Ma (top of the Sicilian substage). As *S. elliptica* also occurs in the mid- to upper part of the studied section, the Marcellino *Arctica* beds can be referred to the later part of Early Pleistocene time. The deposits studied here should therefore be regarded as coeval with those of Ficarazzi (Di Stefano & Rio 1981; Buccheri 1984) and Valle del Belice (Di Stefano *et al.* 1991), both in western Sicily.

Besides some 'classical' BG molluscs, the studied section provides other species with the same, or similar, cold implications. The mollusc *C. rosariae*, the bryozoans *H. clavata*, *P. skenei*, *Amphiblestrum* sp. 1, *Phylactella* sp.1, the spirorbids *S. spirorbis*, *S. corallinae* and *Circeis* sp., and the calcareous alga *P. laevigatum* are reported for the first time from the Mediterranean region, except for the recently recorded *P. skenei* (Rosso & Di Geronimo 1998).

The three faunal groups show a common trend through the section (Fig. 4), with a maximum number of BG, RBG and Aa species in the mid-part. Eight cold-water molluscan species were recorded from sample 9. Molluscs show a better-defined pattern through the section, probably because of the greater diversity of this group, compared with the bryozoans or the spirorbids. Calcareous algae do not provide any trend, but much more work is needed to investigate the contribution of this group to palaeoclimatological investigations (Di Geronimo 1998).

The distribution of cold species through the section clearly points to a cold peak falling within the mid-part of the section i.e. the *Arctica* beds (samples 5–12). A parallel pattern is also shown by the valve size of *T. ovata* and *P. papillosum*, thus suggesting a common cause (i.e. water cooling). Size changes in benthic invertebrates are poorly investigated. However, large specimens could indicate a slowing down of metabolic rates, related to the lowering water temperature. Nicol (1964, 1967) remarked that 'some eurythermal pelecypod species having a wide geographic range grow more slowly; reach sexual maturity later; live longer; and attain a larger maximum size in the colder-water part than in the warmer-water portion of their geographic range'. Similar considerations could also explain the large size of *P. triqueter* and *E. pusillus*. The *Arctica* beds themselves testify to the development of rich communities of cold-water affinity, which also provided biogenic substrates for encrusting bryozoan and serpuloidean communities, among which the cold species developed.

Thanks are due to S. Spinoso for having called our attention to the outcrop, and R. Leonardi (Catania University) for help in fieldwork. H. M. Pedley (University of Hull, UK) and two anonymous referees reviewed and improved the manuscript.

References

ADEY, W. H. 1964. The genus *Phymatolithon* in the Gulf of Maine. *Hydrobiologia*, **24**, 377–420.

BELLOMO, E. & RAFFI, S. 1991. Osservazioni su *Macoma obliqua* (Sowerby, 1817), ospite boreale nel Pleistocene inferiore del Mediterraneo. *Bollettino della Società Paleontologica Italiana*, **30**, 3–6.

BUCCHERI, G. 1984. Pteropods as climate indicators in Quaternary sequences: a lower–middle Pleistocene sequence outcropping in Cava Puleo (Ficarazzi, Palermo, Italy). *Palaeogeography, Palaeoclimatology, Palaeoecology*, **45**, 75–86.

BUCQUOY, E., DAUTZENBERG, P. & DOLLFUS, G. 1887–1898. *Les Mollusques marins du Roussillon. 2. Pelecypodes*. J. B. Balliére, Paris.

CARBONE, S. 1985. I depositi pleistocenici del settore nord-orientale ibleo tra Agnone e Melilli (Sicilia SE): relazione tra facies e lineamenti strutturali. *Bollettino della Società Geologica Italiana*, **104**, 405–420.

DI GERONIMO, I. 1974. Significato paleoecologico di *Cylichna alba* nuovo 'Ospite Nordico' del Pleistocene della Sicilia. *Memorie del Museo Civico di Storia Naturale di Verona*, **20**(1972), 569–572.

——— 1975. La malacofauna siciliana del Ciaramitaio (Grammichele, Catania). *Conchiglie*, **11**(5–6), 101–137.

——— 1979–80. Paleocomunità di Molluschi e profondità di sedimentazione nel Quaternario della Sicilia. *Oceanis*, **5** (Horse Series), 283–299.

DI GERONIMO, R. 1998. *Le Corallinales del Quaternario dell'Italia Meridionale (Sistematica, paleoecologia e paleobiogeografia)*. PhD Thesis, University of Messina.

DI STEFANO, E. & RIO, D. 1981. Biostratigrafia a nannofossili e biocronologia del Siciliano nella località tipo di Ficarazzi (Palermo, Sicilia). *Acta*

Naturalia dell'Ateneo Parmense, **17**, 97–111.

———, SPROVIERI, R. & SCARANTINO, S. 1991. Biostratigrafia e paleoecologia della sezione infrapleistocenica di Case Parrino (Foce del Belice, Sicilia sud-occidentale). *Naturalista siciliano*, serie 4, **15**(3–4), 115–148.

GLIBERT, M. & VAN DE POEL, L. 1971. Mollusques cenozoiques nouveaux ou mal connus. *Bulletin de l'Institut royal des Sciences naturelles de Belgique*, **47**, 1–17.

GRASSO, M. 1993. Pleistocene structures at the Ionian side of the Hyblean Plateau (SE Sicily). Implications for the tectonic evolution of the Malta Escarpment. *In*: MAX, M. P. & COLANTONI, P. (eds) *Geological Development of the Sicilian–Tunisian Platform*. UNESCO Reports in Marine Science, **58**, 49–54.

———, LENTINI, F. & PEDLEY, H. M. 1982. Late Tortonian–Lower Messinian (Miocene) paleogeography of SE Sicily: informations from two new formations of Sortino Group. *Sedimentary Geology*, **32**, 279–300.

IRVINE, D. E. G. & CHAMBERLAIN, Y. M. 1994. *Seaweeds of the British Isles. 1. Rhodophyta, part 2B. Corallinales, Hildebrandiales*. HMSO, London.

KNIGHT-JONES, P., KNIGHT-JONES, E. W. & BUZHINSKAYA, G. 1991. Distribution and interrelationships of northern Spirorbid genera. *Bulletin of Marine Science*, **48**, 189–197.

LA PERNA, R. 1998. *Crenella rosariae* n. sp. (Bivalvia, Mytilidae) from the Mediterranean Pleistocene. *Bollettino dell'Accademia Gioenia di Scienze Naturali di Catania*, **30**(352), 269–283.

MALATESTA, A. & ZARLENGA, F. 1986. Northern guests in the Pleistocene Mediterranean Sea. *Geologica Romana*, **25**, 91–154.

MARS, P. 1963. Les faunes et la stratigraphie du Quaternaire méditerranéen. *Receuil des Travaux de la Station Marine d'Endoume*, **28**, 61–97.

NICOL, D. 1964. An essay on size of marine pelecypods. *Journal of Paleontology*, **38**, 968–974.

——— 1967. Some characteristics of cold-water marine pelecypods. *Journal of Paleontology*, **41**, 1330–1340.

PANETTA, P. & DELL'ANGELO, B. 1977. Il genere *Venerupis* Lamarck, 1818 nel Mediterraneo. *Conchiglie*, **13**(1–2), 1–26.

PARKE, M. & DIXON, P. 1964. A revised check-list of British marine algae. *Journal of the Marine Biological Association, UK*, **44**, 499–542.

PEDLEY, H. M. 1981. Sedimentology and palaeoenvironment of the Southeast Sicilian Tertiary platform carbonates. *Sedimentary Geology*, **28**, 273–291.

RAFFI, S. 1986. The significance of marine boreal molluscs in the Early Pleistocene faunas of the Mediterranean area. *Palaeogeography, Palaeoclimatology, Palaeoecology*, **52**, 267–289.

ROSSO, A. 1996. Valutazione della biodiversità in Mediterraneo: un esempio dai popolamenti a briozoi della Biocenosi del Detritico Costiero. *Biologia Marina Mediterranea*, **3**, 58–65.

——— & DI GERONIMO, I. 1998. Deep-sea Pleistocene bryozoa of Southern Italy. *Géobios*, **30**, 303–317.

RUGGIERI, G., BUCCHERI, G., GRECO, A. & SPROVIERI, R. 1976. Un affioramento di Siciliano nel quadro della revisione della stratigrafia del Pleistocene inferiore. *Bollettino della Società Geologica Italiana,* **94**, 889–914.

HAYWARD, P. J. & RYLAND, J. S. 1998. *Cheilostomatous Bryozoa. Part I. Aeteoidea–Cribrilinoidea. Synopses of the British Fauna No. 10* (second edition). Academic Press, London.

RZHAVSKY, A. V. 1992. Review of Circeinae and Spirorbinae (Polychaeta, Spirorbidae) from seas of USSR with a description of a new species, *Circeis guiganovae. Zoologicheskii Zhurnal,* **71**(7), 5–13.

SANFILIPPO, R. 1998. *Spirorbis spirorbis* (Linnaeus, 1758) and *Spirorbis corallinae* de Silva & Knight-Jones, 1962 (Polychaeta, Spirorbidae), Boreal Guests from the Mediterranean Pleistocene. *Rivista Italiana di Paleontologia e Stratigrafia,* **104**, 279–286.

SGANGA, P. 1963. La sezione stratigrafica calabriana di Naso (Messina). *Rivista Mineraria Siciliana,* **14**, (82–84), 1–23.

TAVIANI, M., BOUCHET, P., METIVIER, B., FONTUGNE, M. & DELIBRIAS, G. 1991. Intermediate steps of southwards faunal shifts testified by last glacial submerged thanatocoenoses in the Atlantic Ocean. *Palaeogeography, Palaeoclimatology, Palaeoecology,* **86**, 331–338.

ZIBROWIUS, H. 1972. *Hydroides norvegica* Gunnerus, *Hydroides azorica* n. sp. et *Hydroides capensis* n. sp. (Polychaeta, Serpulidae), espèces vicariantes dans l'Atlantique. *Bulletin du Muséum national d'Histoire naturelle, Paris, série 3, 39 Zoologie,* **33**, 433–446.

Late Glacial and Post-Glacial pollen records and inferred climatic changes from Lake Balaton and the Great Hungarian Plain

E. NAGY-BODOR[1], M. JÁRAI-KOMLÓDI[2] & A. MEDVE[3]

[1] Geological Institute of Hungary PO 106, H-1442, Budapest, Hungary
[2] Head of the Botanical Department, Hungarian Natural History Museum, PO 222, H-1476, Budapest, Hungary
[3] Geological Institute of Hungary, PO 106, H-1442, Budapest, Hungary

Abstract: In Hungary, the Great Plain and the western part of Transdanubia had different geomorphological and climatic characteristics during the Late Glacial, the Holocene and the present day. As a result, the vegetation in the two areas was also different, as indicated by palynological data. In Late Glacial times the Great Plain had an extremely continental climate whereas the western part of Transdanubia had a somewhat milder one. This clear distinction can be made on the basis of pollen spectra between the palaeo-vegetation of Transdanubia and that of the Great Plain.

A comparison of the pollen data, and the climatic changes indicated by the carbonate content, and $\delta^{18}O_{PDB}$ and $\delta^{13}C_{PDB}$ isotope ratios of the sediments of Lake Balaton, have shown that these data can be correlated beginning from the Bölling, the onset of permanent water coverage.

The comparatively small territory of Hungary (located between latitudes 48°N and 46°N and longitudes 16°E and 22°E; see Fig. 1) has been subjected to transitional climatic influences. Its climate differs from, but has been influenced by, those of adjacent areas. The basically Central European climate is essentially similar to the 'Atlantic–Continental' type (Kordos 1978). It is, however, subject to an arid continental climatic influence from the east, a sub-Mediterranean influence from the south, and a humid Atlantic–Alpine influence from the west (Fig. 2).

According to Köppen's climate classification (Köppen Geiger 1930–1936), there are two major climatic boundaries in the territory of Hungary: the boundary between a warm-moderate and cold-moderate climate, and the boundary between a humid and semi-arid climate. The latter is represented by a boundary between two major vegetation zones – the zone of deciduous forests and the zone of forest steppe – comprising the Little Hungarian Plain and the Great Hungarian Plain (Borhidi 1981; Fig. 2). This difference roughly corresponds to the difference in climate between the Great Hungarian Plain and Transdanubia. It is shown

by our studies that this is detectable from Pliocene time onwards (approximately 2.5 Ma). On the Great Hungarian Plain the climate is arid and continental. In Transdanubia the continental climate is already more humid and moderate. This may be due to its particular basinal position that developed after the uplift of the Alps and the Carpathians, both of which exerted a protective influence (Ziegler 1990).

At the site of the investigated borehole (Tó-31 in Lake Balaton, Fig. 1), the mean annual temperature is 10.8°C and the average annual precipitation is 715 mm. A humid, cool Atlantic (sub-Atlantic) influence from the SW–W (with an annual precipitation of 900–700 mm and mean annual temperature of 8–9°C is also present in the area of Lake Balaton, which belongs to the phyto-geographic region of the Transdanubian Central Range (Pócs 1981).

These differences in climate are reflected in the diverse Recent flora of Hungary and – as proved by our palynological studies – this has also been the situation during the past 15 000 years, in both the Late Glacial and the Holocene. Due to climatic and edaphic factors (less precipitation and the very deep position of the groundwater

From: HART, M. B. (ed.) *Climates: Past and Present*. Geological Society, London, Special Publications, **181**, 121–133, 1-86239-075-4/$15.00

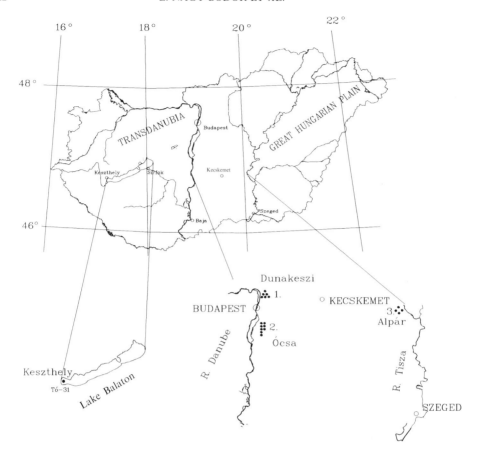

Fig. 1. Map of Hungary with details of the study area and the location of sites and boreholes.

table) there is much less Arbor Pollen in the Great Plain area (Fig. 3) than in the forested area of Transdanubia (Fig. 4). As a result, the history of the Hungarian forest area differs, in certain respects, from that of several other areas in Europe (Lang 1994).

We have attempted to reconstruct the palaeo-climate on the basis of a palynological analysis of borehole sections at Dunakeszi, Ócsa and Alpár in the Great Hungarian Plain (Járai-Komlódi 1966, 1968, 1969, 1987, 1997) and from borehole Tó-31 drilled into the Keszthely Sub-basin in Transdanubia (Cserny, 1993; Bodor & Cserny 1997). We have taken into consideration climate indicator plants (Iversen 1944), the productivity changes of the plants deduced from the oxygen and carbon isotope ratios of carbonates (Talbot 1990; Medve 1996) and the quantitative and qualitative changes in the vegetation under the influence of climate (Kordos Járai-Komlódi 1988). An increase in the

productivity of the vegetation corresponds to a rise in the $\delta^{13}C_{PDB}$ values. The $\delta^{18}O_{PDB}$ value range changes with the intensity of evaporation which, itself, is a function of the temperature. Rising temperatures in the environment result in an increase of $\delta^{18}O_{PDB}$ values of the carbonates, while an increase in precipitation results in a shift of the curve towards negative $\delta^{18}O_{PDB}$ values. An example can be seen in Fig. 5 where, after the warm arid Boreal (V), the curves show negative correlation in the Atlantic (VI–VII) phases. A similar, but less significant phenom-enon, can be observed for Bölling (Ib; Cserny *et al.* 1995).

Our chronology is mainly relative (Lang 1994), but there are also two radiocarbon dates (12 490±300 year BP and 12 210±300 year BP). Our results are supported by these radiocarbon analyses that were carried out on samples taken from peats. The pH of the peat layers was slightly acid.

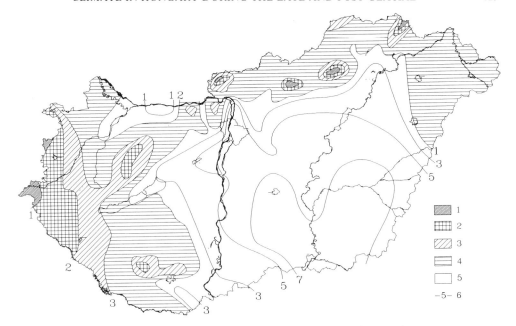

Fig. 2. The main vegetation types of Hungary (after Borhidi 1961) 1, Zone of mountainous *Aconito–Fagetum silvaticae*; 2, zone of submountainous *Melitti–Fagetum silvaticae*; 3, zone of *Querco petraeae–Carpinetum*; 4, zone of *Quercetum petraeae–cerris*; 5, zone of forest steppes. 1,2,3,5,7, Isoxera lines indicating zones of the same aridity on the Great Hungarian Plain, according to the values of the semi-aridity index.

Data for the reconstruction of the Late Glacial palaeoclimate (15–10 ka BP)

Dryas I (Ia)

In Transdanubia the climate was mild, similar to the Dryas II–Bölling climate of the Great Hungarian Plain. Of the thermophilic deciduous trees only *Corylus* and *Ulmus*, that were better able to withstand climatic extremities, occur sporadically. *Corylus* and *Ulmus* may suggest a poor soil and milder microclimate. Of the herbaceous species, *Artemisia* was common. By this time Lake Balaton is supposed to have been filled with water and, as a result, the aquatic vegetation was characterized by the marshy plants *Rorippa amphibia, Hottonia palustris* and *Alisma*. The spread of *Pteridophyta* in the environs of the lake may also have been due to the humidity of the microclimate created by the lake. Consequently, at that time, the western part of Transdanubia may have been characterized by an extremely continental climate; arid summers but with a more humid, cold winter.

From the Dryas I (Ia) to the Atlantic (VI–VII) phase, Lake Balaton was not a single lake with a continuous water surface. It was made up of a number (four) of Sub-basins.

Bölling + Dryas II (Ib, Ic)

The vegetation records, in the western part of Lake Balaton, an increase in temperature and precipitation that, in the Great Hungarian Plain, took place only in the warming period of Allerod (Járai-Komlódi 1966).

Summers were already more humid, as indicated by the higher number of herbaceous plants, (*Ribes, Erica, Sagina*). The first of the lake-shore (riparian) forests also appeared at this time. The persistent occurrence and spread of *Ephedra* may also be due to a rise in temperature. Open-water aquatic plants (*Myriophyllum spicatum, M. verticillatum, Potamogeton natans*) became permanent features of the lake (Fig. 6).

In the profiles studied on the Great Hungarian Plain the Bölling could not be separated (Járai-Komlódi 1966).

Dryas II (Ic)

In Transdanubia, the July mean temperature might have been above 14°C. This is suggested by the presence of the *Nymphaea* sp.–*Typha latifolia* association (Iversen 1954, 1964; Járai-Komlódi 1968). Increasing aridity is indicated

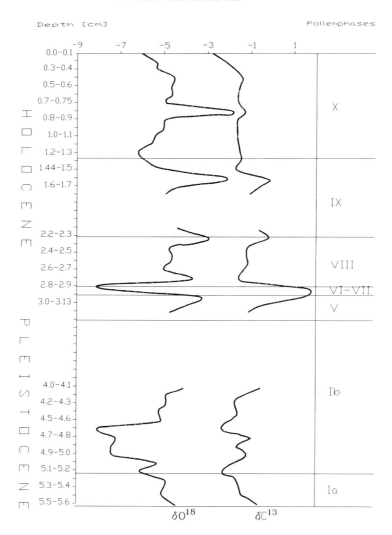

Fig. 3. Combined (arithmetically averaged) pollen diagram of the borehole profiles (Dunakeszi, Ócsa, Alpár-Toserdo) on the Great Hungarian Plain (arbor pollen + non-arbor pollen).

by the disappearance of representatives of the lake-shore (riparian) forests. In these layers the ^{14}C ages were 12 210–12 490±300 year BP.

At the same time, on the Great Hungarian Plain, Arctic meadows with mosses and lichens alternated with (not too dense) *Pinus–Betula* forests and continental, cold steppes. *Artemisia*, which had became dominant in Transdanubia during Dryas I, was significant at that time on the Great Hungarian Plain (Járai-Komlódi 1969; Fig. 3). The climate of the Great Hungarian Plain may have been arid and cold, with a mean temperature around 11° to 13°C in July and *c.* –6 to –8°C in January. This is suggested

by the co-occurrence of *Armeria, Gypsophila and Pleurospermum*, which prefer a colder and more arid climate.

Allerod (II)

A gradual increase in temperature and precipitation in Transdanubia can be deduced from the evidence of mixed deciduous forests (*Tilia, Quercus*). At that time, on the Great Hungarian Plain, the improvement of the climate included a rise in summer temperatures. The July mean temperature may have been around 17 to 18°C,

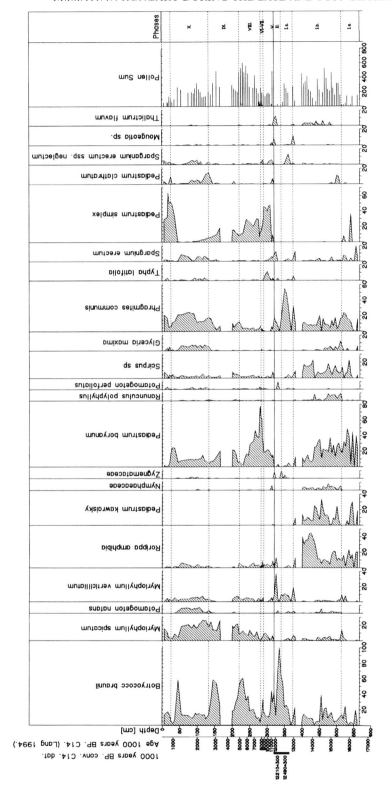

Fig. 4. Arbor pollen percentages (100% = SAP) in the borehole Tó-31 (Lake Balaton) Pollen = 100%.

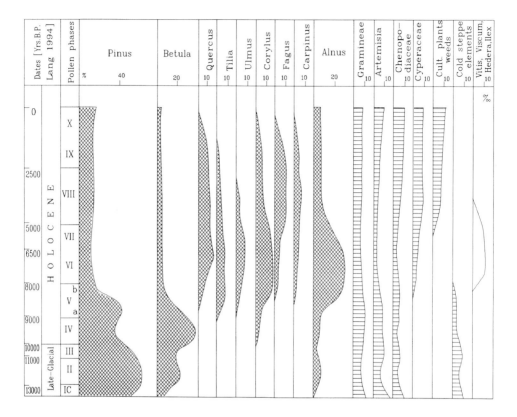

Fig. 5. The δO^{18} PDB and δC^{13} PDB ratios of the calcareous mud in the borehole Tó-31 (Lake Balaton).

(Járai-Komlódi 1968). In addition to the *Pinus–Betula* forests, some thermophilic deciduous trees (*Tilia, Quercus, Ulmus*) also appeared (Járai-Komlódi 1968). On the shores, *Salix–Populus–Alnus* riparian forests were present. In wet conditions *Alnus* peat swamp forests, with *Thelypteris palustris*, are known, while in the stagnant waters, aquatic plants indicate a climate improvement (*Typha latifolia, T. angustifolia, Sparganium* are identified). All of these plants occur only sporadically on the Great Plain during the colder Dryas stages.

Dryas III (III)

It was at this time that the western Sub-basin of Lake Balaton may have dried up (Cserny & Bodor 1996). There is a hiatus (in the vicinity of borehole Tó-31) in the succession and no data are available on the temperature regime at that time.

On the Great Hungarian Plain a brief, and minor, deterioration of climate took place which

did not attain the level of cooling observed in Dryas II. The mean temperature is likely to have been around 13 to 14°C in July and –4 to –6°C in January. According to our pollen analysis, the forests decreased, the riparian forests and the aquatic vegetation were reduced and continental cold steppes rich in heliophytic species (*Sanguisorba minor, Centauera montana*) became more typical (Járai-Komlódi 1966, 1969).

Palynological data for the reconstruction of the Holocene palaeoclimate (10 ka BP present)

Preboreal (IV)

At this time the western part of Lake Balaton was still dry. On the Great Hungarian Plain, in addition to the *Pinus–Betula* forests, broadleaved thermophilic trees (*Tilia, Quercus, Ulmus, Acer*) spread into the area and remain up to the

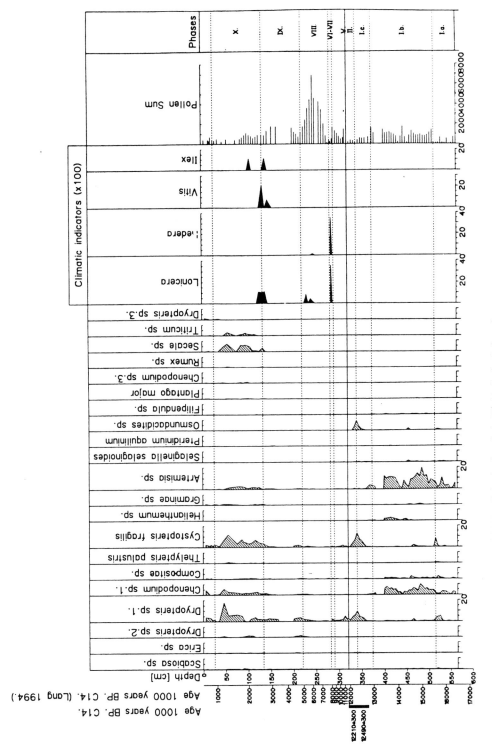

Fig. 6. Water and swamp plants percentages (100% = S) in the borehole Tó-31 (Lake Balaton).

present day in response to a strong warming (Fig. 3). The July mean temperature might have attained 18–20°C.

The landscape of the Great Hungarian Plain was still a *Betula*-bearing forest steppe (*Betula, Salix, Populus*) which was mixed with other thermophilic deciduous trees.

The taxa that had been dominant until that time in the steppe vegetation (*Artemisa, Chenopodium*) were decreasing, being replaced by grass genera of the Family Poaceae (*Festuca, Stipa*). At that time the shallow lakes on the Great Hungarian Plain were populated with open-water aquatic plants, according to pollen analysis (Járai-Komlódi 1966, 1969).

Boreal (V)

In the area of Lake Balaton, there was a marked rise in temperature and, at the beginning of the Boreal, precipitation increased, in comparison to the Great Plain. This is supported by the fact that the part of the lake basin that had previously been dried up (from the Dryas III until the Preboreal) become covered again by water. An increase in precipitation is suggested by the presence in large numbers of various *Pediastrum* algal species, particularly *Pediastrum simplex* (Morzadec-Kerfourn 1988; Fig. 6). At the beginning of the Boreal, *Corylus* and *Fagus* were dominant. Besides *Tilia cordata*, *T. platyphyllos* also appeared in the pollen spectrum. *Pterydophyta* were common. Accordingly, at that time, the climate might have been more humid and uniform over the year than on the Great Hungarian Plain.

In the second part of the Boreal, the climate became warmer and more arid. Much of the lake became shallow and in the Keszthely sub-basin it was subjected to considerable eutrophication (Bodor & Cserny 1997). *Fagus* forests became reduced, whereas *Quercus* forests expanded (Fig. 4). The *Corylus–Quercus* dominance of this period in the Lake Balaton area was also detected by Zólyomi (1995). *Pinus* forests are also likely to have thinned in response to the arid and warming climate (Járai-Komlódi 1968). Only the xerophilic herbaceous vegetation persisted. The number of *Pteridophyta* was reduced. The vegetation became similar to the *Quercus* forest-steppes that today characterize the Great Hungarian Plain.

On the Great Hungarian Plain the climate seems to have been warm and arid during the entire Boreal, with no detectable increase of precipitation, not even in the early Boreal. The Boreal can also be separated into two subphases on the basis of the differences in afforestation

and in the composition of the herbaceous vegetatione (Fig. 3). At the beginning of the Boreal (Phase Va), the Great Hungarian Plain was characterized by the dominance of an eastern-type *Pinus silvestris* forest-steppe, with *Quercus, Tilia, Corylus* and with large patches of steppe. These may have resembled the forest steppes which exist today in the Ukraine.

In the second half of the phase (Phase Vb), mixed *Quercus* forest-steppes developed which were composed of stands of *Tilia* and *Corylus* forests and large warm continental steppe meadows. *Corylus* was of less importance on the Great Hungarian Plain than in Transdanubia. This was due to the more arid and warmer climate of the Great Hungarian Plain which was not conducive to the propagation of the mesophilic *Corylus*. By the end of the Boreal, the smaller stagnant waters became marshy and the composition of the aquatic vegetation became poorer. The area lost its forests and communities of sandy and sodic steppes developed in the treeless sandy areas situated between the Danube and Tisza rivers. On the loess ridges, climatic steppes and loess grasslands are likely to have been dominant. Along the Danube, and in the marginal areas of the Great Hungarian Plain, the forest steppes continued to exist (Fig. 3).

Atlantic (VI–VII)

In Transdanubia and on the Great Hungarian Plain alike, the appearance of the plants suggests a 'climatic optimum'. These plants needed high temperatures, humidity and a moderate climate that were atypical of the other phases of the Holocene. Climatic optimum indicators, appearing in Transdanubia, were *Lonicera* and *Hedera, Vitis* and *Ilex* (Fig. 4). During the Atlantic phase, the lake became increasingly shallow. This is indicated by the diatom flora, which includes species typical of shallow water and a restricted salinity (e.g. *Anomoeneis sphaerophora var. polygramma*; Hajós, pers. comm.).

Lake Balaton, at this time, might have become even marshy at many sites (Erdélyi 1963). This conclusion is also supported by the increasing abundance of *Polygonum amphibium*. This plant was able to withstand shallow water or even drying up for an extended period. This apparent contradiction may result from the fact that, according to the Talbot diagram based on isotope data which shows an open hydrological system, an outflow was established (Talbot 1990) between the Keszthely and Szigliget Sub-basins of Lake Balaton through a channel The presence of a water outflow may have caused the water level to decrease (Cserny *et al.* 1998).

On the Great Hungarian Plain, the terrain – except for the loess ridges – appears to have been covered with forests. The most widespread community was a mixed *Quercus* forest-steppe with *Corylus*. In the first half of the Atlantic phase, *Corylus*, unlike the situation in Transdanubia, was of the same importance as in the Boreal. At more humid sites, a confined *Quercus robur* forest might have been typical. In this forest *Hedera* and *Ilex* grew, perhaps indicating a climatic optimum (Fig. 3). Just as in Transdanubia, the swamp forests and the *Quercus – Fraxinus – Ulmus* riparian forests might also have expanded. Sandy steppes and loess steppe meadows also existed on the Great Hungarian Plain. In both areas, the July mean temperature is likely to have been 24 to 25°C.

Subboreal (VIII)

In the western part of Transdanubia, temperatures probably decreased while humidity probably increased. Within this phase four minor climate changes are detected that essentially coincide with the zoological evidence supplied by 'Avicola humidity' of Kordos (1978). These studies are based on the micro-mammal biostratigraphy developed by Kretzoi (1969) and later by Kordos (1978). The actual geographical distribution of aquatic voles (*Avicola*) is determined first of all by the climate (temperature) and the vegetation. Their succession in a particular area, therefore, provides evidence for palaeoclimatological interpretations. A knowledge of the relative frequency of the aquatic vole can assist in the recognition of changes in humidity (Kordos 1978).

1. At the beginning of the Subboreal the temperature decreased with respect to the previous climatic optimum: *Lonicera* and the lake-shore riparian forests disappeared. According to Kordos, a humidity minimum occurred between 5000 to 4000 years BP, at the beginning of the Subboreal.
2. Following this, an increase in temperature and precipitation took place. The spread of *Abies* shows that winters were mild. Mixed deciduous forests became more and more dominant. The climatic optimum indicators *Lonicera* and *Hedera* also re-appear in the pollen record (Fig. 7). The warming may have coincided with that indicated by Kordos (1978) at 3000 years BP.
3. As a result of a subsequent fall in temperature and increase in precipitation, ferns became widespread. This event can be correlated with that of 2750 years BP, previously identified by Kordos (1978).
4. At the end of the phase, another warming took place, but there was no increase in precipitation. *Fagus silvatica* became less common and *Corylus* and *Quercus* became widespread (Fig. 4). This change in vegetation indicates a warming at 2500 years BP as identified by Kordos (1978) but it does not indicate the increase in precipitation.

At that time on the Great Hungarian Plain, the climate became cooler and slightly arid but was still relatively humid. Cooling is indicated by the fact that there was less *Vitis* and *Hedera* present, while *Ilex* disappeared. The *Quercus* forests became closed and there were no steppe areas present. *Carpinus* also spread into the area and *Fagus* became more common. Large *Quercus–Fraxinus–Ulmus* riparian forests were developed. The most typical forests were probably the *Quercus robur* riparian forest with *Convallaria* and the *Quercus robur–Carpinus betulus* mixed forest.

The strong correlation of the $\delta^{18}O_{PDB}$ and $\delta^{13}C_{PDB}$ values indicate a hydrologically closed lacustrine system from the Dryas I (Ia) until the Subatlantic (X), that is, until the present day. An exception is the Atlantic (VI–VII) which had a warm climate and more abundant precipitation. It was the latter which contributed to the formation of the continuous water surface of Lake Balaton in the Subboreal (VIII) from 5100 years BP.

The negative correlation of the two isotopes may be due to the increased water supply to a south-directed drainage (towards Nagyberek) which rendered the hydrological system open for some time (Cserny 1999).

Subatlantic (IX–X)

Transdanubia was characterized by a temperate climate. The temperature was lower than in the Subboreal and shows increasing precipitation. The July mean temperature could have been around 21–23°C. The predominance of *Fagus silvatica* indicates that the annual mean precipitation must have been over 800 mm. According to our investigations at Lake Balaton, two subphases (IX and X) can be identified within the phase on the basis of an abrupt decrease of *Fagus* (Fig. 4).

Within both subphases, two warming and cooling periods are detected. In the cooling periods, a decrease of 2 to 3°C in the mean annual temperature is probable. The mean

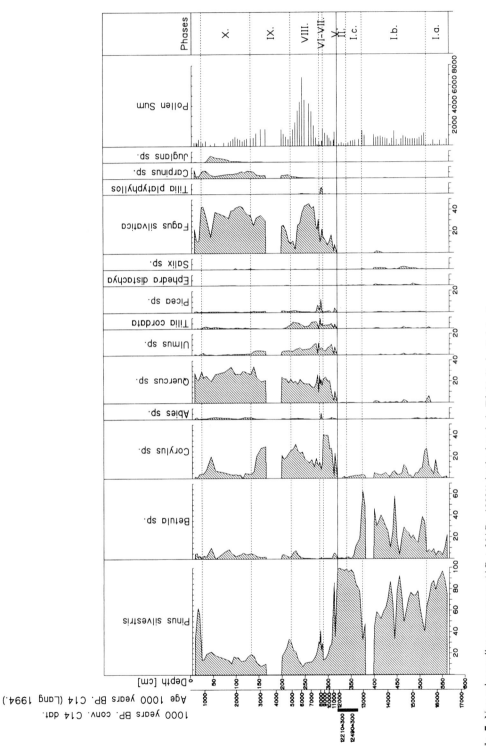

Fig. 7. Non-arbor pollen percentages (AP + NAP = 100%) in the borehole Tó-31 (Lake Balaton).

annual precipitation is unlikely to have fallen below 800 mm because of the extent of the *Fagus silvatica* forests as indicated by the palynological evidence.

On the Great Hungarian Plain the climate became somewhat more arid and cool, showing some similarity to the present day climate. *Fagus* vanished from the Great Hungarian Plain, except for in its northern part. *Carpinus* and *Quercus* became less abundant. The areal extent of the ancient *Quercus* forests also decreased but this may have been the result of agriculture.

Subatlantic IX. Within this subphase two changes in temperature can be detected in the sediments of Lake Balaton.

1. This period began with a fall in temperature and an increase in precipitation. This is indicated by the greater abundance of *Carpinus*, which prefers a higher humidity, and the propagation of *Querco petraeae–Carpinetum* with *Tilia*. The frequent occurrence of *Polygonum bistorta* suggests an increase in the soil humidity. A cooling is also suggested by the re-appearance of *Selaginella*. The zoological evidence (Kordos 1978) also indicates a cooling period between 1750 and 2500 years BP.

2. This period was followed by another period of warming. *Vitis* and even *Ilex* re-appeared (Fig. 7). The mean annual temperature was the warmest at this time. According to zoological evidence (Kordos 1978) this warming took place in the period from 900 to 1750 years BP.

Subatlantic X. Two changes in temperature can be distinguished.

1. In the first period there was a decrease in both temperature and precipitation. *Lonicera and Vitis* disappeared and *Ulmus* became less dominant. Ferns became less dense in the underwood.

2. The second half of this period is characterized by a continued decrease in temperature, and the precipitation regime become more extreme. The differences between summer and winter average temperatures exceeded 20°C. In the vegetation, *Quercus* became more common, *Fagus* became sub-dominant and *Ilex* vanished. (Járai-Komlódi 1991). According to palynological analysis, precipitation could have attained its minimum level at this time during the Subatlantic phase.

The present-day vegetation was developed, including the so-called 'puszta'. This is a kind of anthropogenic semi-desert and is a result of human intervention. In modern times the major part of the territory of Hungary belongs to the zone of European broad-leaved deciduous forest. There are two exceptions. Part of the Great Hungarian Plain is the westernmost extremity of the forest steppe zone of Eastern Europe, while some of the coniferous forests of the western part of Transdanubia are part of the zone of Central European mixed coniferous forest (Jakucs 1981). Since the time when human activity began to have significant impact on the landscape the area of native forests has decreased to about one-quarter of their original area in Hungary. Today only one-half of the existing forests are similar to the primary forests. These small relict areas preserve the remains of the primary vegetation and include the representatives of the ancient flora. Today, the Great Hungarian Plain is virtually treeless; a man-made, cultivated, semi-desert (the Hungarian 'puszta'; Zólyomi & Fekete 1994).

Summary

The present-day transitional climate of Hungary (i.e. the different climatic influences being effective in Transdanubia and on the Great Hungarian Plain) can be recognized in the earlier Holocene vegetation.

- **Dryas I(Ia)** The climate was milder in Transdanubia than on the Great Hungarian Plain. It was similar to the Dryas II–Bölling climate of the Great Hungarian Plain. Lake Balaton is assumed to have been filled with water by this time.
- **Bölling (Ib)** In the western part of Lake Balaton, a noticeable rise in temperature and precipitation occurred; an event that on the Great Hungarian Plain took place only in the warming phase of Alleröd.
- **Dryas II (Ic)** The climate turned more arid and on the Great Hungarian Plain it was even more arid and cold than in Transdanubia.
- **Alleröd** Both the Great Hungarian Plain and Transdanubia record a general warming took place, with less precipitation.
- **Dryas III (III)** The western Sub-basin of Lake Balaton dried up. On the Great Hungarian Plain a brief, and slight, deterioration of climate took place which did not attain the level of the Dryas II cooling.
- **Preboreal (IV)** The western part of Lake Balaton continued to be dry land. On the Great Hungarian Plain there was a strong warming.

- **Boreal (V)** In the Lake Balaton area, the climate of the first part of the Boreal was more humid and temperate than that of the Great Hungarian Plain. In the second part of the Boreal the climate was even warmer, but more arid. On the Great Hungarian Plain, as a consequence of the arid and warm climate, a treeless steppe vegetation, sand meadows, and grassy sodic steppe areas were formed by the end of the Boreal. The climate was arid and warm during the entire Boreal
- **Atlantic (VI–VII)** In Transdanubia and on the Great Hungarian Plain alike, this phase is characterized by the appearance of plants indicating a climatic optimum.
- **Subboreal (VIII)** In the western part of Transdanubia, temperature seems to have decreased and humidity increased. Within the phase, four minor changes in climate could be detected mainly by means of palynological studies. On the Great Hungarian Plain, a cool phase began that became slightly less humid.
- **Subatlantic (IX–X)** Transdanubia was characterized by a temperature lower than in the Subboreal, with an increasing precipitation and a temperate climate. Two subphases (IX, X) could be distinguished, with two warming and cooling periods within each subphase. On the Great Hungarian Plain the climate became somewhat more arid and cool; similar to the climate of today.

We wish to thank Martha Hajós for some valuable data, Kate Jármi for her assistance and Károly Brezsnyánszky, Tibor Cserny and the National Scientific Research Found (OTKA) No. 17321 and T 022371 for financial support.

References

BODOR, E & CSERNY, T. 1997. A Keszthelyi-öböl vízzel borítottságának fejlodéstörténete, (History of water coverage of the Keszthely Bay). *Hidrológiai Közlöny*, **77**, 98–100.

BORHIDI, A. 1981. Magyarország éghajlatának néhány növényföldrajzi vonatkozása, Klimazonális övek (Some phytogeographic implications of the climate of Hungary. Climate zones) *In*: *Növényföldrajz , Társulástan és Ökológia*, Tankönykiadó, Budapest, 365–372.

CSERNY, T. 1993. Lake Balaton, Hungary. *In*: GIERLOWSKI-KORDESCH, E. & KELTS, K. (eds) *A Global Geological Record of Lake Basins*. Cambridge University Press, Cambridge, 397–401.

—— 1999. A Balaton üledékeinek környezetföldtani célú vizsgálata *In*: ANDA, A. (ed.) *Vizalatti talajok*

szerepe a tavak környezetvédelmében. Keszthely, 47–71.

—— & BODOR, E. 1996. The Pre-Quaternary Morphology and Quaternary Evolution of Lake Balaton. *IAG European Regional Geomorphological Conference, Hungary*. Abstract 29.

——, HERTELENDI, E., TARJAN, S. *et al.* 1995. Results of isotope-geochemical studies in the sedimentological and environmental geologic investigations of Lake Balaton. *Acta Geologica Hungarica*, **38**, 355–376.

——, NAGY-BODOR, E. & MEDVE, A. 1998. Linmology Research at the Geological Institute of Hungary. *Geological Survey Annual Report 1996*. Budapest, 31–34.

ERDÉLYI, M. 1963. A Balatonnak és környezetének változása az ember tevékenysége következtében, / Veränderungen am Balaton und Umgebung auf Einfluss der menschlichten Tätigkeit *Hidrológiai Közlöny*, **3**, 219–224.

IVERSEN, J. 1944. Viscum, Hedera, Ilex as climate indicators. *Geologiska Föreningens Förhanlingar*, **66**, 463–483.

—— 1954. The late Glacial flora of Denmark and its relation to climate and soil *Geologischer Undersuchungen II. Report*. **80**, 87–119.

—— 1964. Plant indicators of climate, soil and other factors during the Quaternary. *Report, VIth INQUA, Warsaw, 1961*, vol. II, 421–428.

JAKUCS, P. 1981. Magyarország legfontosabb növénytársulásai. (The important plant communities of Hungary). *In*: *Növényföldrajz, Társulástan és Ökológia*. Tankönyvkiadó Budapest, 225–263.

JÁRAI-KOMLÓDI, M. 1966. Adatok az Alföld negyedkori klíma és vegetációtörténetéhez I. A vegetáció változása a Würm Glaciális és a Holocén folyamán palinológiai vizsgálatok alapján (Quaternary climatic changes and vegetational history of the Great Hungarian Plain). *Botanikai Közlemények*, **53**, 191–201.

—— 1968. The Late Glacial and Holocene flora of the Hungarian Great Plain. *Annales Universitatis Scientiarum Section Biologica*, Budapest, **9–10**, 199–225.

—— 1969. Adatok az Alföld negyedkori klíma és vegetációtörténetéhez. II. A klíma változása a Würm glaciális és a holocén folyamán, palinológiai vizsgálatok alapján (Quaternary climatic changes and vegetational history of the Great Hungarian Plain. Climatic changes during the Würm glaciation and Holocene, as evidenced by palynological investigations). *Botanikai Közlemények*, **56**, 43–55.

—— 1987. Postglacial climate and vegetation history in Hungary. *Holocene Environment in Hungary*. Geographical Research Institute of the Hungarian Academy of Sciences Budapest, 37–47.

—— 1991. Magyarország pleisztocénvégi vegetációtörténete az utolsó interglaciális óta (Vegetation history of the Hungarian Late Pleistocene since the last Interglacial), *Oslénytani Viták (Discussiones Palaeontologicale)*, Budapest, **36-37**, 201–215.

—— 1997. A legutóbbi, azaz a holocén beerdősödés flóratörténetéből (History of the last (Holocene) afforestation in Hungary). *Botanikai Közlemények*,

84, 3–15.

KÖPPEN, W. GEIGER, R. 1930–1936. *Handbuch der klimatologie* I-IV. Berlin.

KORDOS, L. 1978. Changes in the Holocene climate of Hungary reflected by the Vole-Thermometer method. *Földrajzi Közlemények*, **1–3**, 222–230.

——— & JÁRAI KOMLÓDI, M. 1988. Az elmúlt tízezer év klímaváltozásai Közép- Európában (Climate changes in Central Europe during the past ten thousand years). *Journal of the Hungarian Meteorological Service*, **92**, 96–100.

LANG, G. 1994. *Quartäre Vegetationsgeschichte Europas*. Gustav Fischer Verlag, Stuttgart.

MEDVE, A. 1996. *A Balaton vízének és holocén üledékének stabil izotópos vizsgálata*. (Stable isotopic investigation of the water and the Holocene sediments of Lake Balaton). MSc thesis, Lorand Eötvös University, Budapest.

MORZADEC KERFOURN, M. T. 1988. Paleoclimats et Paleoenvironnements du Tardiglaciaire au récent, en Mediterranée orientale, a l'est du delta du Nil: l'apport des microfossiles a membrane organique (Palaeoclimates and Palaeoenviroments from the Lateglacial to recent, in the Eastern Mediterranean, East of the Nile Delta, the contribution of organic-walled microfossils). *Paléoclimats et Paléoenvironnements Quaternaires*, **12**, 267–275.

PÓCS, T.1981. Magyarország növényföldrajzi beosztása (Phytogeographic scheme of Hungary) *In*: *Növényföldrajz, Társulástan és Ökológiai*. Tankönykiadó, Budapest 120–166.

TALBOT, M. R. 1990. A review of the palaeohydrological interpretation of carbon and oxygen isotopic ratios in primary lacustrine carbonates. *Chemical Geology*, **80**, 261–279.

ZIEGLER, A. P. 1990. *Geological Atlas of Western and Central Europe*, Shell International Petroleum Maatschappij, B.V., 143–160.

ZÓLYOMI, B. 1995. Opportunities for Pollen Stratigraphic Analysis of Shallow Lake Sediments: the Example of Lake Balaton, *Geojournal*. **36**, 237–241.

——— FEKETE, G. 1994. The Pannonian loess steppe differentiation in space and time. *Abstracta Botanica*. **18**, 29–41.

Palaeoceanography and numerical modelling: the Mediterranean Sea at times of sapropel formation

E. J. ROHLING[1], S. DE RIJK[1], P. G. MYERS[2] & K. HAINES[2]

[1] *University of Southampton, School of Ocean and Earth Science, Waterfront Campus, Southampton SO14 3ZH, UK (e-mail: ejr@mail.soc.soton.ac.uk)*
[2] *Department of Meteorology, University of Edinburgh, Edinburgh EH9 3JZ, UK*

Abstract: Palaeo-circulation concepts based on (micro-)palaeontological records are compared with results of a general circulation model (GCM), in a review of the eastern Mediterranean conditions during sapropel S1 deposition (7–8ka BP). We discuss two conceptual models, based on proxy data, which integrate bottom water anoxia and high productivity during sapropel formation. One infers a weakened version of the present-day anti-estuarine circulation, whereas the other invokes a reversal of the circulation to an estuarine type. Although both seem reasonable, in view of the available palaeoceanographic proxy data, they imply completely different states of the Mediterranean fresh-water budget. Numerical modelling has been based on three available reconstructions of changes in the west–east salinity gradient compared with present-day conditions (small decrease, intermediate decrease and reversing). The three scenarios give surprisingly similar results. The interface between the surface and intermediate water has shoaled significantly compared with the present, which fuels a deep chlorophyll maximum. All three simulations show stagnant deep waters, which could cause bottom water anoxia. Reversal of circulation is not observed, not even in the reconstruction of most extreme salinity changes. The combined approach of palaeoceanographic and numerical modelling reviewed in this paper shows that modelling provides an indispensable feedback on the viability of the various proxy-based conceptual models.

At the July 1997 meeting of the European Palaeontological Association in Vienna, it became clear that the oldest discipline in palaeo-environmental studies, palaeontology, still plays a pivotal role in the description and understanding of natural environmental changes. Nothing characterizes a past environment as directly as a record of what actually lived in that environment. However, it has also become increasingly obvious that palaeontological information must be combined with that from other palaeo-proxy studies, for example, geochemistry, to allow the development of sensible, cross-validated, palaeo-environmental reconstructions in the form of conceptual models. Next, evaluation of these multi-disciplinary conceptual models with physical oceanographic–climatological models is needed to determine whether the concepts are viable and properly preserve the physics.

To illustrate the problem-solving potential of this cross-disciplinary approach, we review an example concerning Mediterranean palaeo-circulation, which draws on the results to date of EU-Mast3 Project 'Climatic Variability of the Mediterranean Palaeo-circulation (CLIVAMP)'. Where appropriate, the specific contributions to the study made by (micro-)palaeontological disciplines are highlighted and discussed in greater detail.

The present-day Mediterranean

The Mediterranean Sea (Fig. 1) is separated from the Atlantic Ocean by the Strait of Gibraltar (284 m deep), and is subdivided into the western and eastern basins by a sill in the Strait of Sicily (365 and 430 m deep in the two major channels).

A strong excess of evaporation over total fresh-water input into the Mediterranean results in a net water loss of around $0.5 \, \text{m a}^{-1}$ (Bryden *et*

From: HART, M. B. (ed.) *Climates: Past and Present.* Geological Society, London, Special Publications, **181**, 135–149, 1-86239-075-4/$15.00

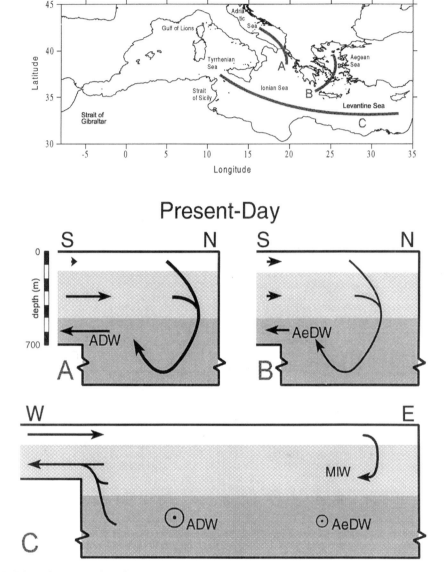

Fig. 1. Schematic presentation of the present-day vertical circulation in the eastern. The map on top shows the locations of the three sections A, B and C. ADW, Adriatic Deep Water; AeDW, Aegean Deep Water; MIW, Mediterranean Intermediate Water.

al. 1994). This evaporation-precipitation (E–P) balance dominates the buoyancy flux and causes anti-estuarine exchange over both major sills (Fig. 1). Relatively low-salinity waters flow eastward at the surface, above a subsurface westward flow of relatively high-salinity (dense) waters.

The strong evaporation causes an eastward increase in the salinity of surface waters, from 36‰ at the Strait of Gibraltar to over 39‰ in the easternmost areas. Winter cooling of these high-salinity waters causes the formation of Mediterranean Intermediate Water (MIW) between Cyprus and Rhodes, which spreads between 150 and 600 m depth throughout the Mediterranean Sea (Wüst 1961; Malanotte-Rizzoli & Hecht 1988; Malanotte-Rizzoli & Bergamasco 1989; POEM Group 1992). There is no equivalent major intermediate water mass formation in the western Mediterranean.

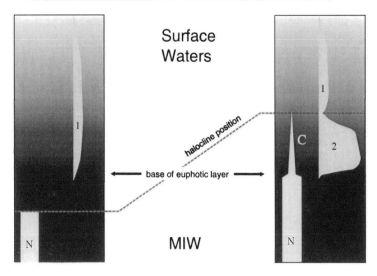

Fig. 2. Schematic representation of the influence of depth variations of the boundary between MIW and surface waters relative to the base of the euphotic layer, on the development of a distinct deep chlorophyll maximum (see Rohling & Gieskes 1989). 1, mixed layer phytoplankton association; 2, deep chlorophyll maximum association; C, *in situ* autotrophic consumption of nutrients, which causes the nutricline to remain at the base of the euphotic layer.

The eastern and western Mediterranean basins each have their own source of deep water, which settles below the MIW. Eastern Mediterranean Deep Water masses are formed in the Adriatic Sea (ADW, Fig. 1) and in the Aegean Sea (AeDW, Fig. 1), through mixing of warm highly saline MIW that penetrates these satellite basins with cooled surface waters of somewhat lower salinities (e.g. Pollak 1951; Wüst 1961; Malanotte-Rizzoli & Hecht 1988; POEM Group 1992; Roether *et al.* 1996). Western Mediterranean Deep Water (WMDW) is not formed in a satellite basin, but in the open Gulf of Lions. Intensive preconditioning of surface waters during late winter, as a result of the influence of the 'Mistral' winds, eliminates the density difference between surface and intermediate waters, and culminates in deep convective overturn that forms WMDW (Wüst 1961; MEDOC Group 1970; Leaman & Schott 1991). The modern overturn in both the eastern and western basins is very efficient, and flushing times are less than a century (Béthoux *et al.* 1990; Schlitzer *et al.* 1991).

MIW contains higher nutrient concentrations than the overlying surface waters (McGill 1961; Miller *et al.* 1970). The nutricline is found more or less in association with the (halo-)pycnocline between intermediate and surface waters (see Minas 1971). Although weaker than the seasonally developing (thermo-)pycnocline, the (halo-)pycnocline is arguably more important in the longer term because of its permanence and its position, which are functions of the general circulation of the basin. Where and when circulation causes shoaling of the (halo-)pycnocline into the euphotic layer (Fig. 2), which extends down to 100 or 120 m, a deep chlorophyll maximum (DCM) may develop, fuelled by advection of new nutrients from below (see Rohling *et al.* 1995).

The Mediterranean environment during sapropel deposition

Throughout the Mediterranean, but in particular in the eastern basin (east of the Strait of Sicily), the normal sequence of hemipelagic muds deposited in a well-oxygenated environment is regularly interrupted by dark-coloured intervals. These layers ('sapropels'), are significantly enriched in organic carbon, often contain well-preserved laminations, and are usually devoid of benthic faunas.

Sapropels regularly developed in the Mediterranean since at least 7 Ma BP, in response to minima in the astronomical cycle of precession with a period of about 21 ka (e.g. Rossignol-Strick 1985; Hilgen 1991; Nijenhuis *et al.* 1996). From Quaternary sapropel occurrences, it appears that the upper depth limit of sapropel formation was near 120 m in the Aegean Sea (Perissoratis & Piper 1992) and around 300 m in the open eastern Mediterranean (Rohling & Gieskes 1989).

The commonly observed absence of benthic foraminiferal faunas from sapropels (Van Straaten 1972; Cita & Podenzani 1980; Vismara-Schilling 1984; Nolet & Corliss 1990; Verhallen 1991; Rohling *et al.* 1993*b*, 1997; Jorissen 1999) provides a vital clue to the environmental conditions in which sapropels were deposited. Benthic Foraminifera are more tolerant of low-oxygen conditions than are most other meio- and macrofauna (Josefson & Widbom 1988). Populations of benthic Foraminifera may be extremely numerous in dysoxic environments (Sen Gupta & Machain-Castillo 1993) and several species are known to survive completely anoxic conditions for at least 24 h (Moodley & Hess 1992). Total absence of benthic Foramini-fera from sapropels therefore indicates that bottom waters were persistently anoxic. The absence of burrowing benthic organisms would in turn favour the preservation of the original sedimentary laminae. Abundance of pyrite with-in sapropels (Van Straaten 1972; Stanley 1983; Thunell *et al.* 1984) corroborates the inference of anoxic bottom water conditions during their formation.

Planktonic foraminiferal and calcareous nan-nofossil associations within sapropels reflect increased levels of productivity, in particular near the base of the euphotic layer in a DCM (Rohling & Gieskes 1989; Castradori 1993). A similar inference has been made on the basis of increased abundances of dinoflagellates that selectively graze on diatoms typical of DCM settings (Targarona 1997). The micropalaeonto-logical indications of DCM formation and related increases in productivity agree with geochemical indications of enhanced productiv-ity during sapropel formation, which consist of strongly enhanced Ba/Al ratios (e.g. Calvert 1983; Van Os *et al.* 1991, 1994; Thomson *et al.* 1995, 1999), Cd/Ca ratios (Boyle & Lea 1989) and nitrogen isotope ratios (Calvert *et al.* 1992), and comparisons with conditions in the present-day Black Sea (Pedersen & Calvert 1990).

Two comprehensive hypotheses have been proposed regarding past eastern Mediterranean circulation patterns that integrate the deductions of bottom water anoxia and increased produc-tivity at times of sapropel formation. One infers a weakened version of the present-day anti-estuarine circulation (Rohling & Gieskes 1989; Castradori 1993; Rohling 1994), and the other invokes a reversal of the circulation to an estuarine type (e.g. Stanley *et al.* 1975; Calvert 1983; Thunell *et al.* 1983; Sarmiento *et al.* 1988; Thunell & Williams 1989). Both hypotheses, discussed in detail below, are reasonable con-ceptual models in view of the available proxy data, but they imply radically different states of the Mediterranean fresh-water budget during sapropel formation.

To corroborate the implied changes in the fresh-water budget, both concepts refer to available evidence that the Mediterranean cli-mate at times of sapropel deposition was more humid than today, and that there were sub-stantial increases in fresh-water input into the basin from remote sources via the Nile and Black Sea (for extensive reviews, see Rohling & Hilgen (1991) and Rohling (1994)). This has been deduced from: (1) common occurrence of oxygen isotope depletions in records based on surface-dwelling planktonic Foraminifera; (2) common occurrence of vegetation typical of humid summers as well as winters; (3) increased discharge rates or sediment transport by Eur-asian rivers as well as the Nile; (4) results of hydraulic control models of Black Sea outflow relevant to sapropels associated with deglacia-tions (among many others: Olausson 1961; Cita *et al.* 1977; Vergnaud-Grazzini *et al.* 1977; Williams *et al.* 1978; Adamson *et al.* 1980; Rossignol-Strick *et al.* 1982, Rossignol-Strick 1985, 1987; Thunell *et al.* 1987; Thunell & Williams 1989; Wijmstra *et al.* 1990; Kallel *et al.* 1997; Lane-Serff *et al.* 1997; Rohling & De Rijk 1999).

To date, however, all this information fails to unequivocally demonstrate whether evaporation from the Mediterranean remained larger than total fresh-water input, or whether the Mediter-ranean fresh-water budget instead changed into one of net fresh-water input. Even the most compelling data, based on mapping of past oxygen isotope values (see below), has been interpreted both in favour of a reduced present-day type fresh-water deficit and of a net fresh-water input. It therefore remains unresolved which of the two proposed circulation patterns characterized the Mediterranean, particularly the eastern basin, during sapropel formation.

Weakened anti-estuarine circulation. In this hypothesis, the reconstructions of bottom water anoxia and increased productivity are combined in a concept of eastern Mediterranean circula-tion during sapropel formation that consists of a present-day type circulation (Fig. 3) that was substantially weakened and active only at shallow and intermediate depths, above a stagnant mass of Old Deep Water (Rohling & Gieskes 1989; Castradori 1993; Rohling 1994).

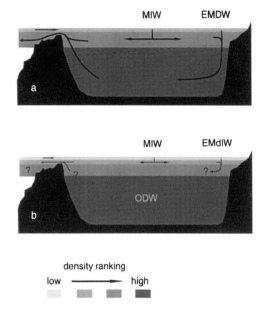

MIW   EMDW

a

MIW   EMdIW

ODW

b

density ranking

low ———————▶ high

Fig. 3. Schematic presentation of the weakened anti-estuarine circulation hypothesis as presented by Rohling (1994). (**a**) The present-day vertical circulation in the eastern Mediterranean Sea. (**b**) The vertical circulation in the eastern Mediterranean Sea at times of sapropel formation. EMDW, East Mediterranean Deep Water; EMdIW, East Mediterranean deep Intermediate Water; ODW, Old Deep Water.

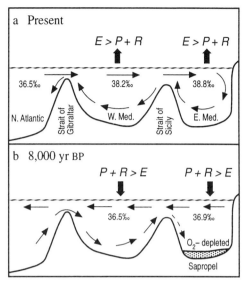

a   Present

$E > P + R$ $E > P + R$

36.5‰   38.2‰   38.8‰

N. Atlantic   Strait of Gibraltar   W. Med.   Strait of Sicily   E. Med.

b   8,000 yr BP

$P + R > E$ $P + R > E$

36.5‰   36.9‰

O_2– depleted

Sapropel

Fig. 4. Schematic presentation of the circulation-reversal hypothesis, modified after Thunell & Williams (1989). This concept constitutes a reversal of the vertical overturn in the Mediterranean from the present-day anti-estuarine (**a**) to an estuarine type during sapropel formation (**b**). E, evaporation; P, precipitation; R, fresh-water runoff.

Such conditions would have developed in response to a substantial reduction of the Mediterranean excess evaporation, which caused a decrease in the W–E salinity gradient.

Even with generally lower surface salinities, localized winter cooling would have continued to cause sinking of water masses that would spread at depth. However, it was proposed that no new water masses were formed of sufficient density to replace the old deep waters that originated from times before the general reduction of surface salinities. As a consequence, the Old Deep Water (ODW) mass became isolated below a very intense (halo-)pycnocline at the base of the newly formed intermediate waters. Continuing oxic degradation of organic matter raining down from the surface then gradually consumed all dissolved oxygen from the ODW over a period of some 500–1000 a (Rohling 1994).

At the same time, shoaling of the interface between surface and intermediate waters, with which the main nutricline is associated, would have brought it close to, or above, the base of the euphotic layer. Continuation of the surface to intermediate water circulation ensured con-

tinuing advection and nutrient (re-)distribution. This resulted in a basin-wide development of a DCM, as witnessed in calcareous nannofossil, dinoflagellate and planktonic foraminiferal records through sapropels (Rohling & Gieskes 1989; Castradori 1993; Targarona 1997).

Circulation reversal. This concept comprises a reversal of the vertical overturn in the Mediterranean from the present-day anti-estuarine type to an estuarine type during sapropel formation (Fig. 4). Reversal is thought to have occurred in response to a change in the fresh-water budget from the present-day type with an excess of evaporation over total fresh-water input to a regime characterized by net fresh-water input (e.g. Stanley *et al.* 1975; Calvert 1983; Thunell *et al.* 1983; Sarmiento *et al.* 1988; Thunell & Williams 1989).

The estuarine circulation pattern consists of surface outflow across the sill of relatively fresh marine waters (diluted by the inferred net fresh-water input into the Mediterranean) above a weak subsurface inflow of relatively saline marine waters from the adjacent ocean. The subsurface inflow is driven by gradual loss of buoyancy from the basin, a result of the admixture of the net fresh-water gain to the marine waters in the basin. This implies that a gross 'upwelling' occurs throughout the basin, even if the replacement of deep waters is relatively slow.

An important side-effect of the general upwelling would be that nutrients released at depth from mineralization of organic matter raining down from the surface are eventually recycled into the euphotic layer, where they fuel new production. This so-called 'nutrient-trap' effect would drive high productivity in this concept of circulation during sapropel formation. The associated high downward flux of organic matter would consume all oxygen in deeper waters in spite of their relatively slow ventilation.

Progress

To break the stalemate, new methods were needed for exploration of the two different concepts of eastern Mediterranean circulation during sapropel formation. Here, we present a simulation of the circulation at that time with a realistic general circulation model (GCM) to identify which concept would be more likely in view of the various surface condition changes inferred from the data. First, an evaluation is presented of methods to define surface conditions in sufficient detail that they may be used to drive a model. Then follows a brief description of the GCM used. Finally, we summarize the GCM results in comparison with the conceptual models.

Definition of surface fields

Numerical simulation provides vital insight into the possible physical states of the circulation in response to sufficiently detailed reconstructions of surface conditions. Ideally, a model should be run coupled with an atmospheric circulation model, but this is a very difficult problem, which has yet to be successfully solved for the present-day Mediterranean. Forcing the model with fixed atmospheric fluxes has its own suite of difficulties associated with it, and has only recently been used successfully in the present-day Mediterranean. Therefore, restoring boundary conditions, where the model's surface tracers are restored towards a fixed value over a given time scale, are used. Consequently, as accurate as possible characterization of surface conditions from palaeo-proxy studies is essential to a successful modelling exercise. The main conditions to be defined are sea surface temperature (SST) and sea surface salinity (SSS). Several techniques are available to estimate palaeo-SST, but only one, oxygen isotope based, technique

exists for derivation of palaeo-SSS values. To date, three studies have applied the various methods to reconstruct Mediterranean surface conditions during the formation of early Holocene sapropel S1 (Thunell & Williams 1989; Kallel *et al.* 1997; Rohling & De Rijk 1999).

Sea surface temperature

Stable oxygen isotope (δ^{18}O) records were among the first palaeoceanographic proxy data to be interpreted in terms of absolute palaeotemperature values (Epstein *et al.* 1953; Emiliani 1966, 1972), but it was soon acknowledged that δ^{18}O records reflect interaction between several main effects (see next section). Interpretations of fossil flora and fauna distributions in terms of palaeotemperature were developed independently. Planktonic foraminiferal distribution patterns in particular were found to be related to SST (e.g. Beard 1969; overview by Phleger 1976; Emiliani & Ericson 1991), although the link may be complex and via hydrographic control (build-up, maintenance and breakdown of thermal stratification) on food availability (e.g. Tolderlund & Bé 1971; Phleger 1976; Rohling *et al.* 1993a; Pujol & Vergnaud-Grazzini 1995).

Advanced statistical methods using planktonic foraminiferal records include the transfer function (TF; Imbrie & Kipp 1971; Kipp 1976) and modern analogue technique (MAT; Hutson 1979). In the TF, the spatial distribution pattern of loadings on the first factor is normally related to the spatial distribution pattern of average SST as determined from contemporary oceanographic surveys. Thiede (1978) and Thunell (1979) presented TF-based SST results for the Mediterranean during the last glacial maximum, but no TF results are available for the early Holocene period of deposition of sapropel S1. The MAT is also used to interpret SST from planktonic foraminiferal associations. In a nutshell, the MAT calculates similarity coefficients to a set of modern (core-top sample) associations for each down-core association, and so selects a subset of most similar modern associations (modern analogues). Average SST values from contemporary oceanographic surveys then give the 'SSTs of the modern analogues'. The SST of each down-core association is subsequently determined by averaging the SSTs of the modern analogues, either in an unweighted fashion or weighted on the basis of the similarity coefficients (Hutson 1979; Overpeck *et al.* 1985). Kallel *et al.* (1997) used a new MAT for detailed determination of early Holocene Mediterranean SST values during formation of sapropel S1.

A recent organic geochemical development uses a calibration between SST and the C_{37} methyl alkenone unsaturation ratio (U^k_{37}; Brassell *et al.* 1986). It allows changes in the U^k_{37} ratio in organic matter within sediment cores to be employed to portray SST variability during the period of deposition of the cores (e.g. Brassell *et al.* 1986; Prahl & Wakeham 1987; Prahl *et al.*, 1988; Sikes *et al.* 1991; Eglinton *et al.* 1992; Zhao *et al.* 1993, 1995). This method has not yet been systematically applied to the Mediterranean early Holocene period, mainly for lack of a suitable local calibration.

The available reconstructions of early Holocene SST patterns in the Mediterranean are essentially based on *a priori* interpretations of plankonic foraminiferal associations or on MAT studies. They show little to no significant difference between the general early Holocene and present-day SST distributions. In fact, Thunell & Williams (1989) in their assessment of the oxygen isotope records to define palaeosalinity decided to use the present-day temperature distribution. Kallel *et al.* (1997) in their comparable exercise used their new MAT-derived SST values, in which the only significant difference consists of 1.5–2.5°C cooler conditions in the early Holocene NW Mediterranean. Rohling & De Rijk (1999) in turn assessed the isotope records using a the present-day SST field, but included a simple NW Mediterranean cooling after Kallel *et al.* (1997).

Sea surface salinity

Although some early studies described some correspondence between loadings on the second factor in transfer functions and salinity (Kipp 1976; Thunell 1979), palaeo-SSS cannot be convincingly determined from direct interpretation of any of the available proxy data. Usually, therefore, palaeo-SSS is estimated following a rather complex, indirect approach (e.g. Thunell & Williams 1989; Maslin *et al.* 1995; Hemleben *et al.* 1996; Labeyrie *et al.* 1996; Kallel *et al.* 1997). First, a planktonic foraminiferal $\delta^{18}O$ record is generated. Only a single species from a narrow size-interval is used, to eliminate vital or metabolic effects on the water to carbonate $\delta^{18}O$ fractionation, which are assumed to be constant through time for individual species in similar ontogenetic stages. The palaeo-SSS calculation then considers the record to be an end result determined by the 'glacial effect', 'temperature effect' and 'salinity or fresh-water budget effect', and isolates the last effect by simple subtraction of the glacial and temperature effects.

The glacial, or ice volume, effect accounts for $\delta^{18}O$ enrichment in the world ocean as a result of preferential uptake of the lighter isotope (^{16}O) with evaporation and its subsequent precipitation and storage for periods of the order of tens of thousands of years in continental ice-sheets. Recent studies show consistent calibrations of the world ocean $\delta^{18}O$ enrichment in relation to glacio-eustatic sea-level lowering of around $0.012 \pm 0.001\,\permil\,m^{-1}$ (Labeyrie *et al.* 1987; Shackleton 1987; Fairbanks 1989). The temperature effect relates to the change with temperature that occurs in the isotopic fractionation during precipitation of carbonate from sea water. The net result of the temperature-dependent change in fractionation is about $0.2\permil$ enrichment of $\delta^{18}O$ in carbonate relative to that in ambient sea water for every 1°C cooling, or $0.2\permil$ depletion for every 1°C warming.

After subtraction of the glacial and temperature effects from the record, the remaining or 'residual' signal may be interpreted in terms of change in the local fresh-water budget (evaporation, precipitation, runoff) that affected the sample site. As the local fresh-water budget affects not only the isotopic composition of sea water but also its salinity, this final conversion is often performed on the basis of a simple linear regression between modern salinity and $\delta^{18}O$ values in the studied basin. This method implies that the relationship between S and $\delta^{18}O$ is constant through time, which assumption has recently been challenged and may introduce serious errors in palaeo-SSS (Rohling & Bigg 1998; Rohling & De Rijk 1999).

All three studies that assessed oxygen isotopic contrasts between the present day and the height of formation of early Holocene sapropel S1 followed the above method to reconstruct general SSS trends through the early Holocene Mediterranean (Thunell & Williams 1989; Kallel *et al.* 1997; Rohling & De Rijk 1999). Only one included a correction for likely non-constant behaviour of the salinity–$\delta^{18}O$ relationship (Rohling & De Rijk 1999).

In their pioneering study, Thunell & Williams (1989) concluded that the W–E $\delta^{18}O$ gradient was very weak to opposite in sign to that of today, and consequently that eastern Mediterranean salinities were somewhat lower than western Mediterranean salinities. This was taken as proof of a 'dilution' basin configuration with reversed circulation. Kallel *et al.* (1997) added many new records, and used their MAT-based SST reconstructions to make corrections in detail for the temperature effect per site. They observed a virtually complete absence of a $\delta^{18}O$ gradient ('flat field') during the formation of

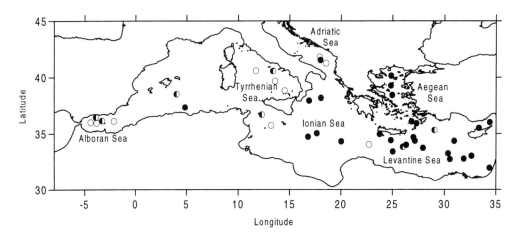

Fig. 5. Locations of isotopic records from the Mediterranean Sea (EU-Mast 3 CLIVAMP palaeo-database).

early Holocene sapropel S1, indicating little or no salinity change from west to east. Rohling & De Rijk (1999) used the extensive dataset compiled from all records available at that time to the palaeo-database sponsored by EU-Mast 3 project CLIVAMP (Fig. 5). Elaborate statistical tests were performed to determine the main significant trends in the early Holocene $\delta^{18}O$ values. The first conclusion was that only records performed on one and the same species should be compared with one another, in contrast to the multi-species approach followed by Thunell & Williams (1989) and Kallel *et al.* (1997). This condition unfortunately reduced the number of suitable records (Fig. 5). Then, it was observed that three main areas may be distinguished within which the early Holocene $\delta^{18}O$ depletions are significantly similar, namely the western Mediterranean, the eastern Mediterranean excluding the Levantine Sea, and the Levantine Sea. Significant differences in the depletions were found between these three areas. Rohling & De Rijk (1999) observed an early Holocene $\delta^{18}O$ gradient of roughly half the present-day magnitude, but of similar sign (eastward increase). The salinity gradient was considered taking into account a likely change in the salinity–$\delta^{18}O$ relationship relative to the present, and was thus determined at roughly 75% of the present gradient.

General circulation model

Model description

To obtain proper simulation of such features as past deep and intermediate water formation, a high-resolution model is needed that adequately reproduces the main circulatory features of the present-day Mediterranean. Even so, model results should be viewed only in terms of their portrayal of past general circulation patterns, as uncertainties and low spatial resolution in the characterization of palaeo-surface conditions inhibit the accurate reproduction of small-scale features such as eddies and jets. Therefore, models should be used primarily to characterize the general state of the palaeo-circulation and its sensitivity to differences between the various reconstructions of surface conditions.

The model Myers *et al.* (1998a) used is the Modular Ocean Model–Array (MOMA). For detailed description of the model and its boundary conditions, the reader is referred to Haines & Wu (1995), Wu & Haines (1996, 1998), and Myers *et al.* (1998a,b). A simulation of the present-day conditions, using the monthly surface temperature and salinity fields from the Mediterranean Oceanic Data Base (Brasseur 1995), served as a control run to verify that the model properly represents all major circulatory

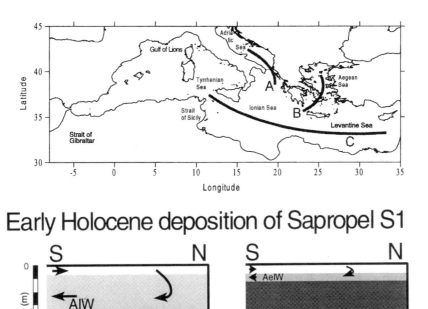

Early Holocene deposition of Sapropel S1

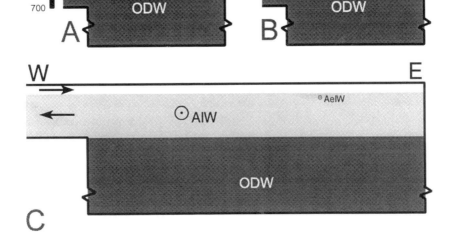

Fig. 6. Schematic presentation of the vertical circulation in the eastern Mediterranean according to the model results. The map shows the locations of the three sections A, B and C. AIW, Adriatic Intermediate Water; ODW, Old Deep Water; AeIW, Aegean Intermediate Water.

features observed in the Mediterranean. Test runs incorporating the somewhat lower sea level at times of S1 formation (about 20 m below the present-day level; Fairbanks 1989) show no significant influences on the simulated circulation patterns. Model runs were continued until statistical steady states were reached, and ranged between 40 and 100 model years.

Early Holocene experiments: surface conditions

In the early Holocene experiments, Myers *et al.* (1998*a*) used the present-day SST values from the Mediterranean Oceanic Data Base (Brasseur 1995), drawing on the above arguments that early Holocene SSTs were comparable. It was noted that this choice might result in slight underestimates of the early Holocene thermally driven overturn in the NW Mediterranean, where 1.5–2.5°C lower SST values were suggested by Kallel *et al.* (1997). Regarding palaeo-SSS, Myers *et al.* (1998*a*) smoothed the three reconstructions (Thunell & Williams 1989; Kallel *et al.* 1997; Rohling & De Rijk 1999) with exponential fits, to obtain values between the coarsely distributed observations through realistic interpolation. A seasonal cycle was superimposed on the annual average values on the basis of modern seasonal fluctuations.

All three scenarios invoke lowered surface salinities, albeit to varying extents. For deeper levels, the model starts with present-day values (Myers *et al.* 1998*a*). This may seem artificial, but also in reality the deep waters would not have been primarily affected by increasing humidity of the climate. In fact, in the case of S1, the basin salinities before the climatically induced freshening of surface waters probably were even higher than the present-day values used in the model. This argument is based on the fact that S1 was formed during the aftermath of the last deglaciation, when peak glacial salinity values in the Mediterranean may have been up to 5 p.s.u. higher than today (e.g. Thunell & Williams 1989; Rohling 1994; Myers *et al.* 1998*a*; Rohling & De Rijk 1999). As it is not clear to what extent this excess salinity would have been mixed up and out of the basin during the deglaciation period, and, for computational reasons, it is not yet possible to run the model over sufficiently long periods to simulate processes during the deglaciation, it seems a reasonable assumption to use deep water values in the early Holocene simulations that are compatible with the present-day conditions.

Early Holocene experiments: model results

Here, we concentrate on the results for the eastern Mediterranean basin only, as sapropel formation seems to have been more common there than in the western basins (for full Mediterranean results, see Myers *et al.* (1998*a*)). The model simulations for the three surface reconstructions give surprisingly similar results (Fig. 6). All three show a stagnant ODW mass at depth in the eastern Mediterranean, separated from actively circulating shallower layers by a very distinct (halo-)pycnocline. The base of this pycnocline resides roughly around 450 m for the least extreme salinity reduction (Rohling & De Rijk 1999), 400 m for the intermediate scenario of surface salinity reduc-

tion (Kallel *et al.* 1997), and 350 m for the most extreme salinity reduction (Thunell & Williams 1989).

The intermediate waters above the ODW in all three scenarios derive from wintertime formation in the Adriatic Sea. This Adriatic Intermediate Water (AIW; Myers *et al.* 1998*a*) ventilates the entire eastern Mediterranean except for the Aegean Sea, which it does not penetrate. There is no equivalent of the modern MIW formation between Cyprus and Rhodes, but there is some intermediate water formation in the Cretan Sea. This water mass spreads in the Aegean down to a maximum depth of about 125 m, below which stagnant ODW prevails in the Aegean.

The interface between intermediate and surface waters in the present-day control run resides at 120–200 m throughout the eastern Mediterranean (Fig. 1), which is well below the base of the euphotic layer, and so is consistent with the observed general absence of a strong DCM (Fig. 2). In the early Holocene experiments, this interface has shoaled dramatically, to 50–80 m or less (Fig. 6). This would be an appropriate change to cause the formation of a DCM fuelled by nutrient advection from the intermediate waters, as has been inferred in general for sapropel periods from planktonic foraminifera (Rohling & Gieskes 1989), calcareous nannofossils (Castradori 1993) and dinoflagellates (Targarona 1997).

Comparison with proxy data based conceptual models

The early Holocene model results of Myers *et al.* (1998*a*) demonstrate a fairly robust overall pattern of circulation; there are no dramatic differences between the circulation patterns for the various reconstructions of surface conditions. The main difference seems to be that the circulation occupies a shallower total depth range, with shoaling of the various interfaces, as more extreme surface salinity reductions are imposed. Even the most extreme salinity reductions (Thunell & Williams 1989) do not result in a reversal of the circulation. In that scenario, surface thermal forcing outweighs the salinity forcing and determines a continuation of net buoyancy loss, and exchange remains anti-estuarine. The anti-estuarine intermediate water circulation, during low surface salinity conditions, is also partly maintained by a negative buoyancy flux coming from the upwards diffusion of salt from the stagnant ODW. The other two reconstructions show very strong agreement in spite of the considerable difference in their surface forcing conditions (Kallel *et al.* 1997; Rohling & De Rijk 1999).

The main characteristics of circulation during deposition of sapropel S1 in all three reconstructions are: (1) continuing mode water formation in the Adriatic, which ventilates the eastern Mediterranean except the Aegean Sea down to 400 m or so, and (weakly) in the South Aegean Sea, which ventilates only the Aegean Sea down to 125 m or so; (2) prevalence of an isolated, stagnant, ODW mass below the intermediate waters; (3) distinct shoaling of the interface between intermediate and surface waters to depths of 50–80 m, which would be well within the euphotic layer, compared with 120–200 m in the present-day control run. Furthermore, the model results suggest that anti-estuarine circulation continued to prevail, albeit considerably weakened relative to the present.

In general, these model conclusions strongly corroborate the weakened anti-estuarine circulation scenario proposed by Rohling (1994) for times of sapropel formation, the only difference being in the more detailed understanding of origin and area–depth of spreading of the two intermediate water masses.

In detail, model conclusions (1) and (2) show close correspondence to the observed upper depth limits of sapropels in these basins. Conclusion (3) agrees well with the observations of basin-wide DCM development during sapropel formation. Here, it needs to be emphasized that for S1, indications that this happened are restricted to the calcareous nannofossils (Castradori 1993). The planktonic foraminiferal record for S1 is at odds with that of almost all other Quaternary sapropels, in that the DCM indicator taxon *Neogloboquadrina* is not abundant (Rohling & Gieskes 1989). It remains to be determined what caused this discrepancy, especially as the model results (Myers *et al.* 1998a) suggest that pycnocline shoaling should have occurred, in agreement with the calcareous nannofossil data. A clue may be found in the fact that sapropel S1, with its relatively low amounts of preserved organic carbon, is also from a geochemical point of view a rather 'weak' sapropel (Cita & Grignani 1982; Pruysers *et al.* 1991; Calvert *et al.* 1992; Fontugne & Calvert 1992). Possibly, the DCM developed, but failed to reach sufficiently high level of productivity to sustain a substantial stock of *Neogloboquadrina*. This would seem to agree with Castradori's (1993) inference that pycnocline shoaling did occur during the formation of S1, but was less pronounced than during the deposition of the other Quaternary sapropels. Finally, model

conclusion (4) endorses proxy data based interpretations that anti-estuarine circulation persisted over the past 140 ka, albeit weakened during periods of sapropel formation (Ross & Kennett 1984; Zahn *et al.* 1987; Vergnaud-Grazzini *et al.* 1989).

Discussion and conclusions: it's palaeontology, but not as we know it

The problem-solving potential of combining proxy-based reconstructions with numerical modelling is obvious from the example presented above. Two equally viable concepts of palaeo-circulation had been elaborated in detail from the plethora of Mediterranean proxy data. Both concepts in places agreed very well with the data, and both had their problems. Different reconstructions of surface conditions were used in arguments to support either concept.

Numerical modelling has now demonstrated that the general circulation of the basin shows a robust basic pattern with relatively minor variations, for all fields of surface conditions within the existing range of reconstructions. This basic pattern is very similar to one of the conceptual models that had been developed from the proxy data (weakened anti-estuarine). The other concept (reversed estuarine) is found to be unlikely with respect to the existing reconstructions of surface conditions, as it does not arise even for the most extreme change in surface conditions.

Palaeontological research has been indispensable in the definition of the proxy-based conceptual models, for which crucial information had been drawn from abundance variations of planktonic and benthic Foraminifera, calcareous nannofossils, pollen and spores, dinoflagellates, diatoms, pteropods, etc. In addition, stable isotope records were performed on the tests of Foraminifera, and palaeontology played an important supporting role to ensure that proper single-species samples were used. For the definition of surface fields used in the numerical model, palaeo-SST and palaeo-SSS patterns were assessed mainly on the basis of planktonic foraminiferal associations and stable isotope records.

The example given in the present paper emphasizes the conclusions reached at the EPA conference, that although the days of descriptive palaeontology may largely be over, applied palaeontology has much to offer to palaeoceanographic–climatological understanding. This is especially relevant when mutual validation of proxy data based interpretations takes place between as many disciplines as possible (including also the various aspects of geochemistry), to define one or more possible integrated conceptual palaeo-environmental models. Sound cross-validated concepts are at least as important to further interpretation as quantitative assessments from, for example, stable isotope data or microfossil transfer functions. The relatively recent liaison with modelling, finally, provides indispensable feedback on the viability of the various proxy-based conceptual models.

This study contributes to EC Mast-3 Project Climatic Variability of the Mediterranean Palaeo-Circulation (CLIVAMP; MAS3-CT95-0043).

References

ADAMSON, D. A., GASSE, F., STREET, F. A. & WILLIAMS, M. A. J. 1980. Late Quaternary history of the Nile. *Nature*, **288**, 50–55.

BEARD, J. H. 1969. Pleistocene palaeotemperature record based on planktonic foraminifers, Gulf of Mexico. *Transactions, Gulf Coast Association of Geological Societies*, **19**, 535–553.

BETHOUX, J. P., GENTILI, B., RAUNET, J. & TAILLIEZ, D. 1990. Warming trend in the western Mediterranean deep water. *Nature*, **347**, 660–662.

BOYLE, E. A. & LEA, D. W. 1989. Cd and Ba in planktonic foraminifera from the eastern Mediterranean: evidence for river outflow and enriched nutrients during sapropel formation. *Transactions of the American Geophysical Union*, **70**, 1134.

BRASSELL, S. C., EGLINTON, G., MARLOWE, I., SARNTHEIN, M. & PFLAUMANN, U. 1986. Molecular stratigraphy: a new tool for climatic assessment. *Nature*, **320**, 129–133.

BRASSEUR, P. 1995. Free data offered to researchers studying the Mediterranean. *EOS Transactions*, **76**, 363.

BRYDEN, H. L., CANDELA, J. & KINDER, T. H. 1994. Exchange through the Strait of Gibraltar. *Progress in Oceanography*, **33**, 201–248.

CALVERT, S. E. 1983. Geochemistry of Pleistocene sapropels and associated sediments from the eastern Mediterranean. *Oceanologica Acta*, **6**, 255–267.

——, NIELSEN, B. & FONTUGNE, M. R. 1992. Evidence from nitrogen isotope ratios for enhanced productivity during formation of eastern Mediterranean sapropels. *Nature*, **359**, 223–225.

CASTRADORI, D. 1993. Calcareous nannofossils and the origin of eastern Mediterranean sapropels. *Paleoceanography*, **8**, 459–471.

CITA, M. B. & GRIGNANI, D. 1982. Nature and origin of Late Neogene Mediterranean sapropels. *In*: SCHLANGER, S. O. & CITA, M. B. (eds) *Nature and Origin of Cretaceous Carbon-Rich Facies*. Academic Press, San Diego, CA, 165–196.

—— & PODENZANI, M. 1980. Destructive effects of oxygen starvation and ash falls on benthic life: a pilot study. *Quaternary Research*, **13**, 230–241.

——, VERGNAUD-GRAZZINI, C., ROBERT, C., CHAM-

LEY, H., CIARANFI, N. & D'ONOFRIO, S. 1977. Paleoclimatic record of a long deep sea core from the eastern Mediterranean. *Quaternary Research*, **8**, 205–235.

EGLINTON, G., BRADSHAW, S. A., ROSELL, A., SARNTHEIN, M., PFLAUMANN, U. & THIEDEMANN, R. 1992. Molecular record of secular sea surface temperature changes on 100-year timescales for glacial terminations I, II and IV. *Nature*, **356**, 423–426.

EMILIANI, C. 1966. Isotopic paleotemperatures. *Science*, **154**, 851–857.

—— 1972. Quaternary paleotemperatures and the duration of the high-temperature intervals. *Science*, **178**, 398–401.

—— & ERICSON, D. B. 1991. The glacial/interglacial temperature range of the surface water of the oceans at low latitudes. *In*: TAYLOR H. P. Jr., O'NEIL, J. R. & KAPLAN, I. R. (eds) *Stable Isotope Geochemistry: a Tribute to Samuel Epstein*. Geochemical Society Special Publication, **3**, 223–228.

EPSTEIN, S., BUCHSBAUM, R., LOWENSTAM, H. A. & UREY, H. C. 1953. Revised carbonate–water temperature scale. *Geological Society of America Bulletin*, **64**, 1315–1326.

FAIRBANKS, R. G. 1989. A 17,000 year glacio-eustatic sea level record: influence of glacial melting rates on the Younger Dryas event and deep-ocean circulation. *Nature*, **342**, 637–642.

FONTUGNE, M. R. & CALVERT, S. E. 1992. Late Pleistocene variability of the organic carbon isotopic composition of organic matter in the eastern Mediterranean: monitor of changes in organic carbon sources and atmospheric CO_2 concentrations. *Paleoceanography*, **7**, 1–20.

HAINES, K. & WU, P. 1995. A modelling study of the thermohaline circulation of the Mediterranean Sea: water formation and dispersal. *Oceanologica Acta*, **18**, 401–417.

HEMLEBEN, C., MEISCHNER, D., ZAHN, R., ALMOGI-LABIN, A., ERLENKEUSER, H. & HILLER, B. 1996. Three hundred eighty thousand year long stable isotope and faunal records from the Red Sea: influence of global sea level change on hydrography. *Paleoceanography*, **11**, 147–156.

HILGEN, F. J. 1991. Astronomical calibration of Gauss to Matuyama sapropels in the Mediterranean and implication for the geomagnetic Polarity Time Scale. *Earth and Planetary Science Letters*, **104**, 226–244.

HUTSON, W. H. 1979. The Agulhas current during the Late Pleistocene: analysis of modern faunal analogs. *Science*, **207**, 64–66.

IMBRIE, J. & KIPP, N. G. 1971. A new micropaleontological method for quantitative paleoclimatology: application to a late Pleistocene Caribbean core. *In*: TUREKIAN, K. K (ed.) *The Late Cenozoic Glacial Age*. Yale University Press, New Haven, CT, 71–181.

JORISSEN, F. J. 1999. Benthic foraminiferal successions across Late Quaternary Mediterranean sapropels. *Marine Geology*, **153** (in press).

JOSEFSON, A. B. & WIDBOM, B. 1988. Differential response of benthic macrofauna and meiofauna to hypoxia in the Gullmar Fjord basin. *Marine*

Biology, **100**, 31–40.

KALLEL, N., PATERNE, M., DUPLESSY, J.-C. *et al.* 1997. Enhanced rainfall in the Mediterranean region during the last sapropel event. *Oceanologica Acta*, **20**, 697–712.

KIPP, N. G. 1976. New transfer function for estimating past sea-surface conditions from sea-bed distribution of planktonic foraminiferal assemblages in the North Atlantic. *In*: CLINE, R. M. & HAYS, J. D. (eds) *Investigation of Late Quaternary Paleoceanography and Paleoclimatology*. Geological Society of America, Memoir, **145**, 3–41.

LABEYRIE, L. D., DUPLESSY, J. C. & BLANC, P. L. 1987. Variations in the mode of formation and temperature of oceanic deep waters over the past 125,000 years. *Nature*, **327**, 477–482.

——, LABRACHERIE, M., GORFTI, N. *et al.* 1996. Hydrographic changes of the Southern Ocean (southeast Indian sector) over the last 230 kyr. *Paleoceanography*, **11**, 57–76.

LANE-SERFF, G. F., ROHLING, E. J., BRYDEN, H. L. & CHARNOCK, H. 1997. Post-glacial connection of the Black Sea to the Mediterranean and its relation to the timing of sapropel formation. *Paleoceanography* *('Currents')*, **12**, 169–174.

LEAMAN, K. D. & SCHOTT, F. A. 1991. Hydrographic structure of the convection regime in the Gulf of Lions: winter 1987. *Journal of Physical Oceanography*, **21**, 575–598.

MALANOTTE-RIZZOLI, P. & BERGAMASCO, A. 1989. The circulation of the eastern Mediterranean. Part I. *Oceanologica Acta*, **12**, 335-351.

—— & HECHT, A. 1988. Large-scale properties of the eastern Mediterranean: a review, *Oceanologica Acta*, **11**, 323–335.

MASLIN, M. A., SHACKLETON, N. J. & PFLAUMANN, U. 1995. Surface water temperature, salinity, and density changes in the northeast Atlantic during the last 45,000 years: Heinrich events, deep water formation, and climatic rebounds. *Paleoceanography*, **10**, 527–544.

MCGILL, D. A. 1961. A preliminary study of the oxygen and phosphate distribution in the Mediterranean Sea. *Deep-Sea Research*, **8**, 259–269.

MEDOC Group 1970. Observation of formation of deep water in the Mediterranean Sea. *Nature*, **277**, 1037–1040.

MILLER, A. R., TCHERNIA, P., CHARNOCK, H. & MCGILL, D. A. 1970. *Mediterranean Sea Atlas of Temperature, Salinity, Oxygen: Profiles and Data From Cruises of R.V.* Atlantis *and R.V.* Chain, *with Distribution of Nutrient Chemical Properties*. MAXWELL, A. E. *et al.* (eds), Alpine, Braintree, MA.

MINAS, H. J. 1971. Résultats préliminaires de la campagne 'Mediprod I' du Jean Charcot (1–15 mars et 4–17 avril 1969). *Investigaçion Pesquera*, **35**, 137–146.

MOODLEY, L. & HESS, C. 1992. Tolerance of infaunal benthic foraminifera for low and high oxygen concentrations. *Biological Bulletin, Woods Hole, MA*, **183**, 94–98.

MYERS, P., HAINES, K. & ROHLING, E. J. 1998a. Modelling the paleo-circulation of the Mediterranean: The last glacial maximum and the Holocene with emphasis on the formation of sapropel S1. *Paleoceanography*, **13**, 586–606.

——, —— & JOSEY, S. 1998b. On the importance of the choice of wind stress forcing to the modelling of the Mediterranean Sea circulation. *Journal of Geophysical Research*, **103**, 15 729–15 749.

NIJENHUIS, I. A., SCHENAU, S. J., VAN DER WEIJDEN, C. H., HILGEN, F. J., LOURENS, L. J. & ZACHARIASSE, W. J. 1996. On the origin of upper Miocene sapropelites: a case study from the Faneromeni section, Crete (Greece). *Paleoceanography*, **11**, 633–645.

NOLET, G. J. & CORLISS, B. H. 1990. Benthic foraminiferal evidence for reduced deep-water circulation during sapropel formation in the eastern Mediterranean. *Marine Geology*, **94**, 109–130.

OLAUSSON, E. 1961. Studies of deep-sea cores. *Reports of the Swedish Deep-Sea Expedition 1947-1948*, **8**, 353–391.

OVERPECK, J. T., WEBB, T. III & PRENTICE, I. C. 1985. Quantitative interpretation of fossil pollen spectra: dissimilarity coefficients and the method of modern analogues. *Quaternary Research*, **23**, 87–108.

PEDERSEN, T. F. & CALVERT, S. E. 1990. Anoxia vs. productivity: what controls the formation of organic-carbon-rich sediments and sedimentary rocks? *AAPG Bulletin*, **74**, 454–466.

PERISSORATIS, C. & PIPER, D. J. W. 1992. Age, regional variation and shallowest occurrence of S1 sapropel in the northern Aegean Sea. *Geomarine Letters*, **12**, 49–53.

PHLEGER, F. B. 1976. Interpretations of Late Quaternary foraminifera in deep-sea cores. *Progress in Micropaleontology*, American Museum of Natural History, 263–276.

POEM Group 1992. General circulation of the eastern Mediterranean. *Earth-Science Reviews*, **32**, 285–309.

POLLAK, M. I. 1951. The sources of deep water of the eastern Mediterranean Sea. *Deep-Sea Research*, **10**, 128–152.

PRAHL, F. G. & WAKEHAM, S. G. 1987 Calibration of unsaturation patterns in long-chain ketone compositions for palaeotemperature assessment. *Nature*, **320**, 367–369.

——, MUELHAUSEN, A. & ZAHNLE, D. L. 1988. Further evaluation of long-chain alkenones as indicators of paleoceanographic conditions. *Geochimica et Cosmochimica Acta*, **52**, 2303–2310.

PRUYSERS, P. A., DE LANGE, G. J., MIDDELBURG, J. J. & HYDES, D. J. 1993. The diagenetic formation of metal-rich layers in sapropel-containing sediments in the eastern Mediterranean. *Geochimica et Cosmochimica Acta*, **57**, 527–536.

PUJOL, C. & VERGNAUD-GRAZZINI, C. 1995. Distribution patterns of live planktic foraminifers as related to regional hydrography and productive systems of the Mediterranean Sea. *Marine Micropaleontology*, **25**, 187–217.

ROETHER, W., MANCA, B. B., KLEIN, B. *et al* 1996. Recent changes in Eastern Mediterranean deep waters. *Science*, **271**, 333–335.

ROHLING, E. J. 1994. Review and new aspects concerning the formation of eastern Mediterranean sapropels. *Marine Geology*, **122**, 1–28.

—— & BIGG, G. R. 1998. Paleosalinity and $\delta^{18}O$: a

critical assessment. *Journal of Geophysical Research*, **103**, 1307–1318.

——— & DE RIJK, S. 1999. The Holocene Climate Optimum and Last Glacial Maximum in the Mediterranean: the marine oxygen isotope record. *Marine Geology*, **153**, 57–75.

——— & GIESKES, W. W. C. 1989. Late Quaternary changes in Mediterranean Intermediate Water density and formation rate. *Paleoceanography*, **4**, 531–545.

——— & HILGEN, F. J. 1991. The eastern Mediterranean climate at times of sapropel formation: a review. *Geologie en Mijnbouw*, **70**, 253–264.

———, JORISSEN, F. J., VERGNAUD-GRAZZINI, C. & ZACHARIASSE, W. J. 1993a. Northern Levantine and Adriatic Quaternary planktic foraminifera; reconstruction of paleoenvironmental gradients. *Marine Micropaleontology*, **21**, 191–218.

———, ——— & DE STIGTER, H. C. 1997. 200 year interruption of Holocene sapropel formation in the Adriatic Sea. *Journal of Micropalaeontology*, **16**, 97–108.

———, DEN DULK, M., PUJOL, C. & VERGNAUD-GRAZZINI, C. 1995. Abrupt hydrographic change in the Alboran Sea (western Mediterranean) around 8000 yrs BP. *Deep-Sea Research I*, **42**, 1609–1619.

———, VERGNAUD-GRAZZINI, C. & ZAALBERG, R. 1993b. Temporary repopulation by low-oxygen tolerant benthic foraminifera within an Upper Pliocene sapropel: evidence for the role of oxygen depletion in the formation of sapropels. *Marine Micropaleontology*, **22**, 207–219.

ROSS, C. R. & KENNETT, J. P. 1984. Late Quaternary paleoceanography as recorded by benthonic foraminifera in Strait of Sicily sediment sequences. *Marine Micropaleontology*, **8**, 315–337.

ROSSIGNOL-STRICK, M. 1985. Mediterranean Quaternary sapropels, an immediate response of the African monsoon to variations of insolation. *Palaeogeography, Palaeoclimatology, Palaeoecology*, **49**, 237–263.

——— 1987. Rainy periods and bottom water stagnation initiating brine accumulation and metal concentrations: 1. The Late Quaternary. *Paleoceanography*, **2**, 333–360.

———, NESTEROFF, V., OLIVE, P. & VERGNAUD-GRAZZINI, C. 1982. After the deluge; Mediterranean stagnation and sapropel formation. *Nature*, **295**, 105–110.

SARMIENTO, J. L., HERBERT, T. & TOGGWEILER, J. R. 1988. Mediterranean nutrient balance and episodes of anoxia. *Global Biogeochemical Cycles*, **2**, 427–444.

SCHLITZER, R., ROETHER, W., OSTER, H., JUNGHANS, H. G., HAUSMANN, M., JOHANNSEN, H. & MICHELATO, A. 1991. Chlorofluoromethane and oxygen in the eastern Mediterranean. *Deep-Sea Research*, **38**, 1531–1557.

SEN GUPTA, B. K. & MACHAIN-CASTILLO, M. L. 1993. Benthic foraminifera in oxygen-poor habitats. *Marine Micropaleontology*, **20**, 183–201.

SHACKLETON, N. J. 1987. Oxygen isotopes, ice volume and sea-level. *Quaternary Science Reviews*, **6**, 183–190.

SIKES, E. L., FARRINGTON, J. W. & KEIGWIN, L. D. 1991. Use of the alkenone unsaturation ratio U^{k}_{37} to determine past sea surface temperatures: core-top SST calibrations and methodology considerations. *Earth and Planetary Science Letters*, **104**, 36–47.

STANLEY, D. J. 1983. Parallel laminated deep-sea muds and coupled gravity flow-hemipelagic settling in the Mediterranean. *Smithsonian Contributions to the Marine Sciences*, **19**, 1–19.

———, MALDONADO, A. & STUCKENRATH, R. 1975. Strait of Sicily depositional rates and patterns, and possible reversal of currents in the Late Quaternary. *Palaeogeography, Palaeoclimatology, Palaeoecology*, **18**, 279–291.

TARGARONA, J. 1997. *Climatic and oceanographic evolution of the Mediterranean region over the last Glacial–Interglacial transition.* PhD thesis, University of Utrecht.

THIEDE, J. 1978. A glacial Mediterranean. *Nature*, **276**, 680–683,

THOMSON, J., HIGGS, N. C., WILSON, T. R. S., CROUDACE, I. W., DE LANGE, G. J. & VAN SANTVOORT, P. J. M. 1995. Redistribution and geochemical behaviour of redox-sensitive elements around S1, the most recent eastern Mediterranean sapropel. *Geochimica et Cosmochimica Acta*, **59**, 3487–3501.

———, MERCONE, D., DE LANGE, G. J. & VAN SANTVOORT, P. J. M. 1999. Recent advances in the interpretation of Eastern Mediterranean sapropel S1 based on geochemical evidence. *Marine Geology*, **153**, 77–89.

THUNELL, R. C. 1979. Eastern Mediterranean Sea during the last glacial maximum; an 18,000 years BP reconstruction. *Quaternary Research*, **11**, 353–372.

——— & WILLIAMS, D. F. 1989. Glacial-Holocene salinity changes in the Mediterranean Sea: hydrographic and depositional effects. *Nature*, **338**, 493–496.

———, ——— & BELYEA, P. R. 1984. Anoxic events in the Mediterranean Sea in relation to the evolution of Late Neogene climates. *Marine Geology*, **59**, 105–134.

———, ———, & CITA, M. B. 1983. Glacial anoxia in the eastern Mediterranean. *Journal of Foraminiferal Research*, **13**, 283–290.

———, ——— & HOWELL, M. 1987. Atlantic-Mediterranean water exchange during the late Neogene. *Paleoceanography*, **2**, 661–678.

TOLDERLUND, D. S. & BÉ A. W. H. 1971. Seasonal distribution of planktonic foraminifera in the western North Atlantic. *Micropaleontology*, **17**, 297–329.

VAN OS, B. J. H., LOURENS, L. J., HILGEN, F. J., DE LANGE, G. J. & BEAUFORT, L. 1994. The formation of Pliocene sapropels and carbonate cycles in the Mediterranean: diagenesis, dilution, and productivity. *Paleoceanography*,9, 601–617.

———, MIDDELBURG, J. J., & DE LANGE, G. J. 1991. Possible diagenetic mobilization of barium in sapropelic sediment from the eastern Mediterranean. *Marine Geology*, **100**, 125–136.

VAN STRAATEN, L. M. J. U. 1972. Holocene stages of oxygen depletion in deep waters of the Adriatic Sea. *In*: STANLEY, D. J. (ed.) *The Mediterranean Sea.* Dowden, Hutchinson & Ross, Stroudsburg, PA, 631–643.

VERGNAUD-GRAZZINI, C., CARALP, M., FAUGÈRES, J. C., GONTHIER, É., GROUSSET, F., PUJOL, C. & SALIÈGE, J. F. 1989. Mediterranean outflow through the Strait of Gibraltar since 18,000 years BP. *Oceanologica Acta*, **12**, 305–324.

———, RYAN, W. B. F. & CITA, M. B. 1977. Stable isotope fractionation, climatic change and episodic stagnation in the eastern Mediterranean during the late Quaternary. *Marine Micropaleontology*, **2**, 353–370.

VERHALLEN, P. J. J. M. 1991. *Late Pliocene to Early Pleistocene Mediterranean Mud-dwelling Foraminifera; Influence of a Changing Environment on Community Structure and Evolution*. Utrecht Micropaleontological Bulletin, **40**.

VISMARA SCHILLING, A. 1984. Holocene stagnation event in the eastern Mediterranean. Evidence from deep-sea benthic foraminifera in the Calabrian and western Mediterranean ridges. *In*: OERTLI, H. J. (ed.) *Proceedings of the 2nd International Symposium on Benthic Foraminifera, 1983, Pau, "Benthosl '83"*, 585–596.

WIJMSTRA, T. A., YOUNG, R. & WITTE, H. J. L. 1990. An evaluation of the climatic conditions during the Late Quaternary in northern Greece by means of multivariate analysis of palynological data and comparison with recent phytosociological and climatic data. *Geologie en Mijnbouw*, **69**, 243–251.

WILLIAMS, D. F., Thunell, R. C. & KENNETT, J. P. 1978. Periodic fresh-water flooding and stagnation of the eastern Mediterranean during the late Quaternary. *Science*, **201**, 252–254.

WU, P. & HAINES, K. 1996. Modelling the dispersal of Levantine Intermediate Water and its role in Mediterranean deep water formation. *Journal of Geophysical Research*, **101**, 6591–6607.

——— & ——— 1998. The general circulation of the Mediterranean Sea from a 100-year simulation. *Journal of Geophysical Research*, **103**, 1121–1135.

WÜST, G. 1961. On the vertical circulation of the Mediterranean Sea. *Journal of Geophysical Research*, **66**, 3261–3271.

ZAHN, R., SARTHEIN, M. & ERLENKEUSER, H. 1987. Benthic isotope evidence for changes of the Mediterranean outflow during the late Quaternary. *Paleoceanography*, **2**, 543–559.

ZHAO, M., BEVERIDGE, N. A. S., SHACKLETON, N. J., SARNTHEIN, M. & EGLINTON, G. 1995. Molecular stratigraphy of cores off northwest Africa: sea surface temperature history over the last 80 ka. *Paleoceanography*, **10**, 661–675.

———, ROSELL, A. & EGLINTON, G. 1993. Comparison of two U^k_{37} sea surface temperature records for the last climatic cycle at ODP site 658 from the subtropical Northeast Atlantic. *Palaeogeography, Palaeoclimatology, Palaeoecology*, **103**, 57–65.

Pollen analysis of the Acerno palaeo-lacustrine succession (Middle Pleistocene, southern Italy)

ELDA RUSSO ERMOLLI

Dipartimento di Scienze della Terra, Università di Napoli Federico II, Largo San Marcellino 10, 80135 Napoli, Italy (e-mail: ermolli@unina.it)

Abstract: Pollen analysis has been performed on a 98 m core recovered in the palaeo-lake of Acerno (Campania, Italy). This study identified four main vegetation zones that can be correlated with Quaternary climatic events. In particular, one interglacial–glacial cycle was recorded in the succession the duration of which was estimated at around 100 ka. The interglacial period (Zone 2) is characterized by large amounts of deciduous and altitudinal forest elements; Mediterranean xerophytes are rather rare because of the distance from the palaeo-sea coast. The interglacial–glacial transition (Zone 3) shows a very rapid decrease of the arboreal taxa and a contemporaneous increase in the frequency of the non-arboreal element. The genus *Pinus*, excluded from the pollen totals, shows its highest percentage and concentration values during the transition. The glacial period (Zone 4) is characterized by the dominance of herbaceous and steppe elements, and by rather large amounts of *Pinus*. The floral composition of the analysed succession allows a late mid-Pleistocene age to be proposed for the Acerno palaeo-lake.

Pollen analysis of thick continental sequences can be considered the most useful method of reconstructing palaeoclimatic changes at very high resolution. The rapid adaptation of vegetation in response to environmental changes also gives an opportunity to detect minor climatic fluctuations and, by the analysis of the whole fossil floral composition, to distinguish them from environmental variations caused by other factors.

The aim of this study is to provide data on the Quaternary continental stratigraphy of southern Italy, in particular that of the mid-Pleistocene period. During that time many lacustrine basins formed in the southern Apennines as a result of the eastward-migrating extension associated with the opening of the Tyrrhenian Sea. There is, however, a lack of suitable outcrops of these lacustrine sediments. Tectonic displacement, coupled with active erosion, means that it is difficult to correlate the sediments. The only way to overcome these problems is to drill through a complete sedimentary succession and recover a continuous core. This has been done successfully in the Vallo di Diano Basin (Russo Ermolli 1994), and in the Acerno palaeo-lake a 98 m succession was recovered.

The pollen data obtained from the Acerno Basin represent a further step in the reconstruction of a palaeoclimatic curve of the Pleistocene period to be used as a reference for the southern Apennine region.

Physical setting

Geography and geology

The Picentini Massif is a Mesozoic carbonate horst that rises several hundred metres above the surrounding hilly landscape where terrigenous units of Miocene age crop out. Tectonic events have strongly influenced the long and complex geomorphological evolution of the Picentini Mountains since Pliocene time. These have alternated with quiescent periods during which erosional processes developed.

The Acerno Basin is located in the central sector of the Picentini Mountains at 655 m above sea level (Fig. 1). The basin represents a tectonic depression initiated during the tectonic activity that affected the southern Apennines in Early Pleistocene time. During the mid-Pleistocene period a lacustrine environment became established in this depression. Erosional capture by the Tusciano River, probably influenced by the general uplift of the Picentini Mountains relative to the Sele Plain, ended lacustrine deposition.

From: HART, M. B. (ed.) *Climates: Past and Present.* Geological Society, London, Special Publications, **181**, 151–159, 1-86239-075-4/$15.00

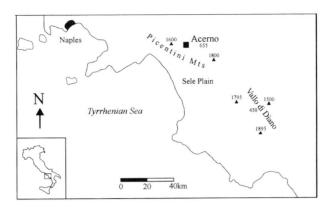

Fig. 1. Location map.

Table 1. *List of taxa and their affiliation to vegetation groups*

Deciduous forest elements	Altitudinal forest elements	Mediterranean xerophytes	Open vegetation	Steppe elements
Quercus deciduous	*Abies*	*Quercus* type *ilex*	Poaceae	*Artemisia*
Carpinus	*Picea*	*Olea*	Asteroideae	*Ephedra*
Acer	*Fagus*	*Phillyrea*	Chicorioideae	*Hippophae*
Ostrya	*Betula*	*Ligustrum*	*Centaurea*	
Ulmus		*Pistacia*	Amaranthaceae–Chenopodiaceae	
Zelkova		*Coriaria mirtifolia*	Caryophyllaceae	
Celtis		*Cistus*	*Plantago*	
Hedera			Cyperaceae	
Alnus			Lamiaceae	
Ericaceae			Papaveraceae	
Tilia			Apiaceae	
Corylus			*Mercurialis*	
Fraxinus			*Armeria*	
Platanus			*Limonium*	
Populus			Dipsacaceae	
Buxus			*Knautia*	
Ilex			*Scabiosa*	
Salix			Brassicaceae	
Juglans			*Geranium*	
			Erodium	
			Linum	
			Ribes	
			Euphorbia	
			Campanulaceae	
			Urticaceae	
			Rumex	
			Polygonum	
			Convolvulus	
			Helianthemum	
			Liliaceae	

Present-day climate and vegetation

The Campanian climate can be considered Mediterranean but, because of the complex physiography of the region, important local variations can be observed. Rainfall, mainly concentrated during the winter and autumn, is more abundant on the southeastern facing slopes, such as the Picentini Mountains, where 1500–2000 mm/year are recorded. Temperatures over the year are also rather variable as a result of the local physiography. In general, the winters are cold and humid and the summers are hot and drought-prone, but in some internal areas the summers can be rather wet. Mean minimum temperatures for the coldest months are 5–6°C but can vary between 1 and −3°C in some internal areas. Mean maximum temperatures for the warmest months are 29–30°C but some variations can be recorded in the internal and more elevated areas.

Vegetation cover is obviously influenced by climatic conditions and its organization ranges from the thermo-Mediterranean belt along the coasts, up to the alti-Mediterranean belt on mountain tops (Ozenda 1975). The Acerno Basin, 655 m high, lies within the supra-Mediterranean belt that is dominated by deciduous oak forest (*Quercus pubescens, Quercus cerris, Carpinus orientalis, Castanea sativa, Tilia platyphyllos,* etc.). The relief surrounding the basin, reaching altitudes of 1600–1800 m, lies within the oro-Mediterranean belt that is now dominated by beech (*Fagus sylvatica*).

Materials and methods

The Acerno lacustrine succession is partly exposed as a result of intense erosion by the Tusciano River but, to analyse the entire record, a borehole was drilled to the base of the succession (98 m). The drill site was located on the top of an erosional terrace in a suitable position for the drilling rig. From the base (98 m) to 94.5 m the core consists of fluvial conglomerates (Fig. 2) that pass rapidly upwards into lacustrine grey silt whose homogeneity is hardly ever interrupted by sandy layers. From 33.2 m to the top of the core, conglomerates alternate with sandy or silty sand levels. The uppermost 33.2 m may represent a regressive trend, which indicates the end of the lacustrine phase.

Pollen analysis was performed on 84 samples taken from the core with a variable sampling interval. Standard chemical treatment was used to remove the inorganic fraction and concentrate the pollen grains as a residue. Tablets containing spores of *Lycopodium clavatum* (Stockmarr 1971) were added to each sample to estimate pollen concentration. A minimum of 300 pollen grains were counted for each sample, allowing 78 taxa to be identified.

Data display

Pollen analysis results are presented in two diagrams (detailed and synthetic) and in some

concentration curves.

The detailed pollen diagram (Fig. 2) includes pollen spectra of the entire sequence and includes almost all the taxa detected in the analysis. Each taxon percentage is calculated on a pollen sum including all the pollen grains except *Pinus* and spores whose percentages are calculated on the same pollen sum increased by *Pinus* and spores.

In the synthetic pollen diagram (Fig. 3) taxa are grouped according to their presumed ecological requirements (Suc 1984). In particular, for the Acerno sequence, five groups were defined (Table 1). The first includes all the elements of the deciduous forest (mesothermic), dominated by *Quercus*, which requires a rather warm and wet climate. The second group includes the elements of the altitudinal forest (microthermic), mainly represented by *Abies* and *Fagus*, which demand high humidity but which can tolerate low temperatures. The third group is formed by the Mediterranean xerophytes, whose rare occurrence in the sequence is probably due to the location of the Acerno lake, which is rather far from the sea. The fourth group includes the herbaceous elements that are mainly represented by the Poaceae, Asteraceae (Asteroideae and Chicorioideae), Chenopodiaceae–Amaranthaceae and Cyperaceae. The fifth group is represented by the steppe elements that are dominated by *Artemisia* and, to a minor extent, by *Ephedra* and *Hippophae*. The last two groups characterize open landscapes with dry conditions.

Concentration curves (Fig. 4) were constructed to clarify some of the problems; e.g. the role of *Pinus* in the vegetation associations.

Description and interpretation of results

Diagram zonation

The main variations of the pollen spectra concern the arboreal taxa (AP; arboreal pollen) percentages in relation to the herbaceous taxa (NAP; non-arboreal pollen) and, to a minor extent, the pollen composition as a whole. For this reason, diagram zonation was based on the AP percentage variations. In particular, in southern Italy, AP percentages higher than 70–80% are typical of forested environments whereas AP percentages lower than 40–50% reflect an open landscape dominated by herbaceous vegetation (Russo Ermolli 1995). The intermediate percentages can be correlated with intermediate environments.

On the basis of this zonation method, four zones were distinguished in the Acerno succes-

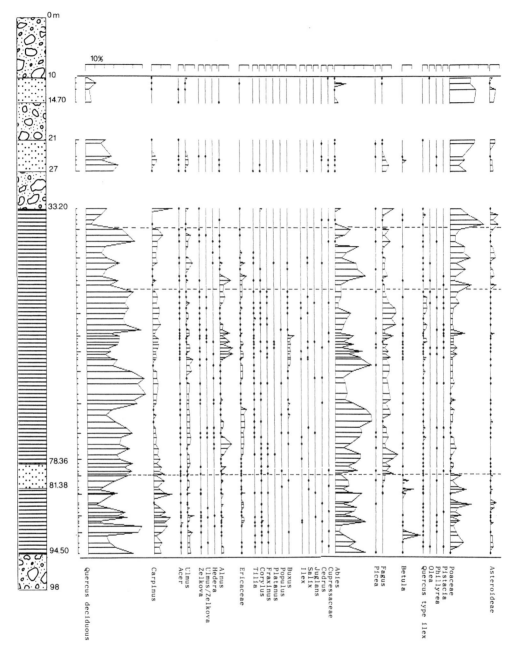

Fig. 2. Pollen percentage diagram plotted against depth.

sion and correlated with the Quaternary climatic cyclicity (Figs 2 and 3). The first zone extends from the base of the succession to 81 m and probably represents the beginning of an interglacial period. The second zone ranges from 81 to 47.6 m and clearly shows an interglacial period in which minor fluctuations are recorded. The third zone, extending from 47.6 to 36 m, represents an interglacial–glacial transition. Finally, the fourth zone, from 36 m to the top of the investigated section (10 m), represents part of a glacial period.

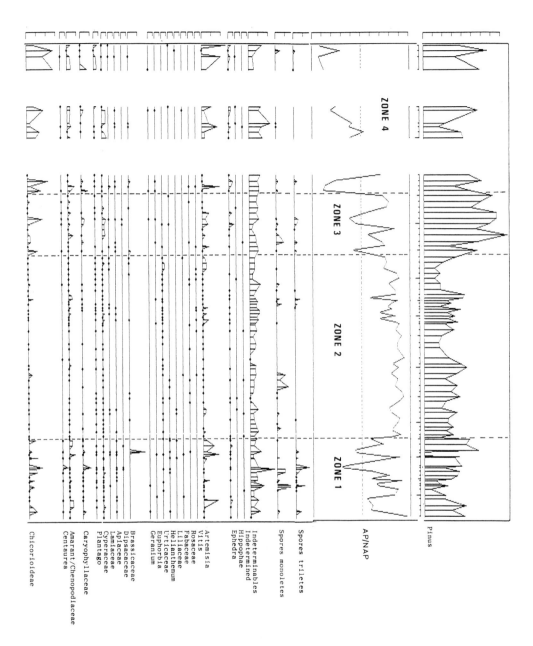

Zone 1

In this zone AP percentages fluctuate from 50 to 90%. *Quercus* is the main element of the deciduous forest, together with significant amounts of *Carpinus* and, to a minor extent, *Acer*, *Ulmus*, *Zelkova* and *Alnus*. *Abies* repre-

sents the main element of the altitudinal forest and its abundance seems to follow the trend of *Quercus*. It is interesting to note that a peak of *Betula* correlates with a decrease in the curves of both *Abies* and *Quercus*. It is possible that *Betula* has the role of a pioneer tree in this less favourable climatic period. The coexistence of

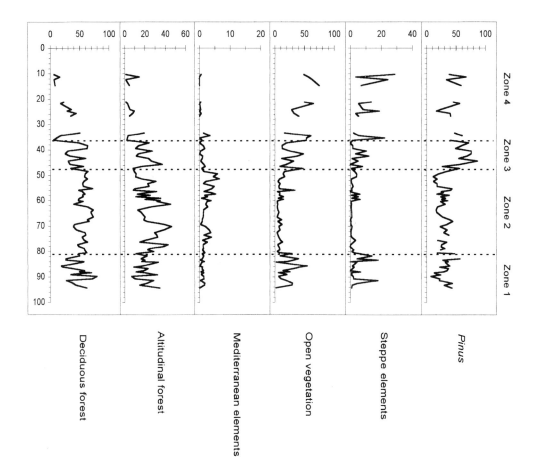

Fig. 3. Synthetic pollen diagram.

significant amounts of herbaceous (mainly Poaceae and Asteroideae) and steppe elements (mainly *Artemisia*), together with the arboreal elements, may indicate that the aridity, which probably characterized the end of the glacial period preceding the Acerno interglacial, was still influencing some vegetation belts by leaving open spaces within the forested landscape. In the Vallo di Diano palaeo-lacustrine succession a similar vegetation evolution can be observed (Russo Ermolli 1994); herbaceous and steppe elements still survive at the beginning of the interglacial periods.

In the Vallo di Diano, the glacial–interglacial transition shows a very rapid increase of arboreal elements (mainly *Quercus*) in response to the rapid rise of temperature that characterizes the cold–warm transition. It has been calculated that a complete transitional phase lasted about 1400 years (Russo Ermolli &

Cheddadi 1997). In Acerno, the starting point of the interglacial was missed because the drilling stopped in barren coarse sediments.

Looking at the concentration curves (Fig. 4) it is interesting to note that during this period both NAP percentages and concentrations show rather high values, a sign of their real representation in the vegetation associations.

Zone 2

In this zone the AP percentages fluctuate from 70 to 90%. The landscape was mainly occupied by an enriched deciduous oak forest and, at higher altitudes, by a mixed fir–beech forest. During this warm and humid interglacial period some climatic fluctuations can be identified by the variable percentages of some elements within the forest cover such as *Ulmus*, *Alnus*, Ericaceae and *Buxus*. Similar oscillations, but with a

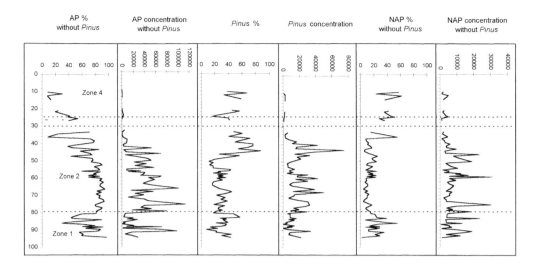

Fig. 4. Comparisons between pollen percentage and concentration curves for selected taxa.

higher amplitude, can be observed in the *Quercus* and *Abies* curves, perhaps as a response to humidity stress periods. The fluctuations of the two main arboreal elements, *Quercus* (more sensitive to temperatures) and *Abies* (more influenced by humidity), do not always seem to follow the same pattern. They are sometimes in opposition and sometimes in phase.

It has already been shown that temperature and humidity follow two different cycles within the Quaternary climatic cyclicity (Van der Hammen *et al.* 1971). The temperature cycle always precedes the humidity cycle so that the beginning of the interglacial and the end of the glacial periods are characterized by aridity whereas the end of the interglacial and the beginning of the glacial periods show high humidity values. The same trend was observed in the Vallo di Diano palaeo-lake located in Campania, south of the Acerno Basin (Russo Ermolli 1994). The vegetation response to such a climatic variation is a rapid development of mixed deciduous oak forest as a consequence of temperature rise and a successive increase of the fir forest as a result of the rise in humidity. This change in the vegetation is not very clearly recorded in the Acerno succession, where *Abies* is rather abundant from the base of the lacustrine phase, probably the result of the general condition of high humidity, which could characterize elevated regions such as the Picentini Mountains.

During this period the concentration values (Fig. 4) are very high for all the taxa, (including the herbaceous), suggesting that the non-arboreal pollen percentages are misleading and that a reliable message is only given by the arboreal pollen.

Zone 3

In the third zone the AP percentages range from 45 to 75%. Arboreal elements are gradually replaced by an open vegetation cover in response to the fall in temperatures that characterize the interglacial–glacial transitions. The available data on the nature and speed of warm–cold transitions indicate that the climate change occurred through a series of short cooling and/ or drying pulses (Reille & Beaulieu 1990) followed by longer periods of relatively stable climate (Woillard 1979; Kukla 1980).

The main herbaceous families (Poaceae and Asteraceae) show increased percentages, along with *Artemisia*, among the steppe elements. *Quercus* remains stable whereas *Abies* decreases its abundance throughout the transitional period.

It is interesting to note that, during the transitional phase, *Pinus* seems to be the main arboreal element surviving during the climatic shift and even increasing its percentages relative to the previous periods. Its concentration shows the highest values, in agreement with its frequencies (Fig. 4). All the other taxa decrease their concentration throughout the period, indicating a vegetation cover reduction.

Zone 4

In this zone, AP percentages vary between 10 and 50%. In response to the incoming cold and dry climatic conditions, all the herbaceous elements increase their percentages to the detriment of the arboreal elements, whose main representative remains *Quercus* with very low values (20% on average). Steppe elements are dominated by *Artemisia*, which comprises about 20% of the total pollen. The lacustrine sequence ends during this glacial period, preventing the reconstruction of the subsequent vegetation evolution.

Concentration values of all the taxa are very low, reflecting the thinning out of the forest cover and the expansion of open spaces where the herbaceous elements dominate, with their lower pollen production and, perhaps, higher clastic sediment flux.

The role of Pinus

The ecological value of *Pinus* is difficult to understand, mainly because it is rather difficult to distinguish *Pinus* species on the basis of pollen morphology. It is abundant throughout the investigated sequence but, during the interglacial–glacial transition as well as during the glacial period, it reaches its highest percentages. The increase of *Pinus* during less forested periods could be linked to different factors such as long-distance transport, high resistance to degradation in soils, a statistical increase owing to the decrease of the other arboreal taxa, or a real representation in the vegetation associations as a consequence of its tolerance of dry conditions. Looking at the concentration curve (Fig. 4) it is noted that the concentration of *Pinus* is very high during the transitional phase, proving that it was really the main arboreal element surviving in the region. However, during the glacial period its concentration values decrease like those of all the other arboreal taxa. The fact that its percentages increase during the same period can only indicate that, in this case, *Pinus* is statistically favoured with respect to the other arboreal taxa.

Discussion and conclusions

The palynological results of the Acerno sequence allow the recognition of one interglacial–glacial climatic cycle and the reconstruction of the vegetation evolution during the investigated period. The composition of the pollen flora seems to confirm the general mid-Pleistocene chronological attribution (Capaldi *et al.*

1989). The absence of 'Tertiary relics' like those still well represented at Camerota in Campania (Brenac 1984) and Crotone in Calabria (Combourieu-Nebout 1990) allows an Early Pleistocene age to be rejected. Moreover, the absence of *Carya*, one of the last relics to disappear from the Italian successions, suggests that the Acerno sequence is younger than the Vallo di Diano sequence, (in Campania) where this element survives in vegetation associations until 450 ka BP (Russo Ermolli 1994; Karner *et al.* 1999). The scarcity of palynological data in southern Italy makes it difficult to be more precise in the dating of the succession. Comparison with other sequences, located far from the Campanian region, do not give more accuracy. In particular, at Valle di Castiglione in Latium (Follieri *et al.* 1988) the mid-Pleistocene interglacial periods (isotopic stages 5, 7 and 9?) are characterized by amounts of *Fagus*, *Carpinus*, *Zelkova* and *Quercus* type *ilex* higher than those recorded in the Acerno interglacial. The Ioanian record, in northern Greece (Tzedakis 1994), shows some similarities with the Acerno sequence. In particular, isotopic stages 7 and 9 are characterized by high amounts of *Abies* and *Fagus*. However, it is dangerous to consider similarities or discrepancies in pollen spectra as chronological constraints for distant regions, especially when modern vegetation associations are involved.

However, on the basis of pollen analysis only, it is possible to propose a late mid-Pleistocene age for the Acerno sequence. The pollen flora is almost the same as the modern one except for the significant presence of *Abies*, which has been replaced completely by *Fagus sylvatica* in the highest vegetation belt of the southern Apennines and today constitutes a relict vegetation on some mountain tops.

The identification of one climatic cycle makes the duration of the lacustrine phase much shorter than previously stated (Capaldi *et al.* 1989). In fact, during mid-Pleistocene time the climatic cyclicity shows a period of about 100 ka (Ruddiman *et al.* 1989) and thus this should be the maximum duration of the lacustrine record analysed here.

References

BRENAC, P. 1984. Végétation et climat de la Campanie du Sud (Italie) au Pliocène final d'après l'analyse pollinique des dépots de Camerota. *Ecologia Mediterranea*, **10**, 207–216.

CAPALDI, G., CINQUE, A. & ROMANO, P. 1989. Ricostruzione di sequenze morfoevolutive nei Picentini Meridionali (Campania, Appennino Meridionale). *Supplementi di Geografia Fisica e Dinamica*

Quaternaria, **1**, 207–222.

COMBOURIEU-NEBOUT, N. 1990. Les cycles glaciaire–interglaciaire en région Méditerranéenne de −2,4 à −1,1 Ma: analyse pollinique de la série de Crotone (Italie Méridionale). *Paléobiologie Continentale*, **17**, 35–59.

FOLLIERI, M., MAGRI, D. & SADORI, L. 1988. 250,000-year pollen record from Valle di Castiglione (Roma). *Pollen et Spores*, **30**, 329–356.

KARNER, D. B., JUVIGNÉ, E., BRANCACCIO, L., CINQUE, A., RUSSO ERMOLLI, E., SANTANGELO, N., BERNASCONI, S., LIRER, L. 1999. A potential early middle Pleistocene tephrostratotype for the Mediterranean basin: the Vallo di Diano, Campania, Italy. *Global and Planetary Change*, **21**, 1–15.

KUKLA, G. 1980. End of the last interglacial: a predictive model of the future? *In*: VAN ZINDEREN BAKKER E. M. & COETZEE, J. A. (eds) *Paleoecology of Africa and the Surrounding Islands*, Balkema, Rotterdam, 395–408.

OZENDA, P. 1975. Sur les étages de végétation dans les montagnes du bassin Méditerranéen. *Documents de Cartographie Ecologique*, **16**, 1–32.

REILLE, M. & DE BEAULIEU, J. 1990. Pollen analysis of a long younger Pleistocene continental sequence in the Velay Maar (Massif Central, France). *Palaeogeography, Palaeoclimatology, Palaeoecology*, **80**, 35–48.

RUDDIMAN, W. F., RAYMO, M. E., MARTINSON, D. G., CLEMENT, B. M. & BACKMAN, J. 1989. Pleistocene evolution: Northern Hemisphere ice sheets and North Atlantic Ocean. *Paleoceanography*, **4**, 353–412.

RUSSO ERMOLLI, E. 1994. Analyse pollinique de la succession lacustre pléistocène du Vallo di Diano (Campanie, Italie). *Annales de la Société Géologique de Belgique*, **117**, 333–354.

—— 1995. *Analyse pollinique des dépots lacustres pléistocènes du Vallo di Diano (Campanie, Italie): cyclicités et quantification climatiques.* PhD thesis, Liège University, Liège.

—— & CHEDDADI, R. 1997. Climatic reconstruction during the Middle Pleistocene: a pollen record from Vallo di Diano (Southern Italy). *Geobios*, **30**, 735–744.

STOCKMARR, J. 1971. Tablets with spores used in absolute pollen analysis. *Pollen et Spores*, **13**, 615–621.

SUC, J.-P. 1984. Origin and evolution of the Mediterranean vegetation and climate in Europe. *Nature*, **307**, 429–432.

TZEDAKIS, P.C. 1994. Vegetation change through glacial–interglacial cycles: a long pollen sequence perspective. *Philosophical Transactions of the Royal Society of London Series B*, **345**, 403–432.

VAN DER HAMMEN, T., WIJMSTRA, T. A. & ZAGWIJN, W. H. 1971. The floral record of the Late Cenozoic of Europe. *In*: TUREKIAN, K. (ed.) *The Late Cenozoic Glacial Ages.* Yale University Press, New Haven, CT, 291–494.

WOILLARD, G. 1979. Abrupt end of last interglacial *s.s.* in NE France. *Nature*, **281**, 558–562.

Large mammal turnover pulses correlated with latest Neogene glacial trends in the northwestern Mediterranean region

B. AZANZA[1,2], M. T. ALBERDI[1] & J. L. PRADO[3]

[1] *Departamento de Paleobiología, Museo Nacional de Ciencias Naturales, CSIC, José Gutiérrez Abascal, 2, 28006 Madrid, Spain (e-mail: mcnaa3j@fresno.csic.es)*
[2] *Departamento de Ciencias de la Tierra, Universidad de Zaragoza, 50009 Zaragoza, Spain*
[3] *INCUAPA–Departamento de Arqueología, Universidad Nacional del Centro (UNC), Del Valle 5737, 7400 Olavarría, Argentina*

Abstract: From the latest Miocene to the Holocene period, three biotic events in the large mammal communities of the western Mediterranean region show significant and robust diversity changes and/or turnover with respect to sampling, and correspond in time to significant pulses of the latest Cenozoic glaciations. The 'Ruscinian mammal turnover pulse' represents a major biotic event close to the 'Messinian salinity crisis'. This can be described more as a turnover than a dispersion event. The pattern of turnover shows a significantly high value of first appearances close to the Mio-Pliocene transition; last appearances were also important but reduced significantly in the following interval. A Southern Hemisphere glaciation event has been related to the Messinian crisis. The second major biotic event corresponds to the so-called '*Equus*–elephant event' during early Villafranchian time. The relevance of this dispersal event is proved by a marked increase in diversity as a result of a high number of first appearances. Between 3.0 and 2.6 Ma the onset of bi-polar glaciations occurred, followed by glacial–interglacial cycles of moderate amplitude sustained at the orbital periodicity of 41 ka. The last major biotic event was the 'Galerian mammal turnover pulse' around 1.0 Ma. This turnover represents an important community reorganization that marks a total rejuvenation of the fauna and is coincident with the arrival of *Homo* species in the western Mediterranean region. At this time, glacial maxima became more extreme and the dominant periodicity of variation shifted to 100 ka, which is attributed to the massive Northern Hemisphere ice sheets.

For some time now, palaeontologists have considered the fossil record to be composed of intervals of ecological and evolutionary stability limited by episodes of relative instability during which fairly rapid changes occurred. The 'co-ordinated stasis model' describes this pattern of linked stability and change (Brett & Baird 1995; Ivani & Schopf 1996). Developmental constraints that restrict change within individual lineages (Eldredge & Gould 1972) or stabilise selection as a result of a lack of environmental change (physical or biotic) that could lead to speciation in multiple lineages have been invoked to explain evolutionary stasis. In turn, the 'equilibrium theory' (MacArthur & Wilson 1967) postulates that extinction and origination (immigration) could both occur independently although giving rise to an ecological balance. Thus, changes in diversity do not necessarily

imply a restructuring of the mammal communities. Consequently, disruptions of this stability require environmental changes (physical or biotic) that provoke fast and concurrent evolutionary changes in multiple lineages. The 'turnover pulse hypothesis' (Vrba 1985, 1992, 1995*a*) questions the role played by biotic factors in clade radiations and extinctions. In consequence, most lineage turnover in the history of life has occurred in pulses that are almost concurrent across widespread groups of organisms, and are clearly in step with changes in the physical environment. Other models, such as the 'Red Queen hypothesis' (Van Valen 1973), which postulates that evolutionary changes would have occurred as a response to the internal dynamics of competitive relationships amongst sympatric populations, would not predict a close relationship between major

From: HART, M. B. (ed.) *Climates: Past and Present.* Geological Society, London, Special Publications, **181**, 161–170, 1-86239-075-4/$15.00

climatic changes and evolutionary events, but would instead propose evolutionary change to continue even during periods of relative stability. A few empirical tests have been made, but the results have been contradictory. The analysis of the record of African bovids seems to corroborate the 'turnover pulse hypothesis' (Vrba 1995b). The analysis of the records of some African omnivores, by Foley (1994) and White (1995) failed to indicate any relationship between biotic pulses and climatic changes. Alternatively, these workers proposed that climatic changes affect evolution primarily through extinction and that further factors, which depend on local competitive conditions, are significant in the appearance of new taxa. On the other hand, the analysis of the Cenozoic fossil record of terrestrial North American mammals (Alroy 1996) indicates that the apparent correlation between origination and extinction rates is a by-product of statistics resulting from the need for a coincidence of the first and last appearances of taxa known during a single temporal interval. Furthermore, although origination rates show many well-defined pulses, most of the observed extinction rates could have been brought about by a single, uniform underlying rate.

The late Cenozoic record from the northwestern Mediterranean region offers remarkable opportunities both for comparative testing and to provide an understanding of the constraining effects of both biotic interaction and abiotic features of the physical environments for mammalian community structure. The region has a good record of large mammals, with chronological and, although debatable, palaeoenvironmental data for this period. It is recognized as a zoogeographic province that has existed from late Miocene time (Bernor 1984, 1986; Fortelius *et al.* 1996), and a number of major biotic crises have been proposed that are related to major pulses of the latest Cenozoic glaciations (Aguirre *et al.* 1976; Azzaroli 1983; Steininger *et al.* 1985; Azzaroli *et al.* 1988; Aguirre & Morales 1990; Sala *et al.* 1992; Torre *et al.* 1992). However, questions concerning species diversity and the dynamics of faunal turnover have not yet been addressed. Moreover, the precocious arrival in the Mediterranean region of man (*Homo antecessor*) indicated by recent discoveries (Bermúdez de Castro *et al.* 1997), poses a challenge to the analysis of this dispersal event within the framework of the large mammal turnover patterns. The earliest evidence of human occupation in Europe has always been controversial. Although recent evidence from the Republic of Georgia indicates that *Homo* was at the gates of

Europe at beginning of Pleistocene time (Gabunia & Vekua 1995), the presence of the earliest traces towards the end of early Pleistocene time places human settlement at around 1.0 Ma (Thouveny & Bonifay 1984; Carbonell *et al.* 1995). Nevertheless, some workers suggest that Europe was not colonized until 0.5 Ma, at least not on a permanent basis (Gamble 1994; Roebroeks & Van Kolfschoten 1994).

If climatic change is the forcing factor for evolutionary changes, then: (1) origination and extinction rates should be correlated; (2) turnover should be a composite function of very low background rates and occasional, dramatic turnover pulses; (3) stasis should result from ecological incumbency (Alroy 1996). Our interest focuses on formulating and testing the hypotheses of community evolution through the analysis of turnover patterns (predictions (1) and (2)); we are not concerned here with its effect on community ecology (prediction (3)).

Methodological approaches

A major problem in explaining the regulation and maintenance of diversity using the fossil record is how to control variations throughout time. The use of previous biochronological scales (MN 'zones', mammal units, biozones) poses a problem, as the boundaries between the units are not defined with regard to faunal turnover and, as a consequence, it is possible to create fictitious events at these boundaries. Two alternative approaches are considered which are based on unequal (biochronological units) and equal time intervals, both being resolved by multivariate methods (Alberdi *et al.* 1997; Azanza *et al.* 1997). For the analysis, previous data compilations (Alberdi *et al.* 1997) have been revised and reformulated (Azanza *et al.* 1999).

Unequal time intervals (UTI) approach

Using multivariate methods based on similarity, Alberdi *et al.* (1997) established nine biochronological units (A–I). The discontinuities between these are assumed to correspond to faunal restructurings associated with major changes in environmental conditions (Alberdi *et al.* 1997; Azanza *et al.* 1999). This assumption is coincident with the concept of *Étage* and *Subétage* of d'Orbigny, which are identified by episodes of relatively rapid change as a result of processes of extinction, adaptive radiations and dispersion, a concept clearly expressed by the Ecological Evolutionary Unit and Subunit (Boucot 1983, 1990). The duration of the units varies between

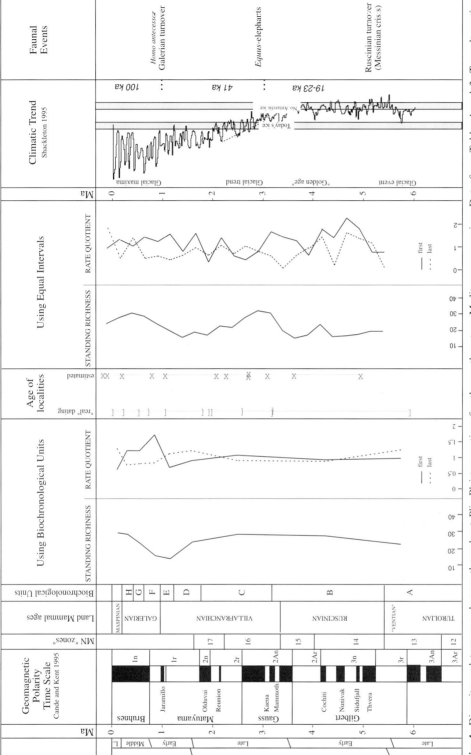

Fig. 1. Diversity and turnover changes throughout Plio–Pleistocene time for the northwestern Mediterranean region. Data from Tables 1 and 2. Two alternative approaches are considered based on unequal (biochronological units) and equal time intervals (see explanation in text). Biochronological units are according to Alberdi *et al.* (1997). The estimated ages of localities were calculated by Azanza *et al.* (1997) from the 'real' ages (radioisotopic and palaeomagnetic datings) of localities following Alroy (1992). The distortion between 'real' and estimated ages of localities is expressed graphically.

Table 1. *Mammalian specific turnover and richness for the unequal time intervals*

Unit	Interval	N	CI	CI$_{bda}$	N$_{sr}$	First obs.	First exp. χ²		RQ		Last obs.	Last exp. χ²		RQ	
A	6.1–5.4	44	100	100	8	25	25.51	0.01	0.98		30	23.945	1.531	1.25	(+)
B	5.4–3.1	50	100	100	10.5	38	40.398	1.142	0.94		37	41.285	0.445	0.89	(++)
C	3.1–1.9	48	100	100	6	35	32.649	0.169	1.07	(+)	29	31.388	0.182	0.92	
D	1.9–1.3	36	97	85	7	17	19.555	0.123	0.92		23	19.623	0.976	1.22	
E	1.3–1.0	17	80	50	3.5	4	5.923	0.624	0.68	(– – –)	9	8.027	0.118	1.12	
F	1.0–0.7	26	92	50	4	17	9.309	4.809	1.72	(++)	9	10.649	0.255	0.84	
G	0.7–0.4	32	82	41	8.5	16	13.072	0.655	1.22		11	13.135	0.347	0.83	(–)
H	0.4–0.2	42	100	100	7	21	17.662	1.065	1.24		11	14.502	0.846	0.76	(– – –)
I	0.2–0	42	100	100	6	11	17.662	2.512	0.62	(–)	19	14.502	1.395	1.31	(–)

Mammalian specific turnover and richness for the nine unequal time intervals (biochronological units) recognized from latest Miocene to Recent time in the northwestern Mediterranean region (Alberdi *et al.* 1997). *N*, number of species. CI, CI$_{bda}$, completeness indices, N_{sr}, standing richness, calculated following Maas *et al.* (1995); obs., observed; Exp., expected. Expected first occurrences were calculated as: Re = exp[0.397ln(duration) + 1.335ln(N_{sr}) −0.962] and expected last occurrences were calculated as: Re = exp[0.456ln(duration) + 0.906ln(N_{sr}) + 0.399] (see explanation in text). RQ, rate quotient; overall significance of the observed turnover pattern is given following Barry *et al.* (1995) by χ² goodness-of-fit, and χ² values are also given for each interval. Significantly high observed turnover is indicated by '+' and significantly low observed turnover is indicated by '–' based on comparison with bootstrapped distribution generated for 10000 simulations (Maas *et al.* 1995). (+ + +) or (– – –) at $P < 0.001$, (+ +) or (– –) at $P < 0.01$, and (+) or (–) at $P < 0.05$. Further data have been given by Azanza *et al.* (1999).

Table 2. *Mammalian specific turnover and richness for the equal time intervals*

Interval	N	N_{sr}	First obs.	First exp.	χ^2	First RQ		Last obs.	Last exp.	χ^2	Last RQ	
5.5–5.25	22	19.5	5	4.863	0.004	1.028		0	0			
5.25–5	27	19.5	5	4.863	0.004	1.028	(−)	12	10.772	0.140	1.114	(−)
5–4.75	27	17.5	12	6.274	5.255	1.913		14	10.837	0.928	1.292	(+)
4.75–4.5	29	16.5	15	6.695	10.322	2.241	(+)	17	10.919	3.399	1.557	(+ +)
4.5–4.25	21	16	9	5.77	1.808	1.560		1	9.777	7.878	0.102	(+ + +)
4.25–4	33	23.5	13	6.422	6.740	2.024		15	10.866	1.578	1.36	(−)
4–3.75	22	17	4	4.557	0.068	0.878		9	10.651	0.256	0.82	(+)
3.75–3.5	21	15	8	5.576	1.055	1.435		6	10.484	1.915	0.55	(+)
3.5–3.25	25	20	10	5.95	2.759	1.681		0	0		1	
3.25–3	38	30.5	13	6.422	6.741	2.024		6	10.484	1.915	0.572	(−)
3–2.75	38	32	6	5.128	0.148	1.170	(− −)	8	10.603	0.638	0.755	(−)
2.75–2.5	34	28	3	4.192	0.338	0.716	(− −)	11	10.735	0.007	1.025	
2.5–2.25	27	21.5	4	4.557	0.068	0.878		7	10.548	1.189	0.664	
2.25–2	30	22.5	10	5.95	2.759	1.681		11	10.735	0.007	1.025	
2–1.75	21	17	2	3.725	0.799	0.537	(−)	6	10.484	1.915	0.572	
1.75–1.5	26	19	11	6.117	3.900	1.798	(+)	10	10.695	0.445	0.935	
1.5–1.25	21	15.5	5	4.863	0.004	0.028		6	10.484	1.915	0.572	
1.25–1	26	20	11	6.117	3.900	1.798		4	10.320	3.862	0.388	(−)
1–0.75	31	24	9	5.77	1.808	1.560	(+)	6	10.484	1.915	0.572	(−)
0.75–0.5	36	29	11	6.117	3.900	1.798		5	10.410	2.812	0.480	(−)
0.5–0.25	39	30.5	8	5.576	1.055	0.435		15	10.866	1.578	1.380	
0.25–0	34	28	10	5.95	2.759	1.681		5	10.410	2.812	0.480	(−)

Mammalian specific turnover and richness for the 22 equal time intervals recognized for the last 5.5 Ma in the northwestern Mediterranean region (from Azanza *et al.* 1997). *N*, number of species; N_{sr}, standing richness, calculated following Maas *et al.* (1995); obs., observed; exp., expected. Expected first appearances were calculated as: $Re = exp [0.291\ln(N_{sr}) + 1.113]$ and expected last appearances were calculated as: $Re = exp[0.74 \ln(N_{sr}) - 0.385]$. RQ, rate quotient; overall significance of the observed turnover pattern is given following Barry *et al.* (1995) by χ^2 goodness-of-fit, and χ^2 values are also given for each interval. Significantly high observed turnover is indicated by '+' and significantly low observed turnover is indicated by '−' based on comparison with bootstrapped distribution generated for 10000 simulations (Maas *et al.* 1995). Further data have been given by Azanza *et al.* (1999). (+ + +) or (− − −) at P=0.001, (+ +) or (− −) at P=0.01, and (+) or (−) at P=0.05.

2.5 and 0.2 Ma, making it necessary to normalize the interval length to avoid computational bias (Van Valen 1984; Foot 1994).

The relative sampling quality among these unequal intervals was estimated by the completeness indices (CI and CIbda) following Maas *et al.* (1995) based on the proportion of range-through taxa; i.e. the inferred occurrences of taxa known from preceding and succeeding intervals. Only the more conservative CIbda index shows significantly low values with respect to the intervals E, F and G. The low value of interval F was expected because it lies between two intervals with a high number of range-through species and in our records there is a low proportion of species present before, during and after (BDA) an interval. Consequently, the richness of the intervals E and G could be underestimated. During Quaternary time scales however, short-term range fluctuations can be picked up in close association with climate oscillations. Thus, the presence or absence of a species may be due merely to shifts in its distribution inside and outside the study area. The absence of some lineages has been evidenced. This is the case, for instance, of the Suidae, with no representative of this family recorded in Europe at the end of Villafranchian time (Van der Made 1989–1990), and also of some hyenids, *Pliocrocuta perrieri*, that disappear during this time and reappear again at the early part of mid-Pleistocene time (Turner 1992). In this situation, the range-through taxa do not always indicate poor quality of the fossil record.

Equal time intervals (ETI) approach

An alternative approach is to use equal intervals, but this requires precise ordering of the first and last appearance events of taxa. This requirement is not achievable directly from the Plio-Pleistocene continental record of the northwestern Mediterranean region, because most data come from isolated localities and age calibrations were not sufficiently precise. The parsimony-based methods of Alroy (1992, 1994) are used to establish the appearance event sequence. The discussion of the results obtained by the analysis has been presented elsewhere (Azanza *et al.* 1997). Calibration of the appearance event sequence is possible from the available radio-isotopic and palaeomagnetic datings of localities (Alroy 1992). Nevertheless, this calibration introduces a temporal distortion that becomes greater as available dating diminishes. The distortion between the 'real' and estimated ages of localities is expressed graphically in Fig. 1. To define an optimal interval length, the order of

magnitude corresponding to the minor length of the biochronological units (0.25 Ma) was selected, so as not to eclipse evolutionary events.

Mammalian diversity and turnover patterns

The changes in large mammal composition throughout time have been explored by two procedures. The analysis of the diversity was measured by the standing richness and the analysis of the turnover by the first and last appearance rate quotients (Azanza *et al.* 1999). The results obtained by both ETI and UTI approaches are compared in Fig. 1. The discrepancies between them are higher during Villafranchian and Ruscinian time, as selected time intervals are very dissimilar, and large mammal locality datings are scarce (Azanza *et al.* 1999). Tables 1 and 2 summarize diversity and turnover analyses following the UTI and ETI approaches, respectively.

Standing richness

Diversity changes throughout time were measured as the standing richness (Harper 1975), which standardizes the number of taxa that actually or potentially occur at each time interval, calculating the species' richness at the midpoint of a time interval. Standing richness (N_{sr}) was calculated following Maas *et al.* (1995).

The standing richness shows similar patterns in both approaches (Fig. 1). Following the UTI approach, the standing richness is high from the latest Miocene to Early Pliocene time, increasing slightly in the 3.1–1.9 Ma interval (interval C), although the maximum number of families was reached during Ruscinian time. After 1.9 Ma it begins to decline, reaching its minimum at the end of Villafranchian time, despite most of the families being recorded. During Galerian time, it increases strongly, later falling to the present diversity. In addition to the Galerian peak, the ETI approach has revealed another pronounced increase in standing richness during Early and Middle Villafranchian time, and the Late Villafranchian drop is less dramatic.

Turnover

Turnover changes through time were measured as rate quotients that normalize the number of first and last appearances in the intervals taking into account both the duration and the differences in species' richness among them (Gingerich 1987). They are calculated as the ratio of observed to expected turnover, where expected turnover is derived from multiple regression of the number of events on richness and duration in the analysis based on the unequal time intervals and from lineal regression on richness in the analysis based on the equal time intervals. The χ^2 goodness-of-fit was used to test deviations in the observed pattern from the expected distribution of first and last appearances (Barry *et al.* 1995). To test the null hypothesis that first and last appearances are randomly distributed among intervals, a bootstrapping method has been used (Maas *et al.* 1995).

The first and last appearance rate quotients (FRQ and LRQ, respectively) track one another closely during latest Miocene to Early Pliocene time, suggesting that a broad range of rates is possible within the equilibrium. Nevertheless, a weak turnover pulse at the beginning of Ruscinian time could be indicated by the high values of both FRQ and LRQ revealed by the ETI approach. During Early Villafranchian time the equilibrium is disrupted, with the FRQ higher in accordance with the acme of standing richness close to 3.0 Ma. The Late Villafranchian drop in richness is coincident in the UTI approach, with higher values of LRQ than of FRQ. In contrast, FRQ and LRQ track one another closely in the ETI approach. This seems to be in agreement with a less dramatic drop in richness shown by this approach. From the beginning of Galerian time, around 1.0 Ma, both approaches reveal significantly high values of FRQ, which is in agreement with the pronounced increase in standing richness. First and last appearances were significantly heterogeneous among intervals following the ETI approach ($\chi^2 = 56.69$ for first appearances and $\chi^2 = 44.11$ for last appearances; $P < 0.01$ and d.f. $= 22$ and 20 respectively), leading us to reject the null hypothesis.

For each interval, the observed numbers of first or last appearances are compared with the expected numbers generated for each of the 10 000 simulations. Significantly high or low observed turnover based on comparison with this bootstrapped test distribution are shown in Tables 1 and 2.

Climate–faunal change correlation

Climatic change is not a simple phenomenon and 'therefore' different patterns of climatic–evolutionary interactions would be expected depending upon which climatic variables are used. The correlation between the diversity and turnover patterns resulting from the ETI approach and the climatic changes was therefore explored using several variables estimated by Foley (1994) from isotopic curves: number of

climatic cycles, maximum $\delta^{18}O$, minimum $\delta^{18}O$, modal temperature, magnitude of climatic variation (maximum temperature minus minimum temperature), and magnitude of climatic change (modal temperature of one period less the modal temperature from the preceding period). The correlation between these variables and the standing richness, and first and last appearance rate quotients were explored. Only the first appearance rate quotient correlated significantly with the number of climatic cycles per interval ($r^2 = -0.88$ with $P = 0.0007$) and with the magnitude of climate change among intervals ($r^2 = -0.74$ with $P = 0.015$).

Discussion and conclusions

Patterns of mammalian species' turnover, richness and faunal composition are assessed in the light of relative sampling quality, and their correlation with climatic changes was evaluated. The results obtained support the argument that climatic change is a forcing factor in first appearances (including migrations) of mammal taxa, in agreement with Alroy (1996) and in contradiction to Foley (1994). Despite the fact that the last appearances do not correlate with any of the climatic variables, we detected several pulses where last appearances are observed higher or lower than expected by comparison with the bootstrapped distribution. This is an empirical test that supports Vrba's 'turnover pulse hypothesis'.

Three biotic events from latest Miocene to Holocene time in the western Mediterranean region show significant and robust diversity changes and/or turnover with respect to sampling. These episodes of faunal change correspond in time to significant changes in the physical environment, as several lines of evidence indicate (Shackleton 1995; Suc *et al.* 1995).

The first event was classically interpreted as a dispersal event between Eurasia and Africa, favoured by the desiccation of the Mediterranean Sea (Aguirre *et al.* 1976; Steininger *et al.* 1985; Cita *et al.* 1995). Nevertheless, some migrations could have occurred before the Messinian crisis (Pickford *et al.* 1995; Agustí & Llanas 1996), and the latter rules out the possibility of interchange with Africa, given that these taxa could have come across from more eastern Eurasia. On the other hand, alternative interpretations are put forward today for evaporites that do not imply desiccation of the deep Mediterranean (Martínez del Olmo 1996; Michalzik 1997). The pattern of turnover shows a significantly high value of first appearances in the interval 4.75–4.5 Ma; the last appearances were also high between 5.25 and 4.5 Ma but drop significantly in the following interval (Table 2). This could indicate a turnover pulse more than a dispersal event at the beginning of Ruscinian time. Near the Mio-Pliocene transition (around 5.4 Ma), there were significant changes in the physical environment, including the Messinian salinity crisis, which is interpreted as a first pulse of the latest Neogene glaciations and also diastrophic events (Hodell *et al.* 1986; Shackleton 1995).

The second major biotic event resulting from our analyses corresponds to the so-called *Equus*–elephant event during Early Villafranchian time (Aguirre *et al.* 1976; Azzaroli 1983; Steininger *et al.* 1985; Azzaroli *et al.* 1988; Aguirre & Morales 1990). The results obtained confirm the relevance of this dispersal event, which signified a marked increase in diversity as a result of the high number of first appearances (Table 1). Around 3.5 Ma a double seasonality was established by the emergence of cool winters generating a thermic seasonality, which was superimposed on the pre-existing hydric seasonality in the north. From 3.5 to 2.6 Ma the Mediterranean region experienced a modern climate, and the modern Mediterranean floral assemblage was developed. Between 3.0 and 2.6 Ma the onset of bi-polar glaciations occurred, followed by glacial–interglacial cycles of moderate amplitude sustained at the orbital periodicity of 41 ka as indicated by isotopic curves (Shackleton 1995; Suc *et al.* 1995).

The 'Galerian mammal turnover pulse' around 1.0 Ma represents the last major biotic event in Europe. This turnover, represented by a major community reorganization, records a total rejuvenation of the fauna. At that time, a great number of lineages appear that constitute the present mammal communities, including humans. Human remains from Atapuerca-TD6 (Carbonell *et al.* 1995; Bermúdez de Castro *et al.* 1997) represent the early occupation of western Europe by *Homo* species older than 0.78 Ma (Parés & Pérez González 1995). However, lithic artefacts and faunal remains included in this analysis were present at Atapuerca-TD4 level (4 m below TD6) and in Le Vallonet and Solilhac dated around 0.9 Ma (Thouveny & Bonifay 1984; Carbonell *et al.* 1995). At that time, the first caballine as *Equus sussenbornensis* (*sensu* Alberdi *et al.* 1995), the first true elaphine cervids, *Cervus elaphus acoronatus*, and the first *Dama, Bos* and modern *Bison* species all appear. Also recorded for the first time as a typical mid–late Pleistocene mammal was *Ursus deningeri* in

the speleoid lineage, *Dolichodoryceros savini* (the first representative of the *Megaloceros* group in the Megacerini), rhinoceros *Stephanorhinus kirchbergensis*, and proboscideans *Elephas antiquus* and *Mammuthus trogontherii*. These are all Eurasian taxa. The taxa of African origin were *Panthera leo, Panthera pardus* and *Crocuta crocuta*. The two first taxa were recorded in Le Vallonet (De Lumley *et al.* 1988; Turner 1995), and the last in several Spanish localities (Aguirre *et al.* 1976; Aguirre & Morales 1990). *Crocuta crocuta* was previously recorded at Ubeidiya (Belfer-Cohen & Goren-Inbar 1994; Bar-Yosef 1995), supporting the hypothesis that *Homo*, like other mammals, migrated to the western Mediterranean through the 'Levantine corridor' and not directly from Africa. Between 1.07 and 0.99 Ma, glacial maxima became more extreme, involving a long alternation of xeric–cool phases (glacials) and relatively humid–warm phases (interglacials), as cycles of 100 ka duration attributed to massive Northern Hemisphere ice sheets (Shackleton 1995; Suc *et al.* 1995). No further turnover pulse is recorded after the turnover around 1.0 Ma. In response to glacial–interglacial oscillations, mammals either migrated or expanded their ecological tolerance to withstand extreme climatic and floral changes. During glacial periods, the Mediterranean region was a refuge for a great variety of taxa that today are dispersed throughout different environments. Consequently, we consider the Galerian mammal communities as disharmonious faunas, which have no analogues in Recent times (in accordance with the individualistic shifts in species' distribution in the Gleasonian model (Graham *et al.* 1996)). The intensity of human occupation in Europe coincides with the local extinction of two of the most important carcass destroyers (*Pachycrocuta brevirostris* and *Pliocrocuta perrieri*), which were already absent in Spain as far back as early Galerian time (Turner 1995). These data agree with a stable human settlement in southern Europe and also support the hypothesis of the anagenetic evolution of humans during mid-Pleistocene time (Carbonell *et al.* 1995; Arsuaga *et al.* 1997; Bermúdez de Castro *et al.* 1997).

The authors wish to express their thanks to M. Maas and J. Alroy for their kind suggestions about their analytical programs, and also to C. Howell and E. Cerdeño for their critical revision of the manuscript and valuable discussions. The present work was made possible by the following research grants: PB91-0082 and PB-94-0071 of DGICYT, and the IberoAmerican Cooperation Program, MEC (Spain), and UNC-grant (Argentina) to J.L.P.

References

AGUIRRE, E. & MORALES, J. 1990. Villafranchian faunal record of Spain. *Quartärpaläontologie*, **8**, 7–11.
──────, LÓPEZ, N. & MORALES, J. 1976. Continental faunas in southeast Spain related to the Messinian. *In: Il Significato Geodinamico della Crisi di Salinità del Miocene Terminale nel Mediterraneo*, Messinian Seminar, Gargano, 5–12 September Vol 2, 62–63.
AGUSTÍ, J. & LLANAS, M. 1996. The late Turolian muroid rodent succession in eastern Spain. *Acta zoologica cracoviensia*, **39**, 47–56.
ALBERDI, M. T., AZANZA, B., CERDEÑO, E. & PRADO, J. L. 1997. Similarity relationship between mammal faunas and biochronology from latest Miocene to Pleistocene in western Mediterranean area. *Eclogae Geologicae Helvetiae*, **90**, 115–132.
──────, PRADO, J. L. & ORTIZ JAUREGUIZAR, E. 1995. Patterns of body size changes in fossil and living Equini (Perissodactyla). *Biological Journal of the Linnean Society*, **54**, 349–370.
ALROY, J. 1992. Conjunction among taxonomic distributions and the Miocene mammalian biochronology of the Great Plains. *Paleobiology*, **18**, 326–343.
────── 1994. Appearance event ordination: a new biochronologic method. *Paleobiology*, **20**, 191–207.
────── 1996. Constant extinction, constrained diversification, and uncoordinated stasis in North American mammals. *Palaeogeography, Palaeoclimatology, Palaeoecology*, **127**, 285–311.
ARSUAGA, J. L., MARTÍNEZ, I., GRACIA, A. & LORENZO, C. 1997. The Sima de los Huesos crania (Sierra de Atapuerca, Spain). A comparative study. *Journal of Human Evolution*, **33**, 219–281.
AZANZA, B., ALBERDI, M. T., CERDEÑO, E. & PRADO, J. L. 1997. Biochronology from latest Miocene to Middle Pleistocene in the western Mediterranean area. A multivariate approach. *In*: AGUILAR, J.-P., LEGENDRE, S. & Michaux. J. (eds) *Actes du Congrès BiochroM'97*. Mémoires et Travaux de l'École Practique des Hautes Études, Institut de Montpellier, **21**, 567–574.
──────, ALBERDI, M. T. & PRADO, J. L. 1999. Mammalian diversity and turnover patterns during Plio-Pleistocene in northwestern Mediterranean area. *Revista de la Sociedad Geológica de España*, **12**, in press.
AZZAROLI, A. 1983. Quaternary mammals and the 'End-Villafranchian' dispersal event – a turning point in the history of Eurasia. *Palaeogeography, Palaeoclimatology, Palaeoecology*, **44**, 117–139.
──────, DE GIULI, C., FICCARELLI, G. & TORRE, D. 1988. Late Pliocene to early Mid-Pleistocene mammals in Eurasia: faunal succession and dispersal events. *Palaeogeography, Palaeoclimatology, Palaeoecology*, **66**, 77–100.
BARRY, J. C., MORGAN, M. E., FLYNN, L. J. et al. 1995. Patterns of faunal turnover and diversity in the Neogene Siwaliks of Northern Pakistan. *Palaeogeography, Palaeoclimatology, Palaeoecology*, **115**, 209–226.
BAR-YOSEF, O. 1995. The role of climate in the interpretation of human movements and cultural transformations in western Asia. *In*: VRBA, E. S., DENTON, G. H., PARTRIDGE, T. C. & BURCKLE, L. H. (eds) *Paleoclimate and Evolution with Emphasis on Human Origins*. Yale University Press, New Haven, CT, 507–522.
BELFER-COHEN, A. & GOREN-INBAR, N. 1994. Cognition and communication in the Levantine lower Palaeolithic. *World Archaeology*, **26**, 144–157.
BERMÚDEZ DE CASTRO, J. M., ARSUAGA, J. L., CARBONELL, E., ROSAS, A., MARTÍNEZ, I. & MOSQUERA, M. 1997. A hominid from the Lower Pleistocene of Atapuerca, Spain: possible ancestor to neanderthals and modern humans. *Science*, **276**, 1392–1395.
BERNOR, R. L. 1984. A zoogeographic theater and a biochronologic play: The time/biofacies phenomena of Eurasian and African Miocene mammal provinces. *Paléobiologie Continentale*, **14**, 121–142.
────── 1986. Mammalian biostratigraphy, geochronology, and zoogeographic relationships of the Late Miocene Maragheh fauna, Iran. *Journal of Vertebrate Paleontology*, **6**, 76–95.
BOUCOT, A. J. 1983. Does evolution take place in an ecological vacuum? II. *Journal of Paleontology*, **57**, 1–30.
────── 1990. Community evolution: its evolutionary and biostratigraphic significance. Paleontological Society, Special Publications, **5**, 48–70.
BRETT, C. E. & BAIRD, G. C. 1995. Coordinated stasis and evolutionary ecology of Silurian to Middle Devonian faunas in the Appalachian Basin. *In*: ERWIN, D. H. & ANSTEY, R. L. (eds) *New Approaches to Speciation in the Fossil Record*. Columbia University Press, New York, 285–315.
CANDE, S. C. & KENT, D. V. 1995. Revised calibration of the geomagnetic polarity timescale for the Late Cretaceous and Cenozoic. *Journal of Geophysical Research*, **100**(B4), 6093–6095.
CARBONELL, E., BERMÚDEZ DE CASTRO, J. M., ARSUAGA, J. L., et al. 1995. Lower Pleistocene hominids and artifacts from Atapouerca-TD6 (Spain). *Science*, **269**, 826–830.
CITA, M. B., RIO, D. & SPROVIERI, R. 1995. The Pliocene series: chronology of the type Mediterranean record and standard chronostratigraphy. *In*: WRENN, J. H. & SUC, J. P. (eds) *Paleoecology, Climate and Sequence Stratigraphy of the Pliocene*. American Association of Stratigraphy and Palynology, Special Volume.
DE LUMLEY, H., KAHLKE, H. D., MOIGNE, A. M. & MOULLE, P. E. 1988. Les faunes de grands mammifères de la Grotte du Vallonnet Roquebrune-Cap-Martin, Alpes Maritimes. *L'Anthropologie*, **92**, 465–496.
ELDREDGE, N. & GOULD, S. J. 1972. Punctuated equilibria: an alternative to phyletic gradualism. *In*: SCHOPF, T. J. M. (ed.) *Models in Paleobiology*. Freeman, Cooper, San Francisco, CA, 82–115.
FOLEY, R. A. 1994. Speciation, extinction and climatic change in hominid evolution. *Journal of Human Evolution*, **26**, 275–289.
FOOT, M. 1994. Temporal variation in extinction risk and temporal scaling of extinctions metrics. *Paleobiology*, **20**, 424–444.

FORTELIUS, M., WERDELIN, L., ANDREWS, P. *et al.* 1996. Provinciality, diversity, turnover, and paleoecology in land mammal faunas of the later Miocene of western Eurasia. *In*: BERNOR, R. L., FAHLBUSCH, V. & MITTMANN, H.-W. (eds) *The Evolution of Western Eurasian Neogene Mammal Faunas*. Columbia University Press, New York, 414–448.

GABUNIA, L. & VEKUA, A. 1995. A Plio-Pleistocene hominid from Dmanisi, East Georgia, Caucasus. *Nature*, **373**, 509–512.

GAMBLE, C. 1994. Time for Boxgrove man. *Nature*, **369**, 275–276.

GINGERICH, P. D. 1987. Extinction of Phanerozoic marine families. *Geological Society of America, Abstracts with Programs*, **19**, 677.

GRAHAM, R. W., LUNDELIUS, E. L., JR. *et al.* 1996. Spatial response of mammals to late Quaternary environmental fluctuations (FAUNMAP working group). *Science*, **272**, 1601–1606.

HARPER, C. W. JR 1975. Standing diversity of fossil groups in successive intervals of geologic time: a new measure. *Journal of Paleobiology*, **49**, 752–757.

HODELL, D. A., ELMSTROM, K. M. & KENNETT, J. P. 1986. Latest Miocene benthic δ^{18}O changes, global ice volume, sea level and the "Messinian salinity crisis". *Nature*, **120**, 411–414.

IVANI, L. C. & SCHOPF, 1996. New perspectives on faunal stability in the fossil record. *Palaeogeography, Palaeoclimatology, Palaeoecology*, special issue, **127**, 1–361.

MAAS, M. C., ANTHONY, M. R. L., GINGERICH, P. D., GUNNELL, G. F. & KRAUSE, D. W. 1995. Mammalian generic diversity and turnover in the Late Paleocene and Early Eocene of the Bighorn and Crazy Mountains Basins, Wyoming and Montana (USA). *Palaeogeography, Palaeoclimatology, Palaeoecology*, special issue, **115**, 181–207.

MACARTHUR, R. H. & WILSON, E. O. 1967. *The Theory of Island Biogeography*. Princeton University Press, Princeton, NJ.

MARTÍNEZ DEL OLMO, W. 1996. Yesos de margen y turbidíticos en el Messiniense del Golfo de Valencia: una desecación imposible. *Revista de la Sociedad Geológica de España*, **9**, 97–116.

MICHALZIK, D. 1997. Sedimentary cycles in the Messinian Tertiary, Late Miocene, of SE Spain. *Neues Jahrbuch für Geologie und Paläontologie, Abhandlungen*, **203**, 89–143.

PARÉS, J. M. & PÉREZ GONZÁLEZ, A. 1995. Paleomagnetic age for hominid fossils at Atapuerca archaeological site, Spain. *Science*, **269**, 830–832.

PICKFORD, M., MORALES, J. & SORIA, D. 1995. Fossil camels from the Upper Miocene of Europe: implications for biogeography and faunal change. *Geobios*, **28**, 641–650.

ROEBROEKS, W. & VAN KOLFSCHOTEN, T. 1994. The earliest occupation of Europe: a short chronology. *Antiquity*, **68**, 489–503.

SALA, B., MASINI, F., FICCARELLI, G., ROOK, L. & TORRE, D. 1992. Mammal dispersal events in the Middle and Late Pleistocene of Italy and western Europe. *Courier Forschungs-Institut Senckenberg*, **153**, 59–68.

SHACKLETON, N. J. 1995. New data in the evolution of

Pliocene climatic variability. *In*: VRBA, E. S., DENTON, G. H., PARTRIDGE, T. C. & BURCKLE, L. H. (eds) *Paleoclimate and Evolution with Emphasis on Human Origins*. Yale University Press, New Haven, CT, 242–248.

STEININGER, F. F., RABEDER, G. & RÖGL, F. 1985. Land mammal distribution in the Mediterranean Neogene: a consequence of geokinematic and climatic events. *In*: STANLEY, D. J. & WEZEL, F. C. (eds) *Evolution of the Mediterranean Basin*. Raimondo Selli Commemoratio, Springer, Berlin, 559–571.

SUC, J.-P., BERTINI, A., COMBOURIEU-NEBOUT, N. *et al.* 1995. Structure of west Mediterranean vegetation and climate since 5.3 Ma. *Acta zoologica cracoviensia*, **38**, 3–16.

THOUVENY, N. & BONIFAY, E. 1984. New chronological data on European Plio-Pleistocene faunas and hominid occupation sites. *Nature*, **308**, 355–358.

TORRE, D., FICCARELLI, G., MASINI, F., ROOK, L. & SALA, B. 1992. Mammal dispersal events in the Early Pleistocene of eastern Europe. *Courier Forschung-institut Senckenberg*, **513**, 51–58.

TURNER, A. 1992. Villafranchian–Galerian larger carnivores of Europe: dispersions and extinctions. *Courier Forschungs Institut Senckenberg*, **153**, 153–160.

—— 1995. Variaciones regionales en la fauna de grandes mamíferos del Pleistoceno inferior y medio de Europa. Una perspectiva Ibérica. *In*: BERMÚDEZ DE CASTRO, J. M., ARSUAGA, J. L. & CARBONELL, E. (eds) *Human Evolution in Europe and the Atapuerca Evidence, Vol. 1*, Junta de Castilla-León, 57–73.

VAN DER MADE, J. 1989–1990. A range-chart for European Suidae and Tayassuidae. *Paleontologia i Evolució*, **23**, 99–104.

VAN VALEN, L. 1973. A new evolutionary law. *Evolutionary Theory*, **1**, 1–30.

—— 1984. A resetting of Phanerozoic community evolution. *Nature*, **307**, 50–52.

VRBA, E. S. 1985. Environment and evolution: alternative causes of the temporal distribution of evolutionary events. *South African Journal of Science*, **81**, 229–236.

—— 1992. Mammals as a key to evolutionary theory. *Journal of Mammalogy*, **73**, 1–28.

—— 1995a. On the connections between paleoclimate and evolution. *In*: VRBA, E. S., DENTON, G. H., PARTRIDGE, T. C. & BURCKLE, L. H. (eds) *Paleoclimate and Evolution with Emphasis on Human Origins*. Yale University Press, New Haven, CT, 24–45.

—— 1995b. The fossil record of African antelopes (Mammalia, Bovidae) in relation to human evolution and paleoclimate. *In*: VRBA, E. S., DENTON, G. H., PARTRIDGE, T. C. & BURCKLE, L. H. (eds) *Paleoclimate and Evolution with Emphasis on Human Origins*. Yale University Press, New Haven, CT, 385–424.

WHITE, T. D. 1995. African omnivores: global climatic change and Plio-Pleistocene hominids and suids. *In*: VRBA, E. S., DENTON, G. H., PARTRIDGE, T. C. & BURCKLE, L. H. (eds) *Paleoclimate and Evolution with Emphasis on Human Origins*. Yale University Press, New Haven, CT, 369–384.

The contribution of Quaternary vertebrates to palaeoenvironmental and palaeoclimatological reconstructions in Sicily

LAURA BONFIGLIO[1], ANTONELLA C. MARRA[1] & FEDERICO MASINI[2]

[1] *Department of Earth Science, University of Messina, Via Sperone 31, casella postale 54, 98166 Messina, Italy (e-mail: laura@labcart.unime.it)*

[2] *Department of Geology and Geodesy, Corso Tükory 131, 90134 Palermo, Italy*

Abstract: In Sicily few studies have been devoted to the climatic–environmental changes of the Pleistocene and Holocene period. Most of the studies on Quaternary vertebrates in Sicily have been focused on the evolutionary–taxonomic aspects of the fauna. Sicily experienced at least four vertebrate dispersal events during Quaternary time, which are of different provenence (African and/or European) and have been controlled by filtering barriers of different intensities. The marked endemism and the extremely low diversity of the fossil assemblages of early and early–mid-Pleistocene time do not allow detailed interpretations. By way at contrast, younger assemblages are more diverse and, although they display some endemic characters, are similar to those of southern peninsular Italy. The late mid-Pleistocene and early Late Pleistocene assemblages (*Elephas mnaidriensis* faunal complex) are characterized by the occurrence of a red deer (*Cervus elaphus siciliae*), a dwarf fallow deer-like endemic megalocerine (*Megaceroides carburangelensis*), auroch (*Bos primigenius siciliae*), bison (*Bison priscus siciliae*), elephant (*Elephas mnaidriensis*), hippopotamus (*Hippopotamus pentlandi*), boar (*Sus scrofa*), brown bear (*Ursus* cf. *arctos*) and three large social carnivores (*Panthera leo, Crocuta crocuta* and *Canis lupus*). Most of these taxa, except for the megalocerine, are characterized by slightly reduced body size compared with the same taxa from mainland Europe. These assemblages are indicative of a climate with temperate, Mediterranean affinity and of landscapes in which forested areas were associated with more open environments. The relatively low abundance of the red deer and the dominance of the megalocerine in several assemblages suggests that a Mediterranean-type forest locally dominated some of the climatic phases. The assemblages of the youngest Late Pleistocene period on Sicily are characterized by a dramatic drop in diversity, with the disappearance of elephant, hippopotamus, bison, the endemic megalocerine, and the largest predators. This would indicate an environmental crisis probably linked to the drier climatic oscillations of late Pleniglacial time, as is suggested also by the spread of the ground vole, which is the dominant small mammal in several assemblages. The Late Glacial period is characterized by the spread of equids (horse and wild ass), which are indicators of open landscapes and of xerophytic steppe-like cover. The beginning of the Holocene period is characterized by the expansion of forested areas and by a more humid climate, as suggested by the abundance of red deer, and by the dispersal of the common dormouse (*Glis glis*) and water vole (*Arvicola* sp.).

Most studies of Quaternary vertebrates in Sicily have focused on evolutionary and taxonomic aspects (Pohlig 1893, 1909; Vaufrey 1929; Accordi & Colacicchi 1962; Aguirre 1968-69; Ambrosetti 1968; Gliozzi & Malatesta 1984; Brugal 1987). Chronological schemes utilizing the Pleistocene mammals of Sicily were based on the assumption of the phyletic derivation of the dwarf elephant *Elephas falconeri* from the middle-sized *Elephas mnaidriensis*, which is in turn a direct descendant of *Elephas antiquus*

(Accordi & Colacicchi 1962; Accordi 1963, 1965; Ambrosetti *et al.* 1980). Vaufrey (1929) was inclined to assume a post-Tyrrhenian age for all the vertebrate faunas. Since the studies of Accordi (1957, 1963, 1965) it was thought that most of the size reductions in the elephants of Sicily took place during the period preceding the Tyrrhenian period. The smallest species (*Elephas falconeri*) was considered to be limited to the early Würm period and to have evolved as a consequence of environmental stress linked to

From: HART, M. B. (ed.) *Climates: Past and Present.* Geological Society, London, Special Publications, **181**, 171–184, 1-86239-075-4/$15.00

Fig. 1. Locations of the Quaternary mammal-bearing deposits in Sicily quoted in the text.

the Würmian climatic cooling (Ambrosetti 1968; Kotsakis 1979). Most of the known vertebrate remains come from cave deposits and little was known about the palaeoenvironmental conditions of the vertebrate-bearing deposits.

Since 1985, a more recent synthesis has incorporated new stratigraphic and aminostratigraphic data (Belluomini & Bada 1985; Bonfiglio 1987, 1991, 1992*a*, *b*; Burgio & Cani 1988; Bada *et al.* 1991; Bonfiglio & Insacco 1992). Taphonomic data show that Pleistocene vertebrates were distributed in both cave environments and broad, open environments (Bonfiglio 1987, 1992*b*, 1995; Bonfiglio *et al.* 1993, 1997*a*, *b*; Chilardi & Gilotti 1996). Four Pleistocene vertebrate faunal complexes have been recognized. They differ as regards their composition and degree of endemism, and correspond to different dispersal events (Bonfiglio *et al.* 1997*b*). Few studies, however, have addressed the climatic–environmental changes during Pleistocene and Holocene time. Studies of pollen and other vegetation remains, which represent one of the most important sources of data for climatic studies, have been rare and of limited value.

The use of fossil mammal assemblages for palaeoecological–palaeoclimatological interpretations is limited in islands by a number of factors. First, insular environments are usually characterized by poorly diversified assemblages that often do not permit detailed environmental reconstructions. Second, the reorganization of mammal communities in response to climatolo-

gical–environmental changes (i.e. migrations of taxa from different climatic zones and long-term migration cycles) is prevented or strongly conditioned by geographical barriers. Furthermore, some insular endemic mammals often underwent dramatic adaptive changes, which may limit their value as environmental indicators.

Moreover, the palaeogeography of Sicily as well as its relationship with peninsular Italy underwent several important changes from Early Pleistocene to late mid-Pleistocene time (Ruggieri & Unti 1974, 1977; Ruggieri et al. 1976; Di Geronimo & Costa 1978; Di Geronimo 1979; Di Geronimo *et al.* 1980; Grasso & Lentini 1982; Sprovieri 1982; Bonfiglio & Burgio 1992; Bonfiglio & Piperno 1996). During late mid-Pleistocene time vertebrate dispersals from peninsular Italy through Southern Calabria were controlled by two palaeogeographic barriers, located in the area of the Straits of Messina and the Catanzaro isthmus (Fig. 1). In fact, Southern Calabria has to be considered a 'fossil island' as suggested by the occurence in early Late Pleistocene deposits of a small-sized elephant (*Elephas* cf. *antiquus*) (Azzaroli 1982; Bonfiglio & Berdar 1986), a dwarf megalocerine (*Megaceroides calabriae*) closely related to the sicilian endemic megalocerine *Megaceroides carburangelensis* (Bonfiglio 1978), and a small hippopotamus (Bonfiglio & Marra, unpublished data).

Given the complexity of the Sicilian case history, we will consider the biochronological framework and the relationships between verte-

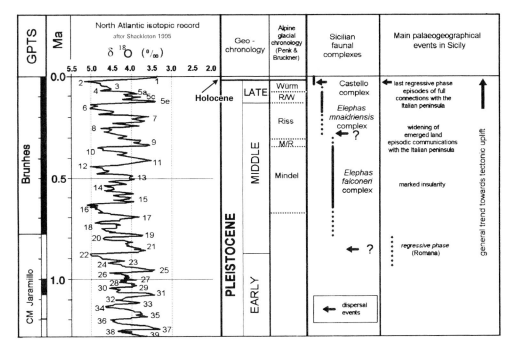

Fig. 2. Relationship between the $\delta^{18}O$ isotope stratigraphy, geochronology, Sicilian faunal complexes and main palaeogeographical events in Sicily.

brate-bearing deposits with evidence of high-stand coastlines and palaeobiogeographical data (route of dispersal). The climatic significance of the faunal assemblages and of the individual taxa have been analysed in comparison with coeval assemblages from the Italian peninsula.

The Quaternary vertebrate complexes of Sicily and their relationships with highstand coastlines

The Pleistocene mammals of Sicily can be divided into four faunal complexes (Bonfiglio et al. 1997b). In each complex the characteristic species are associated with taxa that are common to more than one faunal complex. Some of these common taxa underwent minor evolutionary changes during their existence on the island. For most of these taxa, which actually represent chronomorphs (sensu Martin 1995), detailed anatomical studies are still in progress.

The upper section of Table 1 shows the composition of the four Pleistocene vertebrate complexes of Sicily according to the literature, and in the lower section the common species and chronomorphs are shown. Figure 2 shows the

relationships between the $\delta^{18}O$ isotopic record, geochronology, Sicilian faunal complexes and the main palaeogeographical events in Sicily.

Monte Pellegrino faunal complex

The data used here are taken from De Gregorio (1886), Thaler (1972), Zammit Maempel & de Bruijn (1982), Burgio & Fiore (1988a, 1997) and Masini & Sarà (1999). The genera of the Monte Pellegrino faunal complex were considered indicative of a Pliocene–Early Pleistocene age by Thaler (1972), although Burgio & Fiore (1997) regarded them as of Early Pleistocene age (Late Villafranchian).

Elephas falconeri faunal complex

Data are taken from Accordi & Colacicchi (1962), Ambrosetti (1968), Petronio (1970), Kotsakis (1977, 1986), Kotsakis & Petronio (1981), Esu et al. (1986), Burgio & Cani (1988) and Bonfiglio & Insacco (1992). Esu & Girotti (1991) suggested that the earliest Mid-Pleistocene terrestrial gastropods from the lake deposits of Comiso indicate a wet and cool environment. The gastropods from Comiso are associated with the earliest finds of the Elephas

Table 1. *Composition of the four Pleistocene vertebrate complexes of Sicily according to the literature (upper section), common species and chronomorphs (lower section).*

	Faunal complexes				
	'Monte Pellegrino'	'Elephas falconeri'	'Elephas mnaidriensis'	'Castello'	Holocene
Taxa	Testudo graeca Asoriculus sp. Pannonictis arzilla Apodemus maximus Leithia sp. Maltamys cf. gollcheri Pellegrinia panormensis Hypolagus sp.	Discoglossus cf. pictus Bufo cf. viridis Hyla sp. Emys orbicularis Testudo hermanni Geochelone sp. Lacerta siculomelitensis Lacerta viridis Lacerta sp. Coluber cf. viridiflavus Natrix sp. Several species of bats Crocidura esuae Vulpes sp. Nesolutra trinacriae Elephas falconeri Leithia cartei Leithia melitensis Maltamys gollcheri	Discoglossus cf. pictus Emys orbicularis Testudo hermanni Lacerta siculomelitensis Birds Erinaceus europaeus Crocidura aff. esuae Crocuta crocuta cf. spelaea Panthera leo Canis lupus Nesolutra trinacriae Ursus cf. arctos Elephas mnaidriensis Sus scrofa Hippopotamus pentlandi Cervus elaphus siciliae Megaceroides carburangelensis Bos primigenius siciliae Bison priscus siciliae Leithia cf. melitensis Maltamys cf. wiendincitensis	Birds Erinaceus europaeus Crocidura cf. sicula Canis lupus Vulpes vulpes Equus caballus Equus hydruntinus Sus scrofa Cervus elaphus Bos primigenius Microtus (Terricola) ex gr.savii Apodemus sp. Lepus europaeus	Fishes Bufo bufo Bufo cf. viridis Turtle indet. Emys orbicularis Lacerta viridis Birds Chiroptera indet Erinaceus europaeus Crocidura sp. Canis cf. lupus Vulpes vulpes Felis silvestris Martes sp. Mustela cf. nivalis Ursus sp. Sus scrofa Cervus sp. Bos primigenius, Microtus (Terricola) cf. savii Apodemus sp. Arvicola terrestris Mioxus glis Monachus monachus
Common taxa and/or	Testudo ex gr. hermanni; Maltamys ex gr. gollcheri-wiendincitensis.		Discoglossus cf. pictus, Emys orbicularis, Lacerta siculomelitensis, Natrix sp., Crocidura ex gr. esuae, Leithia melitensis, Nesolutra trinacriae		
chronomorphs			Erinaceus europaeus, Canis lupus, Sus scrofa, Bos primigenius, Cervus elaphus		

falconeri faunal complex. Amino acid racemization dating (Bada *et al.* 1991) yielded an age of 455 ± 90 ka for *Elephas falconeri* from the Spinagallo and Luparello cave deposits. The biochronological events that mark the end of this faunal complex are still poorly defined. The extinction of *E. falconeri* might be the result of predation consequent on the dispersal of the large carnivores of the succeeding *Elephas mnaidriensis* faunal complex.

The marine sedimentary cycle in the Comiso area isolated the Hyblean Plateau from the rest of Sicily during Pliocene and Early Pleistocene time. It ended with lagoonal deposits representing a brackish environment, which are overlain by a palaeosol and by lake deposits containing the first occurrence of the *Elephas falconeri* faunal complex (Conti *et al.* 1980). The lake deposits, in turn, pass upwards into lower Middle Pleistocene calcarenites (Conti *et al.* 1980; Carbone *et al.* 1982). In the Spinagallo cave the *Elephas falconeri* assemblage overlies lower Middle Pleistocene calcarenites (Accordi & Colacicchi 1962; Di Grande & Raimondo 1984; Bonfiglio, 1992*b*). The lake deposits may correspond to the beginning of the 'Roman regression' (Ruggieri *et al.* 1976), which roughly correlates with stage 22 of the $\delta^{18}O$ isotopic record, and the *E. falconeri*-bearing deposits in the Spinagallo cave may correspond to younger oscillations of the oxygen isotope curve.

Elephas mnaidriensis *faunal complex*

Data for this are based on Scinà (1831), Vaufrey (1929), Fabiani (1932, 1934), Burgio *et al.* (1983), Burgio & Fiore (1988*b*), Bonfiglio (1992*a, b*), Bonfiglio & Burgio (1992), Cassoli & Tagliacozzo (1996), Chilardi (1996) and Kotsakis (1996*a, b*). This faunal complex contains elephants of a less reduced size, attributed in the literature to *Elephas antiquus* (Gignoux 1913; Ruggieri 1977), *Elephas* cf. *antiquus* (Ambrosetti 1968), *E. antiquus leonardii* (Aguirre 1968–1969) and *E. mnaidriensis* (Vaufrey 1929).

Bada *et al.* (1991) assigned an age of 200 ± 40 ka to this faunal complex. Electron spin resonance dating for teeth enamel of *Elephas mnaidriensis* and *Hippopotamus pentlandi* from Contrada Fusco (Rhodes 1996) provide an age ranging between 146.8 ± 28.7 and 88.2 ± 19.5 ka.

Where correlated with sediments from littoral environments, the assemblages of the *E. mnaidriensis* faunal complex are associated with upper Middle Pleistocene and/or Upper Pleistocene deposits. At Coste di Gigia (Acquedolci lacustrine basin) in the S. Ciro cave, the *E.*

mnaidriensis assemblages are found in upper Middle Pleistocene terrace deposits overlying a shoreline that probably corresponds to stage 7 of the oxygen isotope curve (Bonfiglio 1991, 1992*b*; Galletti & Scaletta 1991). On the Maddalena peninsula and at Contrada Fusco the *E. mnaidriensis* assemblages are found in deposits underlying lower Upper Pleistocene (Euthyrrenian) calcarenites, or in the same calcarenites (Accordi 1963, 1965; Di Grande & Raimondo 1984; Chilardi & Gilotti 1996). At Rocca Scodonì they are contained in continental deposits overlying an Upper Pleistocene abrasion platform (Bonfiglio 1987). The most likely time interval for this faunal complex is from early stage 6 to early stage 4 of the $\delta^{18}O$ isotopic record. Chronostratigraphical data concerning the last occurrences of the taxa of this faunal complex are still poorly defined.

Contrada Pianetti–Castello complex

Data are based on work by Regalia (1907), Vaufrey (1929), Bidditu (1971), Cardini (1971) and Kotsakis (1979). The Castello faunal complex accompanies the Epigravettian artefacts in Sicily and has to be referred to the Late glacial phase of the climato-stratigraphical scale. Radiometric dating for these human settlements ranges from 15 to 10 ka BP (Leighton 1996).

A new faunal assemblage from a fissure-filling deposit at Contrada Pianetti (Bonfiglio *et al.* 1997*b*) probably represents an intermediate faunal complex that still contains some of the *E. mnaidriensis* complex endemic taxa (*Cervus elaphus siciliae*, *Ursus* cf. *arctos*) as well as new continental taxa dispersed during a Würmian (late Pleniglacial) event, before the Epigravettian colonization of Sicily. This event might be coeval with the spread of the ancient Aurignacian culture in Sicily (Bernabò Brea 1950), which has an age of about 30 ka BP (Guerreschi 1992). This faunal complex roughly correlates with late stage 3 and early stage 2 of the $\delta^{18}O$ isotopic record.

Dispersal routes and comparison with Italian peninsula assemblages

The endemic characteristics of the Monte Pellegrino assemblage, the marked endemism of the *E. falconeri* faunal complex, and the very different palaeogeography of Sicily in Early Pleistocene time make it difficult to compare with any continental assemblage. The features of these faunal complexes suggest that modifications of the endemic taxa occurred during dispersal events over an extensive area (an archipelago perhaps?). Some elements of the first faunal complex have an African origin (*Pellegrinia*) whereas others probably originated by dispersal from European regions (*Mustellercta = Pannonictis* and *Hypolagus*, Thaler 1972; Burgio & Fiore 1997). Some taxa (glirids), on the other hand, may have been derived from older and still undocumented (Pliocene) phases (Masini & Sarà 1999).

The distribution of the *E. falconeri* assemblages in southeastern and western Sicily and the scattered traces of Lower Palaeolithic artefacts on the Hyblean Plateau (southeastern Sicily) and along the area linking it to western Sicily, might suggest that the dispersal events of humans and of some faunal elements followed the same route (Bonfiglio & Piperno 1996). According to Bonfiglio & Piperno, the hypothesis of an African provenance for both *E. falconeri* and human populations through the Sicilian–Tunisian isthmus, as a result of tectonically controlled changes of water depth during Early Pleistocene time (see Vaufrey 1929; Kelling & Stanley 1972; Alimen 1975), should at least merit a fresh and more detailed reconsideration.

The faunal composition of the *Elephas mnaidriensis* complex, although less diverse, is very similar to the coeval fauna of peninsular Italy and suggests the occurrence of relatively easy passages along a land-bridge connection between Italy and Sicily in the area of the Straits of Messina. The Straits of Messina were established in Early Pliocene time and their dynamics and morphology have been determined directly by tectonic activity (Barrier 1987). Uplift during mid-Pleistocene time probably gave rise to a link (Bonfiglio & Burgio 1992), which was emergent during a late mid-Pleistocene lowstand of sea level (stage 6 of the $\delta^{18}O$ isotope curve?). Little is known about the palaeogeography of the Catanzaro isthmus area (see Fig. 1) through which two dispersal events probably occurred during mid-Pleistocene time (Bonfiglio 1978; Azzaroli 1982).

The communities of the Castello faunal complex are also extremely similar to the faunal assemblages of southern Italy, although they are less diverse. This indicates a fully emerged connection. The middle- to small-sized ungulates of mountain environments (ibex, chamois), which characterize several coeval peninsular assemblages, are lacking in Sicily. The fallow deer (which probably became extinct in the Italian peninsula before the Late Pleistocene dispersal events), and *Capreolus* are absent, as are rodents of cold and/or humid evironments (*Microtus arvalis*, *M. agrestis*, *Clethrionomys*,

Arvicola) and of forested areas (*Glis*, *Eliomys*, *Muscardinus*). Among the insectivores commonly found in peninsular Italy, shrews of the Soricini and Soriculini tribes, as well as moles, are lacking.

A similar situation occurs in the Early Holocene fauna. Starting from Neolithic times (about 8000 a BP), humans have introduced domesticated species together with wild and commensal species (e.g. *Hystrix*, *Oryctolagus*).

Ecological and environmental–climatological characteristics of the Sicilian mammal assemblages

Taphonomic data

Most of the vertebrate-bearing deposits of Sicily have been excavated without a scientific aim (collecting for museums), and taphonomic data have not been recorded. Presence or absence of taxa is frequently the only available information. Until recently, small mammal taxa were thought to be absent from the *E. mnaidriensis* assemblages. Data from the Spinagallo cave (Accordi & Colacicchi 1962; Ambrosetti 1968) and recent research in palaeoethnology (Uzzo cave, S. Vito peninsula, Tagliacozzo 1993) and palaeontology (Cape Tindari, Gliozzi & Malatesta 1984; Contrada Pianetti, Bonfiglio *et al.* 1997b; Cimillà, Bonfiglio *et al.* 1993; Acquedolci lacustrine basin, Bonfiglio 1995; Comiso, Bonfiglio & Insacco 1992; Contrada Fusco, Chilardi & Gilotti 1996; K22 site, S. Vito peninsula, Masini unpublished data) provide significant information. The different abundances of the remains may depend on the different sedimentary environments. *E. falconeri* remains and the associated species (e.g. dormice, bats, lizards, and turtles) are densely accumulated in the fissure-filling deposit of the Spinagallo cave, whereas in the lake deposits at Comiso (*E. falconeri*) and at Contrada Fusco the elephant remains as well as the accompanying taxa, are scattered over a broad area. Given the width of the island, and the different palaegeographical conditions, which reflect the complex geology of Sicily, the different compositions of the faunal assemblages belonging to each complex could be due to different environmental conditions as well as to climatic oscillations.

Monte Pellegrino and the *E. falconeri* faunal complex

The occurrence of *Asoriculus* and two species of glirids suggests a forested landscape and temperate conditions for the Monte Pellegrino assemblage.

In the lake and alluvial deposits at Comiso, where the first occurrence of *E. falconeri* is recorded, terrestrial and lacustrine gastropods (*Vertigo substriata, Candidula unifasciata vincae, Discus ruderatus*) indicate a wet and cool climate (Esu & Girotti 1991). According to Kotsakis & Petronio (1981), the bats recovered from the bottom of the deposit in the Spinagallo cave indicate a temperate–hot climate whereas those from its upper levels indicate lower temperatures. According to Kotsakis (1977), amphibians and reptiles from the same cave deposit suggest an arid environment with thin and discontinuous bush cover and a temperature very close to the present temperature of the area. The *Elephas falconeri*-bearing deposits in the Spinagallo cave probably record several climatic fluctuations.

Elephas mnaidriensis *faunal complex*

The assemblages of the *E. mnaidriensis* faunal complex are diverse and balanced from a trophic point of view. Besides two endemic glirids and the Crocidura, large and middle-sized herbivores (elephant of moderately reduced size, hippopotamus, wild boar, ox, bison, two species of deer) are present. The carnivores are represented by bear and otter, but middle- and large-sized gregarious predators, such as wolf, spotted hyena and lion, are also present. This faunal complex does not show very marked endemic characters (most of the herbivores, except for *Megaceroides carburangelensis,* are only moderately reduced in size with respect to the equivalent continental taxa) and, though less diverse, show very marked affinity with the coeval mammal assemblages of the Italian peninsula. Although some assemblages are characterized by the dominance of the cervid *Megaceroides carburangelensis* and of *Hippopotamus pentlandi,* respectively, almost all the taxa are represented in the different assemblages belonging to the *Elephas mnaidriensis* faunal complex (Puntali, Maccagnone, S. Ciro, Cannita, Carburangeli, Capo Tindari caves, Acquedolci lacustrine basin, Contrada Fusco, Messina gravels deposits and S. Teodoro cave at Acquedolci).

The wild ox, *Bos primigenius*, is present in almost all the assemblages of the *E. mnaidriensis* faunal complex, but is never dominant. The body size varies from a slightly reduced form (*Bos primigenius siciliae*) to specimens whose size falls within the range of variability of *Bos primigenius* from southern peninsular Italy. The ox is a ubiquitous species of mid- and Late

Pleistocene time in the Italian peninsula and may be considered generally indicative of open spaces. In Sicily it also occurs in Upper Pleistocene (Pianetti-Castello) and Lower Holocene assemblages.

A bison of reduced size, *Bison priscus siciliae*, has been found in the Puntali, Maccagnone and S. Teodoro caves. The bison is a very rare species and seems to be the only species of marked boreal affinity that reached the island. Its presence suggests that the dispersal of the *E. mnaidriensis* faunal complex in Sicily took place during a cool and arid climatic phase, which produced an expansion of steppe-like or prairie environments in Southern Mediterranean regions. *Bison priscus* is documented in southern Italy at least since stage 6 of the $\delta^{18}O$ oxygen isotope curve (Riss II) and is also the only large bovid of boreal type that reached the Hellenic peninsula (Sala *et al.* 1993). This ungulate, which is considered a taxon indicative of prairie and steppe environments shows, however, a strong phenotypic plasticity that foreshadows considerable ecological flexibility (Sala *et al.* 1993; Masini & Abbazzi 1997).

Deer are more abundant and widespread than bovids in the Sicilian fauna. They are represented by a red deer of reduced size (*Cervus elaphus siciliae*) and by a dwarf, strongly endemic megalocerine (*Megaceroides carburangelensis*). The two species are usually found associated, but with several exceptions (Acquedolci lacustrine basin, Luparello and Monte Gallo caves, Western Sicily) where only the red deer is present. In several localities (e.g. Cape Tindari, Cimillà) *M. carburangelensis* is the dominant species; at sites in which both deer are present, *M. carburangelensis* is always more abundant than the red deer.

According to Gliozzi *et al.* (1993) *Cervus elaphus siciliae* was not a dwarf deer, but merely a small-sized form that tolerated a degree of size reduction (of about 25%) in comparison with peninsular populations. The body was slender and the antlers were slightly modified in comparison with the coeval forms of the Italian peninsula. The red deer is very widespread in the late mid-Pleistocene and in the Late Pleistocene communities of the Italian peninsula, especially on the more humid Tyrrhenian side, where the vegetation cover was more favourable to this ruminant. Although the red deer shows a strong phenotypic plasticity and ecological flexibility (see Masini & Abbazzi 1997) it may be considered a species generally indicative of forested environments with clearings. *C. elaphus siciliae* is present in Sicily up to the Würmian late pleniglacial stage (C. Pianetti site).

Megaceroides carburangelensis, which seems to be limited to the *E. mnaidriensis* faunal complex, is characterized by a strongly reduced size, comparable with that of a small-sized fallow deer, and by well-developed antlers, palmate in their distal portion, and similar in their disposition and orientation to the antlers of the modern fallow deer. The remarkable sturdiness of the limb bones, particularly metapods and acropods, is a distinctive character that was probably inherited from its continental ancestor.

The Sicilian megalocerine is related to *Megaceroides calabriae*, the endemic slightly larger species from the Late Pleistocene period in Calabria (Bonfiglio 1978). The two species are endemic survivors of the genus *Megaceroides*, which became extinct in Europe during mid-Pleistocene time, probably in stage 9 of the $\delta^{18}O$ oxygen isotope curve. The surviving species of *Megaceroides* in the Mediterranean region is not exclusive to the Calabrian–Sicilian region. The type species of the genus, *Megaceroides algericus*, is another endemic form, easily distinguished from *Megaceroides calabriae* and *M. carburangelensis*, and which is present in mid–Late Pleistocene time in Maghreb (North Africa) (see Abbazzi & Masini 1997).

The latest non-endemic species of the genus *Megaceroides* (*M. verticornis* and *M. solilhacus*) are generally thought to be poor climatological–environmental indicators, as they are found in assemblages belonging to both glacial and interglacial phases (von Koenigswald 1988; Abbazzi & Masini 1997). Their very large size suggests they were not adapted to a densely forested environment. In contrast, the much-reduced size of the Sicilian *Megaceroides* might foreshadow a different adaptation and could fit with more densely forested environments. The teeth of *M. carburangelensis* are rather brachyodont, even for a deer, and are more similar to those of the fallow deer than to those of the red deer. The third molars, particularly M/3, are slightly reduced in size with respect to those of other Pleistocene cervids. This suggests that the Sicilian *Megoceroides* fed prevalently on soft vegetable matter, a feature convergent with the fallow deer. Taking into account the above-mentioned features, its abundance in the Sicilian localities and its dominance over the red deer, (which is a species with boreal affinities), it is suggested that the megalocerine was better adapted to the Mediterranean flora than was the red deer. Following these considerations, we regard *M. carburangelensis* as a form very close (ecologically) to the fallow deer and to be indicative of temperate climate and of mixed oak-wood decidous forests (Mediterranean forest).

The wild boar, *Sus scrofa*, is rare but constantly present, as in Italian peninsular communities. It indicates a temperate humid climate and closed forested environments.

The hippopotamus *Hippopotamus pentlandi* is a characteristic form of the *E. mnaidriensis* faunal complex. It is found in various types of deposits but it is prevalent in deposits of coastal plain environments, next to carbonate massifs (Acquedolci lacustrine deposits, S. Ciro, Maccagnone, Cannita caves) where, at present, springs are located. These features suggest that springs discharging from carbonate substrate aquifers supplied water to lakes and ponds on the coastal plain. The living African hippopotamus feeds on the savannah grass and may be considered indicative of open spaces (Sala 1977). However, the phylogenetic relations of the Sicilian form are not yet clear and this species may be related to the European hippopotamus of the *Hippopotamus antiquus* group (Bonfiglio & Marra, unpublished data).

The Sicilian elephant, *Elephas mnaidriensis*, has a reduced size in comparison with the continental ancestor *Elephas antiquus*. It has been recorded from several sites scattered over the island but is very abundant only at the Puntali cave. *Elephas antiquus* is believed to have preferred temperate climates and environments with arboreal covering (see von Koenigswald 1988) and probably had a very wide ecological tolerance. In any case, its presence in Sicily is indicative of a vegetated environment.

The endemic species *Crocidura esuae* is also present in assemblages of the *E. falconeri* faunal complex. This shrew is a medium-sized form (about the same size as the extant *Crocidura russula*) which, according to Kotsakis (1986), may have been adapted to a semi-aquatic environment. Hutterer (1991), however, rejected this interpretation mainly because tendencies towards this adaptation have never been found in any of the many members of the Crocidurinae. The genus *Crocidura* is at present widespread in equatorial to temperate climatic zones of the Palearctic. The present distribution of *Crocidura* does not exceed the 53rd north parallel in Europe, whereas in Eemian time it reached England (Reumer 1984). In the European fossil record *Crocidura* is considered an indicator of temperate or warm–temperate climate .

The two glirids *Leithia* and *Maltamys* which, according to Daams & De Bruijn (1995) belong to the eliomyne group, are highly modified forms. They occur in the *E. falconeri* and *E. mnaidriensis* assemblages but probably reached

Sicily much earlier than the Monte Pellegrino dispersal event. *Leithia* is represented by two very large-sized species, which are about three to four times bigger than the extant *Eliomys quercinus*. Specific morpho-functional studies, aimed at an understanding of the ecological significance of these forms, are lacking. The thick enamel and relatively high crown of the teeth suggest that *Leithia* could feed on more abrasive vegetation. The long survivorship of these two glirids in the Sicily–Malta archipelago may be evidence of relatively stable ecological–environmental conditions. Although the time of the last occurrences of these two glirids is not known in detail, it is probable that their stable ecological niches were disrupted either by the dispersal of new predators (e.g. the fox) and competitors (Muridae, Arvicolidae) or by severe climatic changes during the late Würm phase.

Pianetti–Castello faunal complex

The composition of mammal assemblages during the last Pleniglacial (Castello faunal complex)–ancient Holocene period is demonstrated by the fossil remains collected from cave deposits, where they occur together with lithic artefacts. These remains are generally the products of accumulations derived from hunting by humans, and therefore do not necessarily represent a random sample of the living community.

At Contrada Pianetti a fissure-filling deposit contains *C. elaphus siciliae* associated with *Bos primigenius*, *Ursus* cf. *arctos*, *Canis lupus* and a fully redescribed micromammal assemblage including *Microtus* (*Terricola*) ex gr. *savii*, *Apodemus* sp., *Crocidura* cf. *sicula* (the direct ancestor of the shrew at present living on the island), and *Erinaceus europaeus* (Bonfiglio *et al.* 1997*b*). Large predators and pachyderms are lacking. The Contrada Pianetti assemblage probably represents an intermediate faunal complex composed of some of the endemic *E. mnaidriensis* complex taxa as well as by new continental taxa belonging to a Würmian dispersal event.

The assemblages of the Castello and Holocene faunal complexes are less diverse with respect to those of the *E. mnaidriensis* complex and are composed of taxa that display only slight, if any, endemic characters.

Castello faunal complex

During the Pleniglacial and the Late Glacial period, there was an important faunal exchange between Sicily and the continent. This dispersal event is characterized by the appearance of the

equids, particularly the ass (*Equus hydruntinus*). It is still unclear if the ox, red deer, wild boar and wolf of the Castello faunal complex represent survivors from the *E. mnaidriensis* faunal complex or if they too were newcomers.

Equus hydruntinus, like the extant ass, is considered a species adapted to dry Mediterranean conditions and, unlike other equids, was probably a solitary form that did not depend on an open landscape (Abbazzi *et al.* 1996).

The vole *Microtus* (*Terricola*) ex gr. *savii* is phylogenetically related to the *savii* species, the ground vole widespread in the Italian peninsula since the last interglacial phase. It can be considered as an indicator of a temperate climate and of limited arboreal covering. During the last glacial phase the spread of *M.* (*T.*) *savii* became very reduced in the Italian peninsula and was limited to Southern Italy, whereas during the interglacial and the interstadial phases *M.* (*T.*) *savii* is frequently the dominant form in the Salento region (Southern Italy), which is a limestone plateau with the potential to become arid (San Sidero 3 locality, De Giuli 1983; Fondo Cattie cave, Corridi 1987). In the Castelcivita cave (Southern Italy) *M.* (*T.*) *savii* is a significant presence in levels corresponding to the temperate climatic fluctuations that are correlated with stage 3 of the oxygen isotope curve (Abazzi & Masini 1997).

In Sicily *M.* (*T.*) *savii* is by far the dominant form among the small mammals in several assemblages of the Castello faunal complex (Contrada Pianetti, upper levels of the K 22 site), and in the Holocene communities (Uzzo cave). It is today represented by the endemic subspecies *Microtus* (*Terricola*) *savii nebrodensis*.

Crocidura cf. *sicula* is the second species of the genus represented on the island. This form, which is intermediate in size between *Crocidura russula* and *Crocidura suaveolens*, seems to replace the pre-existing *Crocidura esuae*, probably during the Late Pleistocene Pianetti–Castello dispersal events. The spread of *Crocidura* in Late Pleistocene time in European and Mediterranean regions indicates a temperate and relatively arid climate. The distribution of this shrew was limited to the southern regions of the Italian peninsula during the last glacial phase (Masini & Abbazzi 1997).

The murid *Apodemus* sp. is not found in the *E. mnaidriensis* faunal complex and seems to have arrived in Sicily with *Crocidura* cf. *sicula* and *M.* (*Terricola*) *savii*. *Apodemus* is an opportunistic murid. However, its abundance seems to indicate the presence of a local arboreal covering (Masini & Abbazzi 1997). At present, *Apodemus* remains widespread in Sicily.

Holocene

The early Holocene stratigraphic sequence of the Uzzo cave (S. Vito peninsula) provides the best qualitative and quantitative data concerning the early Holocene environmental evolution available so far (Tagliacozzo 1993). In early Holocene time, at about 10 ka BP, humans frequently hunted red deer and wild boar, both indicators of forested environment, and the wild ox became a very rare animal. The occurrence of wild cat (*Felis sylvestris*) is also indicative of a forested landscape. According to Tagliacozzo (1993), the marked trend towards size reduction observed in the early Holocene red deer of the Uzzo cave is possibly due to selective hunting pressure. However, the extension of a Mediterranean-type forest during the Holocene climatic optimum could also be responsible for the size reduction in this ruminant, which is better adapted to boreal-type forests.

The dispersal of the water vole (*Arvicola*), which has been found in the Uzzo cave (lower Mesolithic levels, 10 ka BP ^{14}C non-calibrated) and at the Partanna Stretto Neolithic site (Upper Belice basin, dated around 5700–6600 a BP, Tusa 1992; Burgio & Di Patti 1994), and of the dormouse (Partanna Stretto only) probably took place during this time. These rodents indicate, respectively, the occurrence of widespread water resources and of wooded landscapes in areas in which a relatively dry climate dominates today (Upper Belice Basin and San Vito Lo Capo Peninsula) (Agnesi *et al.* 1997).

At present, the dormouse lives, together with *Muscardinus*, in the high forested areas of eastern and northwestern Sicily (Mount Etna, and Nebrodi and Madonie mountains). The water vole has not been documented from middle Neolithic levels at the Uzzo Cave and at present is extinct in Sicily.

Discussion and conclusion

The available data for the two older faunal complexes are rather scant. The constant occurrence of glirids, and the occurrence of *Asoriculus* in the Monte Pellegrino faunal complex and of *Crocidura* in the *Elephas falconeri* faunal complex, along with diverse bat and reptile (including a tortoise of very large size) assemblages, are indicative of a generally temperate to warm climate with forest cover. However, some data from terrestrial gastropods and bats indicate cooler climatic oscillations during early mid-Pleistocene time.

High-resolution stratigraphical data are not available for reconstructing the evolution of the climatic environment at the time of the *Elephas mnaidriensis* faunal complex in detail. Nevertheless, autoecology and synecology of the faunal assemblages allow a preliminary picture of the climatic environment during the late mid-Pleistocene, Late Pleistocene and last interglacial period for Sicily.

The assemblages of the *E. mnaidriensis* complex are ecologically well balanced, indicating that the island supported a trophic pyramid that included top predators such as the lion. This indicates that the insular ecological–environmental system was in equilibrium, at least during some phases, and contained different ecological niches. This is in agreement with palaeogeographical reconstructions, which demonstrate that the island was at that time as large, or even larger during sea-level lowstand phases, as Sicily today. It is worth noting the apparent absence of small-sized carnivores (e.g. mustelids) and of non-endemic rodents and insectivores in the Sicilian fossil record. If this absence is real, and not due to collecting bias, it may indicate that a sort of filtering barrier, not fully identifiable with the available data, was acting during the dispersal episodes that formed this faunal complex.

The *E. mnaidriensis* assemblages are indicative of a climate with temperate Mediterranean affinity and of landscapes in which forested areas were associated with open landscapes. Humid environments, rivers, pools and lacustrine basins were locally present. The dominance of the megalocerine in several assemblages suggests that a Mediterranean-type forest locally dominated during some climatic phases, whereas the dominance of the hippopotamus in several other assemblages and the occurrence of the wild ox suggest local open environments. The continuous presence of two forms of deer as well as bear and wild boar suggests that forested areas also extended across the island during the more arid climatic phases, which are identified in western Sicily by extensive aeolian deposits of dune environments. The general tendency towards a tree-covered Mediterranean landscape is indirectly confirmed by the rare, very sporadic, presence of *Bison priscus*, which, probably entered the island during particularly favourable conditions (a cold climatic phase with poorly forested lowlands).

The last glacial phase is marked by a drop in diversity among large mammals, which parallels that observed in coeval faunas from the Italian peninsula and, in general, the faunas of the European regions (extinction or local extinction of elephants, hippopotamus, lion and hyena). We are inclined to relate this decreased diversity to climatic environmental changes rather than to a disequilibrium in the insular faunal system. As an alternative hypothesis, one can suggest a direct influence of humans on extinctions but, in the case of Sicily, this possibility appears to be poorly corroborated by reliable data, as Late Pleistocene human occupation sites are not known before the spread of Ancient Aurignacian human culture in the later part of Late Glacial times.

The late part of the last glacial cycle was also a favourable time for faunal exchanges between Calabria and the Sicily–Malta insular complex, because of a eustatic lowstand. During this time faunal assemblages are characterized by the diffusion of *Microtus* (*Terricola*), which became the dominant species among small mammals, and by *Crocidura* cf. *sicula* and *Apodemus*. Deer, bear and other forest indicators are, however, still represented. The ecological significance of the *savi* ground vole, along with geomorphological data (spread of aeolian dunes and of talus deposits in western Sicily; Agnesi *et al.* 1997), suggests a degraded environment, with reduced arboreal cover, in which frutescent and steppe-like vegetation tended to prevail. This reconstruction is also in general agreement with the vegetation distribution for the last glacial maximum in the Mediterranean based on pollen analysis (Zagwin 1992) and by palynological data collected from the sedimentary sequences of volcanic lakes in the south of the Italian peninsula (Lakes of Monticchio and Basilicata, Watts *et al.* 1996). The tendency towards a strong climatic aridification has been demonstrated for the late Pleniglacial–Late Glacial time also for North Africa and the whole Saharan belt by Petit Maire (1993). During this time, or slightly later, the horse (*Equus caballus*) and the ass (*Equus hydruntinus*) entered Sicily. These equids are much represented in the Pleniglacial (horse) and Late Glacial (wild ass) in the south of the Italian peninsula (Apulia, Paglicci cave, Salento Peninsula) (Torre *et al.*, 1996), and are further confirmation of a rather arid climate. The spread of the horse, a social ungulate, which depends mainly on flat terrain and open landscapes, may have occured along emerged coastal plains.

The early Holocene period is characterized by a general reforestation and by a more humid climate as indicated by the disappearance of horses (a local extinction that occurred over the whole peninsula), by the very rare documentation of the wild ox, by the spread of deer and boar, and by the occurrence of dormice and of the water vole.

This research was supported by MURST 40% to Laura Bonfiglio MURST and 60% to Federico Masini. Helpful suggestions by two anonymous reviewers are gratefully acknowledged.

References

ABBAZZI, L. & MASINI, F. 1997. *Megaceroides solilhacus* and other deer from the Middle Pleistocene site of Isernia La Pineta (Molise, Italy). *Bollettino della Società Paleontologica Italiana, Modena*, **35**, 213–227.

——, DELGADO HUERTAS, A., IACUMIN, P., LONGINELLI, A., FICCARELLI, G., MASINI, F. & TORRE, D. 1996. Mammal changes and isotopic biogeochemistry. An interdisciplinary approach to climatic–environmental reconstructions at the last Pleniglacial/Late Glacial transition in the Paglicci Cave section (Gargano Apulia, SE Italy). *Il Quaternario*, **9**, 573–580.

ACCORDI, B. 1957. Nuovi resti di ippopotamo nano nel Pleistocene dei dintorni di Siracusa. *Atti dell'Accademia Gioenia di Scienze Naturali in Catania*, **11**, 99–109.

—— 1963. Rapporti tra il 'Milazziano' della costa iblea (Sicilia sud-orientale) e la comparsa di *Elephas mnaidriensis*. *Geologica Romana*, **2**, 295–304.

—— 1965. Some data on the Pleistocene stratigraphy and related pigmy mammalian faunas of eastern Sicily. *Quaternaria*, **6**(1962), 415–430.

—— & COLACICCHI, R. 1962. Excavations in the pygmy elephants cave of Spinagallo (Siracusa). *Geologica Romana*, **1**, 217–230.

AGNESI, V., MACALUSO, T. & MASINI, F. 1997. L'ambiente e il clima della Sicilia nell'ultimo milione di anni. *In*: TUSA, S. (ed.) *Prima Sicilia, alle origini della società siciliana*, Ediprint, Palermo, 31–53.

AGUIRRE, E. 1968–1969. Revision sistematica del los Elephantidae por su morfologia y morfometria dentaria. *Estudios Geologicos*, **24**(1968), 109–168; **25**(1969), 123–177, 317–367.

ALIMEN, M. H. 1975. Les isthmes hispano-marocain et siculo-tunisien aux temps acheuléens. *L'Anthropologie*, **79**, 399–436.

AMBROSETTI, P. 1968. The Pleistocene dwarf elephants of Spinagallo (Siracusa, south eastern Sicily). *Geologica Romana*, **7**, 277–398.

——, AZZAROLI, A. & KOTSAKIS, T. 1980. Mammiferi del Plio-Pleistocene delle isole italiane. *In*: *Catalogo della Mostra 'I Vertebrati fossili italiani'*. Verona, 243–248.

AZZAROLI, A. 1982. Insularity and its effects on the terrestrial vertebrates: evolutionary and biogeographic aspects. *'Palaeontology, essential of Historical Geology', First International Meeting, Venice*, 2–4 June 1981. Mucchi, Modena, 193–213.

BADA, J. L., BELLUOMINI, G., BONFIGLIO, L., BRANCA, M., BURGIO, E. & DELITALA, L. 1991. Isoleucine epimerization ages of Quaternary mammals of Sicily. *Il Quaternario*, **4**(1a), 5–11.

BARRIER, P. 1987. Stratigraphie des dépôts pliocènes et quaternaires du Détroit de Messine. *Documents et Travaux, IGAL, Paris*, **11**, 59–81.

BELLUOMINI, G. & BADA, J. L. 1985. Isoleucine epimerization ages of the dwarf elephants of Sicily. *Geology*, **13**, 451–452.

BERNABÒ BREA, L. 1950. Yacimientos paleoliticos del sudest de Sicilia. *Ampurias*, **12**, 115–143.

BIDDITTU, I. 1971. Considerazioni sull'industria litica e la fauna del riparo della Sperlinga di S. Basilio. *Bullettino di Paleontologia Italiana*, **80**(n.s. 22), 64–75.

BONFIGLIO, L. 1978. Resti di Cervide (Megacero) dell'Eutirreniano di Bovetto (RC). *Quaternaria*, **20**, 87–108.

—— 1987. Nuovi elementi faunistici e stratigrafici del Pleistocene superiore dei Nebrodi (Sicilia nord-orientale). *Rivista Italiana di Paleontologia e Stratigrafia*, **93**, 145–164.

—— 1991. Correlazioni tra depositi a Mammiferi, depositi marini, linee di costa e terrazzi medio e tardopleistocenici nella Sicilia orientale. *Il Quaternario*, **4**, 205–214.

—— 1992a. Campagna di scavo 1987 nel deposito pleistocenico a *Hippopotamus pentlandi* di Acquedolci (Sicilia nord-orientale). *Bolletino della Società Paleontologica Italiana*, **30**, 157–173.

—— 1992b. Middle and Upper Pleistocene mammal-bearing deposits in south-eastern Sicily: new stratigraphical records from Coste di Gigia (Syracuse). *Geobios*, **14**, 189–199.

—— 1995. Taphonomy and depositional setting of Pleistocene mammal-bearing deposits from Acquedolci (North-Eastern Sicily). *Geobios*, **18**, 57–68.

—— & BERDAR, A. 1986. Gli elefanti del Pleistocene superiore di Archi (R.C.): nuove evidenze di insularità della Calabria meridionale durante il ciclo Tirreniano. *Bollettino della Società Paleontologica Italiana*, **25**, 9–34.

—— & BURGIO, E. 1992. Significato paleoambientale e cronologico delle mammalofaune Pleistoceniche della Sicilia in relazione all'evoluzione paleogeografica. *Il Quaternario*, **5**, 223–234.

—— & INSACCO, G. 1992. Palaeoenvironmental, palaeontologic and stratigraphic significance of vertebrate remains in Pleistocene limnic and alluvial deposits from South Eastern Sicily. *Palaeogeography, Palaeoclimatology, Palaeoecology*, **95**, 195–208.

—— & PIPERNO, M. 1996. Early faunal and human populations. *In*: LEIGHTON, R. (ed.) *Early Societies in Sicily*. Accordia Research Centre, University of London, London, 21–29.

—— DI GERONIMO, I. S., INSACCO, G. & MARRA, A. C. 1997a. Large mammal remains from late Middle Pleistocene deposits of Sicily: new stratigraphic evidence from the western edge of the Hyblean Plateau (South-Eastern Sicily). *Rivista Italiana di Paleontologia e Stratigrafia*, **102**, 375–384.

—— DI STEFANO, G., INSACCO, G. & MARRA, A. C. 1993. New Pleistocene fissure-filling deposits from the Hyblean Plateau (South Eastern Sicily). *Rivista Italiana di Paleontologia e Stratigrafia*, **98**, 523–540.

—— INSACCO, G., MARRA, A. C. & MASINI, F. 1997b. Large and small mammals, amphibians, reptiles from a new fissure-filling deposit of the Hyblean Plateau (South-Eastern Sicily). *Bollettino della Società Paleontologica Italiana*, **36**, 97–122.

BRUGAL, J. P. 1987. Cas de 'nanisme' insulaire chez l'aurochs. *In*: *112th Congrès National des Sociétés Savantes, Lyon, Vol. 2*, 53–66.

BURGIO, E. & CANI, M. 1988. Sul ritrovamento di elefanti fossili ad Alcamo (Trapani, Sicilia). *Il Naturalista Siciliano*, s. 4, **12**, 87–97.

—— & DI PATTI, C. 1994. La fauna del fossato/trincea di contrada Stretto (Partanna). *In*: TUSA, S. (ed.) *La preistoria del basso Belice e della Sicilia meridionale nel quadro della preistoria sicilaina e mediterranea*, Palermo, 201–109.

—— & FIORE, M. 1988a. La fauna vertebratologica dei depositi continentali di Monte Pellegrino (Palermo). *Il Naturalista Siciliano*, **12**, 91–18.

—— 1988b. *Nesolutra trinacriae* n.sp. lontra quaternaria della Sicilia. *Bollettino della Società Paleontologica Italiana*, **27**, 259–275.

—— 1997. *Pannonictis arzilla* (De Gregorio 1886) a 'Villafranchian' element in the fauna from Monte Pellegrino (Palermo, Sicily). *Il Quaternario*, **10**, 65–74.

——, OLIVA, N. & SCALONE, E. 1983. La collezione vertebratologica della Grotta dei Puntali presso Carini (Palermo). *Il Naturalista Siciliano*, s.4, **7**, 67–79.

CARBONE, S., DI GERONIMO, I., GRASSO, M., IOZZIA, S. & LENTINI, F. 1982. I terrazzi marini quaternari dell'area iblea (Sicilia sud-orientale). *Contributi conclusivi per la realizzazione della Carta Neotettonica d'Italia, Progetto Finalizzato Geodinamica, Napoli*, **506**, 1–35.

CARDINI, L. 1971. Rinvenimenti paleolitici nella grotta Giovanna (Siracusa). *Atti 23a riunione scientifica dell'Istituto Italiano di Preistoria e Protostoria, Siracusa, Malta*, 22–26 October 1968.

CASSOLI, P. F. & TAGLIACOZZO, A. 1996. L'avifauna. IN: BASILE, B. & CHILARDI, S. (eds) *Le ossa dei Giganti. Lo scavo paleontologico di Contrada Fusco*. Arnaldo Lombardi, Siracusa, 61–67.

CATALANO, R. & D'ARGENIO, B. 1982. Schema geologico della Sicilia. *In*: *Guida alla Geologia della Sicilia occidentale*. Società Geologica Italiana, Palermo, 9–41.

CHILARDI, S. 1996. I macromammiferi. *In*: BASILE, B. & CHILARDI. S. (eds) *Le ossa dei Giganti. Lo scavo paleontologico di Contrada Fusco*. Arnaldo Lombardi, Siracusa, 73–80.

—— & Gilotti, A. 1996. Stratigrafia e sedimentologia. *In*: BASILE, B. & CHILARDI, S. (eds) *Le ossa dei Giganti. Lo scavo paleontologico di Contrada Fusco*. Arnaldo Lombardi, Siracusa, 27–34.

CONTI, M. A., DI GERONIMO, I., ESU, D. & GRASSO, M. 1980. Il Pleistocene in facies limnica di Vittoria (Sicilia meridionale). *Geologica Romana*, **18**, 93–104.

CORRIDI, C., 1987. Le faune Pleistoceniche del Salento. *2 La fauna di Fondo Cattie*. Maglie, Lecce.

DAAMS & DE BRUIJN, H. 1995. A classification of the Gliridae (Rodentia) on the basis of dental morphology. *Hystrix*, n.s. **6**, 3–50.

DE BRUIJN, H. 1966. On the Pleistocene Gliridae (Mammalia, Rodentia) from Malta and Mallorca. *Proceedings, Koninklike Nederlandse Akademie van Wetenschappen*, **69**, 480–496.

DE GIULI, C. 1983. Le faune pleistoceniche del Salento.

La fauna di S. Sidero 3. *Quaderni del Museo Paleontologico di Maglie*, **1**, 45–84.

DE GREGORIO, A. 1886. Intorno a un deposito di roditori e di carnivori sulla vetta di Monte Pellegrino con uno schizzo sincronografico del calcare postpliocenico della vallata di Palermo. *Memorie della Società Toscana di Scienze Naturali*, **8**, 217–253.

DI GERONIMO, I. 1979. Il Pleistocene in facies batiale di Valle Palione (Grammichele, Catania). *Bollettino di Malacologia*, **15**, 85– 156.

—— & COSTA, B. 1978. Il Pleistocene di Monte dell'Apa (Gela). *Rivista Italiana di Paleontologia e Stratigrafia*, **84**, 1121–1158.

——, GHISETTI, F., GRASSO, M., LENTINI, F., SCAMARDA, G. & VEZZANI, L. 1980. Dati preliminari sulla tettonica della Sicilia sud-orientale. Fogli 273 (Caltagirone), 274 (Siracusa), 275 (Scoglitti), 276 (Ragusa) e 277 (Noto). *Contributi preliminari alla realizzazione della Carta Neotettonica d'Italia, Progetto Finalizzato Geodinamica, Napoli*, **356**, 747–773.

DI GRANDE, A. & RAIMONDO, W. 1984. Linee di costa plio-pleistoceniche e schema litostratigrafico del Quaternario siracusano. *Geologica Romana*, **21**, 279–309.

ESU, D. & GIROTTI, O. 1991. Late Pliocene and Pleistocene assemblages of continental molluscs in Italy. A survey. *Il Quaternario*, **4**, 137–150.

——, KOTSAKIS, T. & BURGIO, E. 1986. I vertebrati e i molluschi continentali pleistocenici di Poggio Schinaldo (Palermo, Sicilia). *Bollettino della Società Geologica Italiana*, **105**, 233–241.

FABIANI, R. 1932. Risultati di alcuni scavi nella Grotta della 'Za' Minica' presso Capaci (Palermo). *Atti della Reale Accademia di Scienze, Lettere e Belle Arti, Palermo*, **17**, 1–8.

—— 1934. Notizie preliminari sui risultati di uno scavo paleontologico nella Grotta della Cannita (Palermo). *Bollettino di Scienze Naturali ed Economiche di Palermo*, n.s. **16**, 3–7.

GALLETTI, L. & SCALETTA, C. 1991. Descrizione di una sequenza del Pleistocene superiore con fauna continentale a San Ciro-Maredolce (Palermo). *Il Naturalista siciliano*, **15**, 3–10.

GIGNOUX, M. 1913. Les formations marines pliocènes et quaternaires de l'Italie du Sud et de la Sicile. *Annales de l'Université de Lyon*, n.s. **36**, VII–XXIV.

GLIOZZI, E. & MALATESTA, A. 1984. A megacerine in the Pleistocene of Sicily. *Geologica Romana*, **21**, 311–389.

——, —— & SCALONE, E. 1993. Revision of *Cervus elaphus siciliae* Pohlig, 1893, Late Pleistocene endemic deer of the Siculo-Maltese district. *Geologica Romana*, **29**, 307–353.

GRASSO, M. & LENTINI, F. 1982. Sedimentary and tectonic evolution of the eastern Hyblean Plateau (Southeastern Sicily) during late Cretaceous to Quaternary time. *Palaeogeography, Palaeoclimatology, Palaeoecology*, **39**, 261–280.

GRAZIOSI, P. & MAVIGLIA, C. 1946. La grotta di S. Teodoro (Messina). *Rivista di Scienze Preistoriche*, **1**, 227–283.

GUERRESCHI, A. 1992. La fine del Pleistocene e gli inizi dell'Olocene. *In*: GUIDI, A. & PIPERNO, M. (eds)

Italia preistorica. Laterza, 198–237.

HUTTERER, A. 1991. Variation and evolution of the Sicilian shrew: taxonomic conclusions and description of a possibly related species from the Pleistocene of Morocco (Mammalia: Soricidae). *Bonner zoologische Beitrage*, **42**, 241–251.

KELLING, G. & STANLEY, D. J. 1972. Sedimentation in the vicinity of the Strait of Gibraltar. *In*: STANLEY, D. J. (ed.) *The Mediterranean Sea. A Natural Sedimentation Laboratory*. Dowden, Hutchinson & Ross, Stroudsburg, PA, 489–519.

KOTSAKIS, T. 1977. I resti di Anfibi e Rettili pleistocenici della grotta di Spinagallo (Siracusa, Italia). *Geologica Romana*, **6**, 211–229.

—— 1979. Sulle mammalofaune quaternarie siciliane. *Bollettino del Servizio Geologico d'Italia*, **99**, 263–276.

—— 1986. *Crocidura esui* n. sp. (Soricidae, Insectivora) du Pléistocène supérieur de Spinagallo Sicilia orientale, Italie). *Geologica Romana*, **23**, 51–64.

—— 1996a. Anfibi e rettili. *In*: BASILE, B. & CHILARDI, S. (eds) *Le ossa dei Giganti. Lo scavo paleontologico di Contrada Fusco*. Arnaldo Lombardi, Siracusa, 56–60.

—— 1996b. I micromammiferi. *In*: BASILE, B. & CHILARDI, S. (eds) *Le ossa dei Giganti. Lo scavo paleontologico di Contrada Fusco*. Arnaldo Lombardi, Siracusa, 68–72.

—— & PETRONIO, C. 1981. I Chirotterri del Pleistocene superiore della grotta di Spinagallo (Siracusa, Sicilia). *Bollettino del Servizio Geologico d'Italia*, **101**, 49–76.

LEIGHTON, R. 1996. Research traditions, chronology and current issues, an introduction. *In*: LEIGHTON, R. (ed.) *Early Societies in Sicily. New Developments in Archaeological Research*. Accordia Research Centre, University of London, London, 1–19.

MARTIN, R. A. 1995. A new Middle Pleistocene species of *Microtus* (*Pedomys*) from the Southern United States, with comments on the taxonomy and early evolution of *Pedomys* and *Pitymys* in North America. *Journal of Vertebrate Paleontology*, **15**, 171–186.

MASINI, F. & ABBAZZI, L. 1997. L'associazione di mammiferi della Grotta di Castelcivita. *In*: GAMBASSINI, P. (ed.) *Il Paleolitico della Grotta di Castelcivita, cultura e ambiente, 5. Materiae*. Electa, Napoli, 33–57.

—— & SARÀ, M. 1999. *Asoriculus burgioi* sp. nov. (Soricidae, Mammalia) from the Monte Pellegrino faunal complex (Sicily). *Acta Zoologica cracoviensia*. (in press).

——, FICCARELLI, G., ROOK, L. & TORRE, D. 1992. Mammal dispersal events in the Middle and Late Pleistocene of Italy and Western Europe. *In*: KOENIGSWALD, W. & WEDERLIN, L. (eds) *CFS 153, 'Mammalian Migration and Dispersal Events in the European Quaternary, Frankfurt*, 59–68.

PETIT-MAIRE, N. 1993. Past global climatic changes and the tropical arid/semi-arid belt in the North of Africa. *In*: THORWEIHE, U. & SCHANDELMEIER, H. (eds) *Geoscientific Research in Northeast Africa*, 551–560.

PETRONIO, C. 1970. I roditori pleistocenici della Grotta

di Spinagallo (Siracusa). *Geologica Romana*, **9**, 149–194.

POHLIG, H. 1893. Eine Elephantenhohle Siciliens und der erste Nachweis des Cranialdomes von Elephas antiquus. *Abhandlungen der Bayerischen Akademie der Wissenschaften*, **18**, 73–100.

—— 1909. Über zwei neue altpleistocäne formen von Cervus. *Zeitschrift der Deutschen Geologische Gesellschaft*, **61**, 250–253.

REGALIA, E. 1907. Sulla fauna della 'Grotta del Castello' di Termini Imerese (Palermo). *Archivi per l'Antropologia e l'Etnologia*, **37**, 339–374.

REUMER, J. W. F. 1984. Ruscinian and early Pleistocene Soricidae (Insectivora, Mammalia) from Tegelen (The Netherlands) and Hungary. *Scripta Geologica*, **73**, 11–73.

RHODES, E. J. 1996. ESR dating of tooth enamel. *In*: BASILE, B. & CHILARDI, S. (eds) *Le ossa dei Giganti. Lo scavo paleontologico di Contrada Fusco*. Arnaldo Lombardi, Siracusa, 39–44.

RUGGIERI, G. 1977. Rettifiche a due note su malacofaune quaternarie. *Bollettino della Società Paleontologica Italiana*, **16**, 271–272.

—— & UNTI, M. 1974. Pliocene e Pleistocene nell'entroterra di Marsala. *Bollettino della Società Geologica Italiana*, **93**, 723–733.

——, —— 1977. Il Quaternario del Pianoro di S. Margherita di Belice (Sicilia). *Bollettino della Società Geologica-Italia*, **96**, 803–812.

——, ——, UNTI, M. & MORONI, M. A. 1976. La calcarenite di Marsale (Pleistocene inferiore) e i terreni contermini. *Bollettino della Società Geologica d'Italia*, **94**, 1623–1627.

SALA, B. 1977. L'ippopotamo nel Pleistocene superiore in Italia. Considerazioni paleoecologiche. *Rivista di Scienze Preistoriche*, **32**, 283–286.

—— 1983. Variation climatiques et séquences chronologique sur la base des variations des associations faunistiques à grands mammifères. *Rivista di Scienze Preistoriche*, **38**, 161–180.

——, MASINI, F., FICCARELLI, G., ROOK, L. & TORRE, D. 1993. Mammal events in the Middle and Late Pleistocene of Italy and Western Europe. *Courier Forschungs-Institut Senckenberg*, **253**, 59–68.

SCINÀ, D. 1831. *Rapporto sulle ossa fossili di Mardolce e degli altri contorni di Palermo*. Reale Tipografia di Guerra, Palermo, 1–64.

SHACKLETON, N. J. 1995. New data on the evolution of Pliocene climatic variability. *In*: VRBA, E. S., DENTON, G. H., PARTRIDGE, T. C. & BURCKLE, L. H. (eds) *Paleoclimate and Evolution with Emphasis on Human Origins*. Yale University Press, New Haven, CT, 241–248.

SPROVIERI, R. 1982. Considerazioni sul Plio-Pleistocene della Sicilia. *In*: *Guida Geologica della Sicilia Occidentale*. Società Geologica Italiana, Palermo, 115–118.

TAGLIACOZZO, A. 1993. Archeozoologia della Grotta Dell'Uzzo, Sicilia. Da un'economia di caccia ad un'economia di pesca ed allevamento. *Supplemento al Bullettino di Palentologia Italiana*, **84**, n.s. II, 1–278.

THALER, L. 1972. Les rongeurs (Rodentia et Lagomorpha) du Monte Pellegrino et la question des anciens isthmes de la Sicilie. *Comptes rendus de l'Académie des Sciences série 2*, **274**, 188–190.

TORRE, D., ABBAZZI, L., FICCARELLI, G., MASINI, F., MEZZABOTTA, C. & ROOK, L. 1996. Changes in mammal assemblages during the Late-Glacial earliest Holocene. *Il Quaternario*, **9**, 551–560.

TUSA, S. 1992. *La Sicilia nella Preistoria*. Sellerio, Palermo.

VAUFREY, R. 1929. Les éléphants nains des iles mediterranèennes et la question des isthmes pléistocènes. *Archives de l'Institut de Paléontologie Humaine*, **6**, 1–220.

VON KOENIGSWALD, W. 1988. Paläoklimatische Aussage letztinterglazialer Säugetiere aus der nördlichen Oberrheinebene. *Paläoklimaforschung*, **4**, 206–314.

WATTS, W. A., ALLEN, J. R. M., HUNTLEY, B. & FRITZ, S. C. 1996. Vegetation history and climate of the last 15,000 years at Laghi di Monticchio, Southern Italy. *Quaternary Science Review*, **15**, 113-132.

ZAGWIJN, W. H. 1992. Migration of vegetation during the Quaternary in Europe. *Courier Forschungs-Institut Senckenberg*, **153**, 9–20.

ZAMMIT MAEMPEL, G. & BRUIJN, H. 1982. The Plio-Pleistocene Gliridae from the Mediterranean islands reconsidered. *Proceedings, Koninklijke Nederlandse Akademie van Wetenschappen*, **85**, 113–128.

Were climatic changes a driving force in hominid evolution?

J. CHALINE[1], A. DURAND[1], A. DAMBRICOURT MALASSÉ[2], B. DAVID[1], F. MAGNIEZ-JANNIN[1] & D. MARCHAND[1]

[1] *UMR CNRS 5561 et Laboratoire de Préhistoire et Paléoécologie du Quaternaire de l'EPHE, Université de Bourgogne, Centre des Sciences de la Terre, 6 bd. Gabriel, 21000 Dijon, France*

[2] *Institut de Paléontologie humaine du Muséum National d'Histoire Naturelle de Paris (UMR CNRS 9948), 1 rue R. Panhard, 75013 Paris, France*

(e-mail: jchaline@u-bourgogne.fr)

Abstract: A comparison of externalist and internalist approaches in hominid evolution shows that the externalist approach, with its claim that climate was responsible for the appearance of bipedalism and hominization, now seems to be ruled out by the biological, palaeogeographical, palaeontological and palaeoclimatic data on which it was based. Biological data support the embryonic origin of cranio-facial contraction, which determined the increase in cranial capacity and the shift in the position of the *foramen magnum* implying bipedalism. In the internalist approach, developmental biology appears as the driving force of hominid evolution, although climate exerts a significant influence and was involved in the following ways: (1) in the prior establishment of ecological niches that allowed the common ancestor to become differentiated into three subspecies; (2) by dividing up the area of distribution of species, resulting in the present-day subspecies of gorillas and chimpanzees; (3) by facilitating relative fluctuations of the geographical areas of distribution of the various species, particularly the spread of australopithecines across the African savanna from north (Chad, Ethiopia) to south (South Africa); (4) by determining adaptive geographical differentiations among *Homo erectus* and *Homo sapiens* (pigmentation, haemoglobin, etc.).

Climate, as a factor of the environment, exerts a major influence over the geographical distribution of species and subspecies, and over their adaptation to their environment, although there are wide variations in the way it affects different groups. The living world is organized on a hierarchical basis, and the gearing ratios between the different levels may vary substantially. This is the case with hominids, where changes are small at the molecular level (King & Wilson 1975; Hixson & Brown 1986; Miyamoto *et al.* 1987, 1988; Hayasaka *et al.* 1988; Ueda *et al.* 1989; Gonzales *et al.* 1990; Sibley *et al.* 1990; Bailey *et al.* 1991, 1992; Horai *et al.* 1992; Perrin-Pecontal *et al.* 1992; Goodman *et al.* 1994), larger at the chromosomal level (Chiarelli 1962; De Grouchy *et al.* 1972; Stanyon & Chiarelli 1981, 1982, 1983; Yunish & Prakash 1982; Marks 1985, 1993; Dutrillaux & Couturier 1986; Dutrillaux *et al.* 1986; Godfrey & Marks

1991; Matera & Marks 1993) and greatly amplified at the morphological level (Dambricourt Malassé 1987, 1988, 1993, 1996; Chaline 1994, 1998; Deshayes, 1997). These differences constitute the human paradox.

The questions raised are about the precise role of climate in distinct aspects of hominid evolution and the way climate affects the different levels of organization of the living world. There are currently two contrasting views of the origin and evolution of the higher apes and of humankind as related to the role of climate. One is an externalist approach (Coppens 1986, 1994), whereas the other is an internalist approach (Chaline 1994, 1998; Chaline *et al.* 1998).

The aim of this paper is: (1) to present the externalist and internalist approaches; (2) to put forward explanatory models of evolution within a palaeoclimatic and palaeoecological frame-

From: HART, M. B. (ed.) *Climates: Past and Present.* Geological Society, London, Special Publications, **181**, 185–198, 1-86239-075-4/$15.00

work; (3) to test the exact role of climate as a driving force behind human evolution (when, where and how?), and thus to test the two opposing hypotheses, the externalist v. the internalist approach.

The externalist approach: climate as the major driving force behind human evolution

The externalist approach (Coppens 1986, 1994) is a theory in which climate is seen as the principal driving force behind human evolution, alongside tectonics. The theory is based on two observations: (1) the present-day distribution of chimpanzees and gorillas covers all of the tropical forest regions of Africa, but terminates with virtually no overlap at the great tectonic trough of the East African Rift Valley; (2) the sites of hominid fossils that are more than 3 Ma old lie to the east of the Rift Valley.

Coppens deduced from this that panids and hominids were segregated geographically by the tectonic crisis some 8 Ma ago that caused the Rift Valley floor to collapse and raised the western lip of the Western Rift, creating an 8000 m barrier between the summits and the bottom of Lake Tanganyika. 'The great initial province was split in two. The west remained humid and the forests and woodlands were preserved, while the east became increasingly arid with ever more sparse savanna. The common ancestor population of Panids and Hominids was therefore divided into a larger western and a smaller eastern population. It is tempting to imagine that the segregation was the cause of the divergence between the two groups: the western descendants continued to adapt to life in an arboreal environment and formed the Panids while the eastern descendants of the same ancestors "invented" new mechanisms suitable for life in an open environment and gave rise to the Hominids' (Coppens 1994, p. 67). Coppens considered that the earliest forms of australo-pithecines were more confined to arboreal environments than the more recent species (Coppens 1986, p. 232). Man, for his part, is without doubt a pure product of aridity, which Coppens termed the 'Omo event' (Coppens 1994, p. 71).

This theory asserts that the diversity currently observed among hominids results from tectonic and subsequently climatic events. Climate is seen therefore as a major force in human evolution. The great weakness of the externalist approach is that it provides no biological mechanism to account for the substantial changes in body shape.

The internalist approach: the ontogenetic mechanism

This is a radically different theory based upon biological data, especially upon the ontogenetic mechanism. This theory was later explored by Delattre & Fenart (1954, 1956, 1960), who analysed the ontogenesis of higher ape and human skulls and showed that they underwent occipital and facial contraction. These major stages of cranio-facial contraction are a well-known phenomenon in primates, alternatively termed 'flexure of the skull base' or 'occipital shift' (Deniker 1885; Anthony 1952).

The internalist approach is supported by the data of ontogenetic development, which was completely overlooked in the so-called Synthetic Theory of Evolution during the 1940s (Devillers & Chaline 1993). Such an approach links genetics to morphogenesis through developmental biology. It has given rise to stimulating research (Gould 1977; Raff & Kauffman 1983; Devillers & Chaline 1993; Raff 1996 (for synthesis); McGinnis & Kuziora 1997; Pennisi & Roush 1997).

This approach has been taken up more recently by Deshayes (1986, 1991), Dambricourt Malassé (1987, 1988, 1996) and Deshayes & Dambricourt Malassé (1990). Dambricourt Malassé (1987) discovered and proved that cranio-facial contraction is a embryonic phenomenon that is clearly visible in the mandible, and that the living and fossil lower jaws retain the range of contraction. She also demonstrated that primate evolution can be summarized as six 'fundamental ontogenies', because of their embryological origin, corresponding to successive body plans or bauplans, resulting from six major phases of cranio-facial contraction: (1) prosimians; (2) monkeys apes; (3) great apes; (4) australopithecines; (5) *Homo*; (6) *Sapiens*. Recent morphometric studies on the skull (Chaline *et al.* 1998) have shown that the modern human form is not significantly different from the earlier forms of the genus *Homo*. Thus we do not retain here the '*Sapiens*' bauplan (Fig. 1).

The morphological areas defined by these bauplans are exploited by more-or-less numerous speciation events (Fig. 2). Obviously, comparisons between the great ape skull plan, which is contemporary with *Homo sapiens*, and the two fossil hominids skull plans (*Australopithecus* and *Homo*), which extend over time, are a working hypothesis and not an evolutionary model in the strict sense, insofar as we assimilate the living great apes to their Tertiary ancestors.

This internalist approach views development

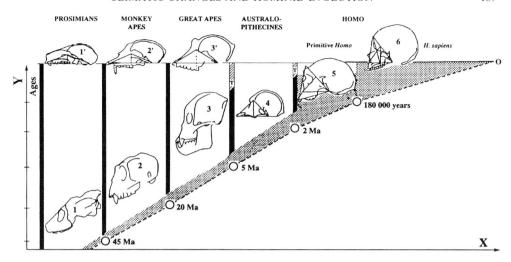

Fig. 1. Phylogeny of embryonic basicranial–facial contraction in primates (after Dambricourt Malassé 1996). X, range of contraction; Y, geological time. 1, Adapiform; 1', Lemuriform; 2, *Mesopithecus*; 2', *Cercopithecus*; 3, *Proconsul africanus*; 3', *Gorilla*; 4, *Australopithecus africanus*; 5, *Homo erectus*; 6, *Homo sapiens*. T, extinction. Explanation: the range of embryonic contraction is unchanged from Adapiforms (1) to living lemurs (1' including lorises and tarsiers). However, it increases between Adapiforms and *Homo*. Each column represents one fundamental ontogeny. Diversity occurs in each: in terms of range *Tarsius* fits in with the prosimians and not the simians, and similarly *Hylobates* fits in with the simians and not the great apes. The names above the figure correspond to the five bauplan names.

heterochronies and mutations of regulator genes as the engines of morphological change, and especially of the cranio-facial contractions that punctuate the evolution of the higher apes and hominids. From great apes to modern man numerous heterochronies (hypermorphosis, hypomorphosis and post-displacements) have occurred during ontogeny (Chaline *et al.* 1998), allowing (1) the acquisition of permanent bipedalism of *Australopithecus* and *Homo*, (2) the increased cranial capacity of primitive forms of *Homo* (*habilis, ergaster, rudolfensis, erectus, heidelbergensis* and *neanderthalensis*) and (3) the disappearance of simian characters associated with renewed increase in cranial capacity in *Homo sapiens*.

Analysis of the neotenic mechanism in the axolotl (*Ambystoma mexicanum*) provides an eloquent example linking genetic regulation (Humphrey 1967; Voss 1995) with environmental conditions (Norris & Gern 1976). In this instance, environment, through temperature, is instrumental during early larval stages in activating or inhibiting genes that control the selection of the most effective morphologies within a fairly broad spectrum of possibilities.

The genes responsible for heterochronies, and more particularly *Hox* genes (Dollé *et al.* 1993; Carroll 1995; McGinnis & Kuziora 1997; Pennisi & Roush 1997; Meyer 1998), or other regulatory genes, may explain the origin of cranio-facial contractions in hominids and certain human deformities, and thus provide a biological solution to the human paradox of the 1% genetic divergence versus the large morphological difference between humans and chimpanzees.

The internalist approach reflects the pre-eminent role of the ontogenetic mechanism (Chaline 1994, 1998). The next step is to examine the climatic data that reflect the external constraints on biological systems.

Climatic patterns through time

The pattern of climates in Africa is controlled primarily by the position of the equator and the tropics (climatic gradients perpendicular to the equator). Ascending air masses in the equatorial zone lead to cloud formation. The next factor in climate distribution is the presence of oceans on either side of the continental landmass, which provide the water vapour necessary for cloud formation (gradients parallel to the equator). Only then does orography come into play by locally altering the gradients parallel and perpendicular to the equator. The present-day situation has arisen very gradually since the time the South Atlantic opened up some 130 Ma ago and Africa wheeled round as it drifted more

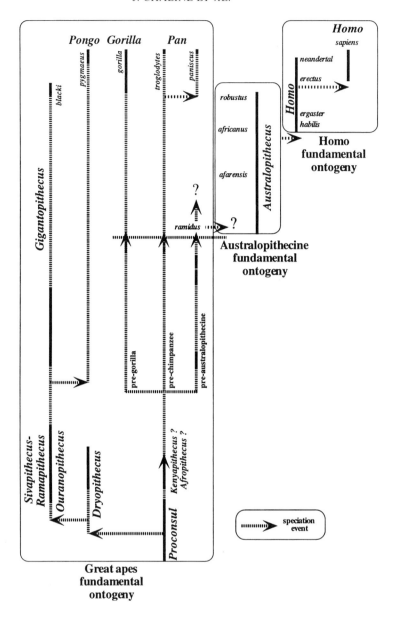

Fig. 2. Phylogenetic diagram showing relationship patterns with regard to all the palaeontological and ontogenetic data. Speciations occur by punctuation, and several lineages, such as australopithecines and early human, seem to undergo gradual evolutionary changes (continuous lines represent the fossil record). The last three fundamental ontogenies of primates ('great apes', 'Australopithecines', and 'Homo') are shown. The 'great apes' bauplan has been diversified into a great many species.

than 3000 km northward (Scotese *et al.* 1988). At 15–20 Ma the situation was already very similar to the present-day situation and was in no way related to the formation of the Rift Valley.

This geographical pattern of climates underwent many large changes, first, because Africa has drifted relative to the equator, which lay 5° further north about 10 Ma ago (Scotese *et al.* 1988), but also because of changes in atmospheric circulation. In Late Miocene time tropical North Africa experienced at least four episodes of substantial climatic deterioration

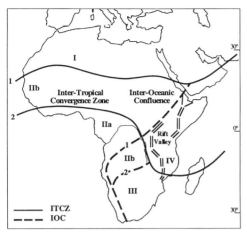

Fig. 3. Schematic and hypothetical geographical distribution of climates and environments in Africa between 23 Ma and Present. ITCZ, Inter-Tropical Convergence Zone; IOC, Inter-Oceanic Confluence; 1, when the sun is over the Tropic of Cancer; 2, when it is over the Tropic of Capricorn; I, area of continental trade winds; II, area of Atlantic monsoon; IIa, area of permanent Atlantic monsoon; IIb, area of seasonal Atlantic monsoon; III, area of Indian Ocean trade winds; IV, area of East African monsoons. The equator was 7° more northerly at 23 Ma, 5° at 10 Ma and 2° at 6 Ma (after Scotese *et al.* 1988).

leading to aridity, the last at the end of Miocene time (about 6–5.3 Ma) coinciding with a very distinct cooling of the Atlantic (Diester-Haass & Chamley 1982; Sarnthein *et al.* 1982). Schematically, since the work by Chudeau (1921), it has been accepted that extensions of the arid Sahara zone are related in part to a reduced summertime advance of the front of the Inter-Tropical Convergence Zone (ITCZ; Fig. 3), which may even remain blocked in the Southern Hemisphere. The inter-Oceanic Confluence (N–S) coincides to some extent with the northern part of the Rift Valley (IOC; Fig. 3).

From Late Pliocene time (*c.* 3.2 Ma) general atmospheric circulation changed radically, for geodynamic reasons. Events such as the closure of the Panama isthmus, the opening of the Bering Straits and mountain building in northwest America and Asia are thought to have been decisive factors, in conjunction with variations in planetary orbit, in triggering the Northern Hemisphere ice ages (Ruddiman & Raymo 1988; Berger 1992; de Menocal 1995). For recent times, when chronology is relatively precise, it is now known that changes occur suddenly. After the last glacial maximum, it is thought that the arid zone extended several hundred kilometres further south on three occasions between

about 19 and 15 ka ^{14}C BP, each episode lasting only 500–1000 a (Durand 1995). The dense humid forest became fragmented and mountain biotopes spread as a result of the fall in temperatures (Maley 1987).

Against this general context we will examine below the impact of climate on: the diversification of the common ancestor, the formation of subspecies and species, the geographical distribution of the different species and the diversity among present-day humans. Chromosomal data must be introduced here that have not yet been included in the general scheme of hominid evolution.

Chromosomal data

A major advance in chromosome research was the development of R-banding and G-banding techniques. Opinions differ about the role of heterochromatin. Dutrillaux & Couturier (1986) considered that the chromosome data from primates provide no evidence either for positional effects or that alterations in heterochromatin influence gene expression. This idea is materialized at the phylogenic level by two equally feasible dichotomic hypotheses entailing acceptance of three convergences or reversions, but a third hypothesis, where each rearrangement is regarded as unique, implies complex populational evolution (Dutrillaux & Richard, 1997).

By contrast, Stanyon & Chiarelli (1982), as well as Marks (1993), have argued that genes can be (1) repressed by placing them in, or near, blocks of heterochromatin, or (2) activated by a shift in their position away from heterochromatin regions. They set great store by active regions detected by Ag-NOR (silver nitrate) type methods and this has repercussions for phylogeny. They concluded that common derived karyological features indicate that *Gorilla* and *Pan* shared a common descent after the divergence of *Homo*. The distribution of heterochromatin at the tips of the chromosomes of gorillas and chimpanzees was confirmed by Marks (1993), suggesting a phylogenetic association between those two taxa.

A trichotomic chromosomal model

The common ancestral chromosome formula of all these species can be reconstructed by deduction. The reconstruction is based on the principle that if two, three or four species share a common chromosome, it is likely that some common ancestor transmitted it to all of them. This is the simplest relationship. We may also

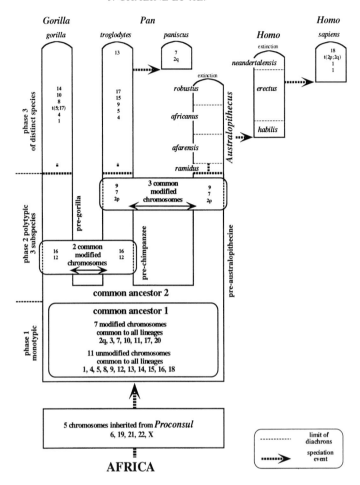

Fig. 4. Trichotomic chromosomal model incorporating palaeontological and ontogenetic data. This model explains the distribution of chromosome formulae of living species and proposes a solution for shared *Pan–Homo, Gorilla–Pan* apomorphic rearranged chromosomes and the absence of *Gorilla–Homo* apomorphic chromosomal features. After a monotypic phase accounting for chromosomal apomorphies 2, 3, 7, 10, 11, 17 and 20, the common ancestor must have become polytypic and divided into three subspecies: pre-gorilla, pre-chimpanzee and pre-australopithecine. Some 5 Ma ago the three subspecies produced the three extant and fossil genera: *Gorilla, Pan* and *Australopithecus*. About 2 Ma ago the *Homo* lineage derived from *Australopithecus* to give primitive man (*Homo erectus*), and around 0.18 Ma ago modern humans (*Homo sapiens*). The figures refer to chromosomes rearrangements.

take the view that two identical chromosomal changes occur in the same way on the same chromosome in two related species derived from an ancestral form. This is statistically far less likely, although it is possible by convergence.

We have attempted to reconstruct the chronology of the formation of the various chromosomes and the successive events that marked the history of the family (Chaline *et al.* 1991, 1996) (Fig. 4).

Among anthropoid apes, the group with the most primitive chromosomal formula is logically the one that branched off earliest in the family history. This is the orang-utan group, which does not share the seven derived specific mutant chromosomes (2q, 3, 7, 10, 11, 17 and 20) of the common ancestor because it emigrated to Asia, where it has been cut off from the African branch for more than 10 Ma (Andrews & Cronin 1982; Lipson & Pilbeam 1982; Simons 1989). It has been relatively stable from the chromosomal point of view, only two new chromosome mutations having occurred since that time.

After the separation of the orang-utan lineage,

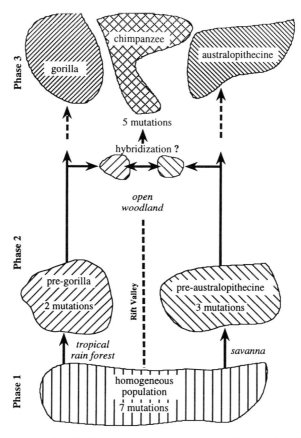

Fig. 5. Dichotomic model explaining the distribution of chromosomes in living species. This most parsimonious model suggests that after the monotypic phase, the common ancestor split into a polytypic phase with two subspecies (pre-gorilla and pre-australopithecine), and that the populations from the two subspecies interbred to produce the third, hybrid, subspecies (the pre-chimpanzees).

the common ancestral lineage of the three extant genera, chimpanzees, gorillas and humans, remained in Africa. This phase corresponds to the 'common ancestor' implied by genetic similarities. That there was a common ancestor is irrefutably shown by the formation of seven specific mutant chromosomes (2q, 3, 7, 10, 11, 17 and 20) and by the retention of 11 non-mutated common chromosomes (1, 4, 5, 8, 9, 12, 13, 14, 15, 16 and 18) inherited by the three living species.

The occurrence of seven characteristic mutant chromosomes found in the gorilla, chimpanzee and human descendants implies that the common ancestor was a single monotypic species, not divided into subspecies. Genetic pooling as a result of interbreeding produces a fairly even spread of chromosome variety, including chromosomal variations that appear in isolated individuals. This first part of the history of the common ancestor corresponds to what we shall term the 'first homogeneous common ancestor phase' (Fig. 4).

Thereafter, the common ancestor diversified and the lineages separated, leading to the gorilla, chimpanzee and humans. This divergence raises a complex populational problem related to the distribution of five new mutated chromosomes.

It is reported that chimpanzees and humans share three mutated chromosomes (2p, 7 and 9) that gorillas do not have. Conversely, chimpanzees and gorillas share two other specific mutated chromosomes (12 and 16) not found in humans. In other words, chimpanzees share three common mutated chromosomes with humans and two common mutated chromosomes with gorillas. But gorillas share no special rearrangements with humans! This position is inexplicable by the standard hypothesis of a split into two branches (dichotomic model) leading to the gorilla–chimpanzee on one side and to humans on the other, as suggested by Marks (1993).

There is, however, a possible solution to the puzzle: the 'second heterogeneous common ancestor phase' (Fig. 4). To explain this major paradox, it must be supposed that at a certain time in their history, after the first undifferentiated phase, the ancestors of chimpanzees and gorillas were able to interbreed and acquire two new chromosome mutations they alone possessed, and not humans. This accords with the suggestion of Smouse & Li (1987) 'that the ancestors of all three taxa were still conspecific' (see also Pruvolo *et al.* 1991).

However, it must also be accepted that for some time, perhaps different from the first period, the ancestors of chimpanzees and humans were able to interbreed and to incorporate three new chromosomal mutations into their genetic constitution, possessed by them alone and not gorillas. The fact that the ancestors of gorillas and humans do not share exclusive mutations in their chromosomal formulae implies that they did not at that time have contacts allowing hybridization. They were geographically isolated.

This model introduces australopithecines as a missing link between the common ancestor and the human lineage proper. They are never considered by chromosome specialists for the obvious reason that, being extinct, their chromosomes are unknown. But as they formed a necessary intermediate stage, they must be included in the model. By deduction, the chromosomal formula of the extinct pre-human form *Australopithecus* can be evaluated fairly accurately, except for the four chromosomes that were rearranged after genetic isolation of the species (ch 1/1/ fusion of 2p–2q/18).

A dichotomic chromosomal model

A further possible dichotomic model suggests that after the first common phase, two subspecies, pre-gorillas and pre-australopithecines became geographically separated and chromosomally differentiated (Fig. 5). Only then did two small populations of each subspecies meet and form the pre-chimpanzee stock. Chimpanzees could be hybrids of pre-gorillas and pre-australopithecines.

To account for the separation into distinct subspecies, a new decisive factor must be incorporated here, that of climate change.

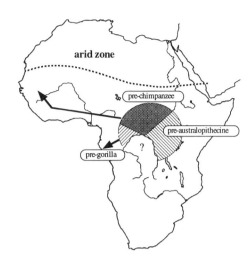

Fig. 6. Distribution of the common ancestor at the start of the polytypic phase. Pre-gorillas must have occupied the wet Atlantic monsoon zone (rain forest), pre-chimpanzees the less humid Atlantic monsoon zone (open forest) and pre-australopithecines the Indian Ocean monsoon zone (acacia savanna).

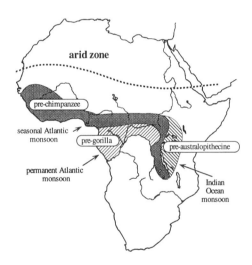

Fig. 7. Distribution of the common ancestor during the maximum extension of the polytypic phase. The pre-chimpanzee group must have been geographically centred around the present-day Lake Victoria region. It stretched to northwest Africa across the area north of the Zaire River in a forest–savanna mosaic or open woodland environment. The pre-gorilla group must have been located on the western edge of the former, still north of the Zaire River, in the very wet tropical rainforest zone. The pre-australopithecine group must have been located further east, in the East African Rift Valley, and further north in the savanna beyond the open forests inhabited by the pre-chimpanzees.

The internalist approach in the climatic context

Climate as the driving force of diversification of the common ancestor

Climatic changes (Durand 1995) would have induced changes in the environment that must have been instrumental in the geographical isolation of the subspecies and species. Figures 6 and 7 explain the logic behind these changes.

The original common area of hominids in central Africa is currently a privileged climatic zone where the ITCZ (east–west) corresponding to the thermal and meteorological equator crosses the IOC (north–south) (Leroux 1983). Very schematically, we find the following features.

- The permanent Atlantic monsoon domain and tropical rainforest occur in the west; forest is documented in the Congo, Gabon and Cameroon basins from the onset of Miocene time (Boltenhagen et al. 1985).
- To the north lies the seasonal Atlantic monsoon domain and the savanna; in Nigeria a seasonal subtropical climate is reported to have existed since the Oligocene–Miocene boundary (Takashi & Jux 1989).
- Further north still is the Sahara zone; evidence of a large arid zone in North Africa from mid-Miocene times is provided by rodents (Jaeger 1975) and by flora (Boureau et al. 1983).
- In the south and east is the domain of the Indian Ocean monsoons and trade winds. Tropical rainforest, open woodlands and mountain forests are reported to have coexisted since 19 Ma, and savannas and grasslands since 15–12 Ma (Bonnefille 1987; Retallack et al. 1990; Kingston et al. 1994). Local reliefs of the Rift Valley are superimposed on this background pattern to form an environmental mosaic that is sensitive to climatic fluctuations (White 1983; Pickford 1990).

Our hypothesis is that differentiation occurred in connection with the environment as determined by climate. Starting from a monotypic common stock (acquisition of seven chromosomal changes) (Figs 6 and 7), the pre-gorillas split away in the permanent monsoon domain north of the Zaire River barrier. The pre-chimpanzees spread widely east–west, on the northern boundary of the permanent Atlantic monsoon domain, ranging from Central to West Africa. The pre-australopithecines developed on the eastern margin under the influence of the Indian Ocean monsoons and trade winds, and spread north–south along the inter-oceanic confluence that roughly coincides with the eastern arm of the Rift Valley and westward across the savanna zone skirting the north of the tropical rainforest range. Such variations in climate could have segregated the three subspecies in a yet unknown history, as fossil remains dating from 12 to 5 Ma are very sparse. Few finds have been made to date: a molar from Ngorora (Kenya) older than 10 Ma, one from Lukeino (7 Ma) (Pickford 1975), a fragment of jaw with a molar from Lothagam, again in Kenya, dated to 5.5 Ma, and the *Samburupithecus* mandible (9.5 Ma).

A further implication concerns the shape of the pelvis. As gorillas and chimpanzees are quadrupeds, and as bipedalism is the apomorphic feature characteristic of australopithecines and their human descendants, the three components of the common ancestor must therefore still have walked on all fours.

Climate as the driving force of diversification of australopithecine, gorilla and chimpanzee species

Allopatric break-up of the ancestral form would have allowed gorillas, australopithecines and chimpanzees to become isolated (Fig. 8), as evidenced by the discovery of the earliest australopithecines (*Ardipithecus ramidus*), dated to some 4.4 Ma in the Pliocene deposits of the Middle Awash (White et al. 1994, 1995).

Fossil finds of gorillas and chimpanzees will probably be few and far between, perhaps because of the lack of appropriate sedimentary environments in their tropical forest habitat, where remains are more rarely preserved than in open sedimentary environments of savannas. However, since becoming separate species, gorillas and chimpanzees have respectively acquired six autapomorphic chromosomal rearrangements (ch 1/4/5–17/8/10/14 for gorillas; ch 4/5/9/15/17/13 for chimpanzees). It should also be pointed out that the bonobo (*Pan paniscus*) has autapomorphic rearrangements on chromosomes 2q and 7. These autapomorphies prove the early isolation of gorillas and chimpanzees, and the probably more recent separation of the bonobo. It is interesting to see that the teeth of *Ardipithecus ramidus* of the Awash (size, morphology, enamel thickness) and

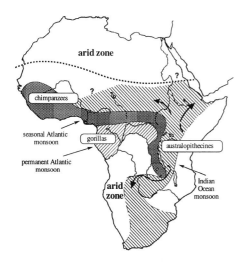

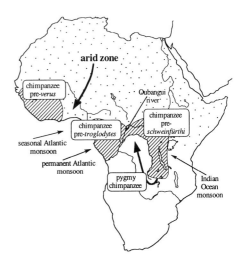

Fig. 8. Distribution of genera *Pan, Gorilla* and *Australopithecus* between 5 and 2 Ma. The chimpanzee lived north of the Zaire River in a forest–savanna mosaic or open woodland environment. The gorilla group must have been located on the western edge of the former, still north of the Zaire River, in the very wet tropical rainforest zone. The australopithecines group must have been located further east, in the East African Rift Valley, and further north in the savanna beyond the open forests inhabited by the chimpanzees, as suggested by the discovery of Koro Toro (Chad).

Fig. 9. Distribution of the three subspecies of common chimpanzees. The westernmost subspecies, the black chimpanzee (*Pan troglodytes verus*), inhabits Guinea whereas the others are found north of the Zaire or Congo River: the typical common chimpanzee (*Pan troglodytes troglodytes*) in the west in Congo, Gabon, Equatorial Guinea and Cameroon, and Schweinfürthi chimpanzee (*Pan troglodytes schweinfurthi*) further east in northern Zaire (after Collet 1988).

even features of the cranium have a strikingly chimpanzee-like morphology very similar to the bonobo, suggesting a possible filiation.

Climate and the environmental diversification it entails acted as controls over the area of distribution of the various species. Australopithecines were able to migrate from Central Africa northward to Ethiopia and the Afars, and southward, skirting round the Zambezi River barrier, to South Africa, where their fossils were first unearthed (Taung, Sterkfontein, Swartkrans) (Fig. 8).

The externalist approach implies that the earliest australopithecines could be found only to the east of the Rift Valley (Coppens 1986, 1994), and not to the west. The discovery of an early australopithecine, named 'Abel', (*Australopithecus bahrelghazali*), 2500 km west of the Rift Valley (Chad) and thought to be 3–3.5 Ma old (Brunet *et al.* 1995, 1996) challenges this part of the theory. This discovery was, however, predictable by the trichotomic theory of the common ancestor in the internalist approach, as the acquisition of bipedal gait provided the potential to colonize all of the African savanna from north to south and east to west (Fig. 8).

Fig. 10. Distribution of the three subspecies of gorillas. Gorillas are divided into western lowland gorillas (*Gorilla gorilla gorilla*) in Congo, Gabon, Equatorial Guinea and southern Cameroon, and eastern mountain gorillas (*Gorilla gorilla graueri* (Burundi and Rwanda); the third subspecies is the mountain gorilla of Rwanda and Uganda (*Gorilla gorilla beringei*) (after Collet 1988).

Climate as the driving force of diversification of present gorilla and chimpanzee subspecies

The repetition of Pleistocene fluctuations is thought to be responsible for the current diversification of chimpanzees and gorillas, each characterized by three subspecies (Figs 9 and 10).

Starting from the initial chimpanzee distribution, it is thought that common chimpanzees then diversified into three subspecies. One subspecies (*Pan troglodytes verus*) is isolated in Guinea (Collet 1988), probably as a result of a break in the forest caused perhaps by a southward shift of the Sahara desert zone in Quaternary times engendering aridity in what is now Nigeria (Fig. 9). A second area is located in the Central African Republic and northern Zaire, along the Oubangui River, which already forms a natural barrier between the areas of distribution of western gorillas and chimpanzees (*Gorilla gorilla gorilla* and *Pan troglodytes troglodytes*) and eastern subspecies (*Gorilla gorilla beringei* and *graueri* as well as *Pan troglodytes schweinfürthi*) and which must have been reinforced by the southward extension of arid conditions.

This mechanism of separation of species into western and eastern populations may have been compounded because the gorilla and chimpanzee populations were unable to cross the great natural barrier of the marshland basin of the Zaire River.

It is impossible at present to provide details and a chronology of what might have happened, but this mechanism is plausible. It can, furthermore, be tested, because if it is correct, fossils of chimpanzees should be found in Tertiary deposits in areas where chimpanzees no longer occur from the Ivory Coast to Nigeria.

Climate as the driving force of diversification of primitive and modern Homo.

Climate plays a role in development of the species *Homo erectus*, which survived for at least one and a half million years and differentiated geographically, as indicated by the two opposing morphological trends observed in *Homo erectus* of Asia and of Europe, the latter giving rise to Neandertals. Climate also plays a further role in development of *Homo sapiens*, by determining clinal differences affecting characters related to climate, in particular pigmentation and various selected physiological factors (adaptations to altitude, humidity, aridity, etc.) (Lewontin 1982;

Langaney 1988). These differentiations must have been the same for the various geographical forms of *Homo erectus*.

Such adaptive differentiation can have occurred only in the last 180 ka for *Homo sapiens*, which derived by cranio-facial contraction from a *Homo erectus* population. It is because this differentiation occurred so recently on the evolutionary scale that modern humans are alone in not being subdivided into geographically quantifiable and identifiable races.

Conclusion: climate as a driving force of hominid evolution?

The foregoing discussion shows that climate was instrumental in hominid evolution, but was it really a driving force behind it?

The externalist approach, with its claim that climate was responsible for the appearance of bipedalism, now seems to be ruled out by biological, palaeogeographical, palaeontological data and palaeoclimatic data on which it was based. The pattern of different environments in relation to the main climatic domains (permanent or seasonal Atlantic monsoon and Indian Ocean monsoon and trade winds) seems to have been comparable with the present-day situation since about 15 Ma, i.e. well before the formation of the Rift Valley (8 Ma).

Biological data do, however, support the embryonic origin of cranio-facial contraction, which determined the increase in cranial capacity and, through the position of the foramen magnum, brought about bipedalism. These processes are certainly partly under the control of so far unidentified homeobox genes (*Hox* genes) or other regulatory genes. The story of human evolution may be more appropriately termed an internalist history rather than an externalist one.

In the internalist approach, developmental biology is the major driving force behind hominid evolution but climate exerts a significant influence at crucial moments in that history: (1) in the prior development of ecological niches allowing the common ancestor (polytypical phase 2) to differentiate into three subspecific lineages foreshadowing the three present-day genera; (2) by facilitating relative fluctuations of the geographical areas of distribution of the various species, especially the extension of australopithecines towards the Sahara in the northwest, Ethiopia in the northeast, and into southern Africa; (3) by breaking up the area of distribution of extant species of gorillas and chimpanzees, which led to the three present-day subspecies of each; (4) by determining slight adaptive differences in pigmentation, which must have been the same for the various geographical forms of *Homo erectus* and *Homo sapiens*.

This work was supported by the UMR CNRS 5561 (Biogeosciences) and by the Laboratoire de Paléobiodiversité et Préhistoire de l'Ecole Pratique des Hautes Etudes (Université de Bourgogne, Dijon). We are grateful to two anonymous reviewers and to M. Hart for helpful comments, to A. Bussière for the final drawings, and to C. Sutcliffe for help with translation.

References

ANDREWS, P. & CRONIN, J. E. 1982. The relationships of *Sivapithecus* and *Ramapithecus* and the evolution of the orangutan. *Nature*, **297**, 541–546.

ANTHONY, J. 1952. La brisure de la base du crâne chez les Primates. *Annales de Paléontologie*, **38**, 71–79.

BAILEY, W. J., FITCH, D. H. A., TAGLE, D. A., CZELUSNIAK, J., SLIGHTOM, J. L. & GOODMAN, M. 1991. Molecular evolution of the w-globin gene locus: gibbon phylogeny and the hominoid slowdown. *Molecular Biology and Evolution*, **8**, 155–184.

——, HAYASAKA, K., SKINNER, C. G., KEHOE, S., SIEU, L. C., SLIGHTOM, J. L. & GOODMAN, M. 1992. Reexamination of the African hominoid trichotomy with additional sequences from the primate β-globin gene cluster. *Molecular Phylogenetics and Evolution*, **1**, 97–135.

BERGER, A. 1992. *Le climat de la Terre*. De Boeck Université, Brussels.

BOLTENHAGEN, E., DEJAX, J. & SALARD-CHEBOLDAEFF, M. 1985. Évolution de la végétation tropicale africaine du Crétacé à l'actuel d'après les données de la palynologie. *Bulletin de la Section des Sciences*, **8**, 165–194.

BONNEFILLE, R. 1987. Evolution des milieux tropicaux africains depuis le début du Cénozoïque. *Mémoire des Travaux de l'Ecole Pratique des Hautes Etudes (EPHE)*, **17**, 101–110.

BOUREAU, E., CHEBOLDAEFF-SALARD, M., KOENIGUER, J. C. & LOUVET, P. 1983. Évolution des flores et de la végétation tertiaires en Afrique, au nord de l'équateur. *Bothalia*, **14**, 355–367.

BRUNET, M., BEAUVILAIN, A., COPPENS, Y., HEINTZ, E., MOUTAYEA, H. E. & PILBEAM, D. 1995. The first autralopithecine 2,500 kilometres west of the Rift Valley (Chad). *Nature*, **378**, 273–274.

——, ——, ——, ——, —— & —— 1996. *Australopithecus bahrelghazali*, une nouvelle espèce d'hominidé ancien de la région de Koro Toro (Tchad). *Comptes Rendus de l'Académie des Sciences, Série IIa*, **322**, 907–913.

CARROLL, S. B. 1995. Homeotic genes and the evolution of arthropods and chordates. *Nature*, **376**, 479–485.

CHALINE, J. 1994. *Une famille peu ordinaire*. Le Seuil, Paris.

—— 1998. Vers une approche globale de l'évolution

des Hominidés. *Le Point sur–Comptes Rendus de l'Académie des Sciences, Serie II*, **326**, 307–318.

——, DAMBRICOURT MALASSÉ, A., MAGNIEZ-JANNIN, F., MARCHAND, D., DAVID, B., MILLET, J.-J. & COURANT, F. 1998. Quantification de l'évolution morphologique des Hominidés et hétérochronies. *Comptes Rendus de l'Académie des Sciences, Série II*, **326**, 291–298.

——, DURAND, A., MARCHAND, D., DAMBRICOURT MALASSÉ, A. & DESHAYES, M. J. 1996. Chromosomes and the origins of Apes and Australopithecines. *Human Evolution*, **11**, 43–60.

——, DUTRILLAUX, B., COUTURIER, J., DURAND, A. & MARCHAND, D. 1991. Un modèle chromosomique et paléobiogéographique d'évolution des primates supérieurs. *Geobios*, **24**, 105–110.

CHIARELLI, B. 1962. Comparative morphometric analysis of primate chromosomes of the anthropoid apes and man. *Caryologia*, **15**, 99–121.

CHUDEAU, R. 1921. Les changements de climat du Sahara pendant le Quaternaire. *Comptes Rendus de l'Académie des Sciences*, **172**, 604–607.

COLLET, J. Y. 1988. *Les petits des grands singes*. Terre sauvage. Nuit et Jour, 6–12.

COPPENS, Y. 1986. Evolution de l'homme. *Comptes Rendus de l'Académie des Sciences*, **3**(3), 227–243.

—— 1994. Une histoire des Hominidés. *Pour la Science*, **201**, 64–71.

DAMBRICOURT MALASSÉ, A. 1987. *Ontogenèses, paléontogenèses et phylogenèse du corps mandibulaire catarhinien. Nouvelle interprétation de la mécanique humanisante (théorie de la foetilisation, Bolk, 1926)*. Thèse de Doctorat, Muséum National d'Histoire Naturelle, Paris.

—— 1988. Hominisation et foetalisation. *Comptes Rendus de l'Académie des Sciences, Série II*, **307**, 199–204.

—— 1993. continuity and discontinuity during Hominization. *Quaternary International*, **29**, 86–98.

—— 1996. Nouveau regard sur l'origine de l'homme. *La Recherche*, **286**, 46–54.

DE GROUCHY, J., TURLEAU, C., ROUBIN, M. & KLEIN, M. 1972. Évolutions caryotypiques de l'homme et du chimpanzé. Étude comparative des topographies de bandes après dénaturation ménagée. *Annales de Génétique*, **15**, 79–84.

DELATTRE, A. & FENART, R. 1954. Rotation occipitale positive et négative. *Comptes Rendus de l'Académie des Sciences*, **239**, 676–678.

—— & —— 1956. L'hominisation du crâne est-elle terminée? *Comptes Rendus de l'Académie des Sciences*, **243**, 429–431.

—— & —— 1960. *L'hominisation du crâne*. CNRS, Paris.

DE MENOCAL, P. B. 1995. Plio-Pleistocene African climate. *Science*, **270**, 53–59.

DENIKER, J. 1885. Recherches anatomiques et embryologiques sur les singes anthropoïdes. Foetus de gorille et de gibbon comparés aux foetus humains et aux anthropoïdes jeunes et adultes. *Archives de Zoologie expérimentale et générale*, **2**(3 bis), 1–265.

DESHAYES, M. J. 1986. *Croissance cranio-faciale et orthodontie*. Masson, Paris.

—— 1991. Reconsidération de la croissance cranio-faciale au cours de l'ontogenèse et de l'évolution. Implications pour les traitements orthopédiques. *Revue d'Orthopédie dentofaciale*, **25**, 353–365.

—— 1997. A new ontogenetic approach to craniofacial growth. The basis of Projet Télécrane International. *Journal of Mastication and Health Society*, 7(2), 59–76.

—— & DAMBRICOURT MALASSÉ, A. 1990. Analyse des différents types architecturaux cranio-faciaux par l'approche ontogénique de l'hominisation. *Revue de Stomatologie et de Chirurgie Maxillofaciale*, **91**, 149–258.

DEVILLERS, C. & CHALINE, J. 1993. *Evolution, an Evolving Theory*. Springer, Berlin.

DIESTER-HAASS, L. & CHAMLEY, H. 1982. Oligocene and post-Oligocene history of sedimentation and climate off Northwest Africa (DSDP site 369). *In*: VON RAD, U., HINZ, K., SARNTHEIN, M. & SEIBOLD, E. (eds) *Geology of the Northwest African Continental Margin*. Springer, Berlin, 529–544.

DOLLÉ, P., DIERICH, A., LE MEUR, M., SCHIMMANG, T., SCHUBAUR, B., CHAMBON, P. & DUBOULE, D. 1993. Disruption of the *Hoxd-13* gene induces localized heterochrony leading to mice with neotenic limbs. *Cell*, **75**, 431–441.

DURAND, A. 1995. Sédiments quaternaires et changements climatiques au Sahel central (Niger et Tchad). *Africa Geoscience Review*, **2**, 323–614.

DUTRILLAUX, B. & COUTURIER, J. 1986. Principes de l'analyse chromosomique appliquée à la phylogénie: l'exemple des Pongidae et des Hominidae. *Mammalia*, **50**, 22–37.

—— & RICHARD, F. 1997. Notre nouvel arbre de famille. L'analyse des chromosomes permet de réécrire l'histoire des primates. *La Recherche*, **298**, 54–61.

——, VIEGAS-PEQUIGNOT, E. & COUTURIER, J. 1986. Les méthodes cytogénétiques. *Mammalia*, **50**, 11–20.

GODFREY, L. & MARKS, J. 1991. The nature and origins of primate species. *Yearbook of Physical Anthropology*, **34**, 39–68.

GONZALEZ, I. L., SYLVESTER, J. E., SMITH, T. F., STAMBOLIAN, D. & SCHMICKEL, R. D. 1990. Ribosomal RNA gene sequences and hominoid phylogeny. *Molecular Biology and Evolution*, **7**, 203–219.

GOODMAN, M., BAILEY, W. J., HAYASAKA, K., STANHOPE, M. J., SLIGHTOM, J. & CZELUSNIAK, J. 1994. Molecular evidence on primate phylogeny from DNA sequences. *American Journal of Physical Anthropology*, **94**, 3–24.

GOULD, S. J. 1977. *Ontogeny and Phylogeny*. Belknap Press of Harvard University Press, Cambridge, MA.

HAYASAKA, K., GOJORI, T. & HORAI, S. 1988. Molecular phylogeny and evolution of primate mitochondrial DNA. *Molecular Biology and Evolution*, **5**, 626–644.

HIXSON, J. E. & BROWN, W. M. 1986. A comparison of the small ribosomal RNA genes from the mitochondrial DNA of the great apes and humans: sequence, structure, evolution, and phylogenetic implications. *Molecular Biology and Evolution*, **3**, 1–18.

HORAI, S., SATTA, Y., HAYASAKA, K. *et al.* 1992. Man's place in Hominoidea revealed by mitochondrial DNA genealogy. *Journal of Molecular Evolution*, **35**, 32–43.

HUMPHREY, R. R. 1967. Albino axolotls from an albino tiger salamander through hybridization. *Journal of Heredity*, **58**, 95–101.

JAEGER, J. J. 1975. *Les Muridae (Mammalia, Rodentia) du Pliocène et du Pléistocène du Maghreb. Origine; évolution; données biogéographiques et paléoclimatiques*. Thèse de Doctorat, Université des Sciences et Techniques du Languedoc, Montpellier.

KING, M. C. & WILSON, A. C. 1975. Evolution at two levels in humans and chimpanzees. *Science*, **188**, 201–203.

KINGSTON, J. D., MARINO, B. D. & HILL, A. 1994. Isotopic evidence for Neogene hominid paleoenvironments in the Kenya Rift Valley. *Science*, **264**, 955-959.

LANGANEY, A. 1988. *Les Hommes*. Colin, Paris.

LEROUX, M. 1983. *Le climat de l'Afrique tropicale*. Champion, Paris.

LEWONTIN, R. 1982. *La diversité des hommes*. Belin, Paris.

LIPSON, S. & PILBEAM, D. 1982. *Ramapithecus* and hominid evolution. *Journal of Human Evolution*, **11**, 545–548.

MALEY, J. 1987. Fragmentation de la forêt dense humide africaine et extension des biotopes montagnards au Quaternaire récent: nouvelles données polliniques et chronologiques. Implications paléoclimatiques et biogéographiques. *Palaeoecology of Africa*, **17**, 307–334.

MARKS, J. 1985. C-band variability in the common chimpanzee, *Pan troglodytes*. *Journal of Human Evolution*, **14**, 669–675.

—— 1993. Hominoid heterochromatin: terminal C-band as a complex genetic trait linking chimpanzee and gorilla. *American Journal of Physical Anthropology*, **90**, 237–246.

MATERA, A. G. & MARKS, J. 1993. Complex rearrangements in the evolution of hominoid chromosome XVII. *Journal of Human Evolution*, **24**, 233–238.

McGINNIS, W. & KUZIORA, M. 1997. Les gènes du développement. *Pour la Science, hors série*, **7614**, 126–133.

MEYER, A. 1998. *Hox* gene variation and evolution. *Nature*, **391**, 225–228.

MIYAMOTO, M. M., KOOP, B. F., SLIGHTOM, J. L., GOODMAN, M. & TENNANT, M. R. 1988. Molecular systematics of higher primates: genealogical relations and classification. *Proceedings of the National Academy of Sciences of the USA*, **85**, 7627–7631.

——, SLIGHTOM, J. L. & GOODMAN, M. 1987. Phylogenetic relations of humans and African apes from DNA sequences in the ψn globin region. *Science*, **238**, 369–373.

NORRIS, D. O. & GERN, W. A. 1976. Thyroxine-induced activation of hypothalamo-hypophysal axis in neotenic salamander larvae. *Science*, **194**, 525.

PENNISI, E. & ROUSH, W. 1997. Developing a new view of evolution. *Science*, **277**, 34–37.

PERRIN-PECONTAL, P., GOUY, M., NIGON, V. M. & TRABUCHET, G. 1992. Evolution of the primate β-globin gene region: nucleotide sequence of the δ–β globin intergenic region of gorilla and phylogenetic relationships between African apes and man. *Jour-*

nal of Molecular Evolution, **34**, 17–30.

PICKFORD, M. 1975. Late Miocene sediments and fossils from the Northern Kenya Rift Valley. *Nature*, **256**, 279–284.

—— 1990. Uplift of the Roof of Africa and its bearing on the evolution of mankind. *Human Evolution*, **5**, 1–20.

PRUVOLO, M. T., DISOTELL, T., AMMARD, M. W., BROWN, W. M. & HONEYCUTT, R. L. 1991. Resolution of the African hominoid trichotomy using a mitochondrial gene sequence. *Proceedings of the National Academy of Sciences of the USA*, **88**, 1570–1574.

RAFF, R. A. 1996. *The Shape of Life*. University of Chicago Press, Chicago, IL.

—— & KAUFMAN, T. C. 1983. *Embryos, Genes, and Evolution*. MacMillan, New York.

RETALLACK, G. J., DUGAS, D. P. & BESTLAND, E. A. 1990. Fossil soils and grasses of a middle Miocene east African grassland. *Nature*, **247**, 1325–1328.

RUDDIMAN, W. F. & RAYMO, M. E. 1988. Northern Hemisphere climatic regimes during the past 3 Ma, possible tectonic connections. *Philosophical transactions of the Royal Society of London, Series B*, **318**, 411–430.

SARNTHEIN, M., THIEDE, J., PFLAUMANN, U. *et al.* 1982. Atmospheric and oceanic circulation patterns off Northwest Africa during the past 25 million years. *In*: VON RAD, U., HINZ, K., SARNTHEIN, M. & SEIBOLD, E. (eds) *Geology of the North West African Continental Margin*. Springer, Berlin, 545–604.

SCOTESE, C. R., GAHAGAN, L. M. & LARSON, R. L. 1988. Plate tectonic reconstructions of the Cretaceous and Cenozoic ocean basins. *Tectonophysics*, **155**, 27–48.

SIBLEY, C. G., COMSTOCK, J. A. & AHLQUIST, J. E. 1990. DNA hybridization evidence of hominoid phylogeny: a reanalysis of the data. *Journal of Molecular Evolution*, **30**, 202–236.

SIMONS, E. L. 1989. Human origins. *Science*, **245**, 1343–1350.

SMOUSE, P. E. & LI, W. H. 1987. Likelihood analysis of mitochondrial restriction–cleavage patterns for the human–chimpanzee–gorilla trichotomy. *Evolution*, **41**, 1162–1176.

STANYON, R. & CHIARELLI, B. 1981. The chromosomes of man and apes: evolution and comparison. *Bionature*, **1**, 11–27.

—— & —— 1982. Phylogeny of the Hominoidea: the chromosome evidence. *Journal of Human Evolution*, **11**, 493–504.

—— & —— 1983. Chromosomal phylogeny in great apes and humans. *Antropologia Contemporanea*, **6**, 157–159.

TAKASHI, K. & JUX, U. 1989. Palynology of Middle Tertiary lacustrine deposits from the Jos Plateau. *Bulletin Nagasaki University of Natural Sciences*, **29**, 181–367.

UEDA, S., WATANABE, Y., SAITOU, N. *et al.* 1989. Nucleotide sequences of immunoglobulin-epsilon pseudogenes in man and apes and their phylogenetic relationships. *Journal of Molecular Biology*, **205**, 85–90.

VOSS, S. R. 1995. Genetic basis of paedomorphosis in the axolotl, *Ambystoma mexicanum*: a test of the single-gene hypothesis. *Journal of Heredity*, **86**, 441–447.

WHITE, F. 1983. *The Vegetation of Africa*. UNESCO, Paris.

WHITE, T. D., SUWA, G. & ASFAW, B. 1994. *Australopithecus ramidus*, a new species of early hominid from Aramis, Ethiopia. *Nature*, **371**, 306–312.

——, —— & —— 1995. *Australopithecus ramidus*, a new species of early hominid from Aramis, Ethiopia. Corrigendum. *Nature*, **375**, 88.

YUNISH, J. J. & PRAKASH, O. 1982. The origin of man: a chromosomal pictorial legacy. *Science*, **215**, 1525–1530.

Late Quaternary vertebrate taphocoenoses from cave deposits in southeastern Austria: responses in a periglacial setting

FLORIAN A. FLADERER

*Institut für Paläontologie, Geozentrum, Universität Wien, Althanstrasse 14, 1090 Wien,
Austria (e-mail: florian.fladerer@univie.ac.at)*

Abstract: Forty-seven species of birds and at least 78 mammalian taxa from caves in the
extant submontane vegetational zone close to the periphery of the Middle and Late
Pleistocene Alpine maximum glacial expansion are comprehensively summarized for the first
time. The communities of the last glacial period show sympatries of today's boreal–tundra
species with those of East European to Central Asian steppe-communities and to a lesser
extent with those of temperate woodlands. Shifts in the individual species' ranges during the
last glacial period (Würm Glacial) can be partly correlated with the climatic fluctuations in
the Eastern Alps as they were observed sedimentologically and glaciologically. Analyses of
high-resolution uppermost Pleniglacial/Late Glacial taphocoenoses (*c.* 14–12 ka BP) in the
Styrian foothills demonstrate a distinctly greater faunal diversity and reflect more diversified
vegetational biota than in the Holocene. The evidence of Middle and Upper Palaeolithic
human populations within the region is outlined. The presented results provide a foundation
for subsequent and comparative studies in diversity and community changes in the adjacent
regions.

The Grazer Bergland is situated north and west
of Graz, the capital of the Austrian Province of
Styria (Fig. 1). The first scientific excavations in
caves of this region with more or less stratified
recoveries and documentation of profiles were
carried out in the early 1920s (Abel & Kyrle
1931), although investigations under the leader-
ship of biologists, prehistorians or geologists
extend back more than 150 years. The first
evidence for Pleistocene humans in the eastern
margin of the Alps is a pointed antler fragment
which was excavated as early as 1837 in the
Grosse Badlhöhle. However, it was not recognized as an Upper Palaeolithic artefact until
1870. From 1946 to the 1960s, Mottl (1951,
1953, 1975), a former student of the Hungarian
palaeontologist, T. Kormos, carried out a basic
survey which involved more than 20 caves in the
Central Styrian karst area and in Upper Styria.

In 1984, a small excavation at the entrance of
the Grosse Badlhöhle marked the beginning of
the recent study period. In 1986–92, small-scale
excavations in five caves (Tropfsteinhöhle, Tunnelhöhle, Grosse and Kleine Peggauerwandhöhle, Rittersaal) were carried out for
assessment purposes or for the construction of
fences against uncontrolled digging (Fuchs 1989;

Fladerer 1993). Partly published samples from
1976 (Grosse Ofenbergerhöhle), 1984 and 1986
(Knochenhöhle), stored at the Natural History
Museum in Vienna, and one collection by the
author in 1994 (Knochenhöhle) were added with
special reference to the palaeoclimatic development of the Styrian foothills. In June 1997
excavations in the fore-cave of the Lurgrotte,
near Peggau, were initiated by the author
together with G. Fuchs (Graz) for documentation and display purposes. The first results of
these investigations are included in this paper.

The geological and morphological setting

The Grazer Bergland (Fig. 1), which has a
maximum extent of 50 km from SW to NE,
consists mainly of carbonates and clastic sediments of Silurian to Carboniferous age (Flügel
1975). Devonian rocks, mostly carbonates, are
the most widespread. Para-metamorphic base-
ment rocks form the fringe and also outcrop in
the southeast of the area. In the west, the
'Palaeozoic of Graz' is covered by Cretaceous
molasse sediments. To the south, the pre-
Tertiary formations dip below Neogene marine
and terrestrial sediments. In the east, the small

From: HART, M. B. (ed.) *Climates: Past and Present.* Geological Society, London, Special
Publications, **181**, 199–213, 1-86239-075-4/$15.00

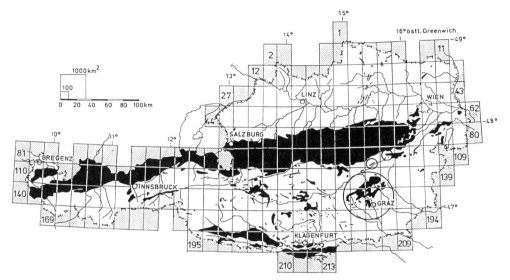

Fig. 1. Karstified (carbonate) rocks in Austria (after Stummer, in Trimmel (1978), modified). Black areas indicate karstified rocks; large circle; Central Styrian karst; small circles; locations of the Grosse Ofenbergerhöhle and Knochenhöhle near Kapellen in the Mürzalpen. Tetragonal grid encloses serial parts of the official Austrian federal map (Österreichische Karte). Pleistocene vertebrate taxa evidence is restricted to cave sediments in the shaded regions, to loess deposits in the northeast (east of c.13°, north of c.48°), and to isolated finds in river sands and gravels.

Basin of Semriach and the larger Basin of Passail contain Upper Tertiary sediments.

The Grazer Bergland is a hilly and mountainous region that is extensively covered by deciduous forests and, as a modern modification, by coniferous plantations. The mountains reach an altitude of 1720 m. The valleys are generally narrow and steep sided. The southwards orientated breakthrough valley of the main River Mur is a Late Tertiary feature (Maurin 1994). The tops of the hills, as well as the higher cave-systems in the carbonate rocks, occur at around 500–1100 m. They correspond to Late Tertiary–Early Pleistocene river terraces and karstification levels. The karstification process started in the Late Miocene and is still active today. The region was never connected to the Pleistocene Alpine glacier systems (Fig. 2). During the last glacial cycle a fluvial accumulation of up to more than 30 m of gravels and sands covered the Eemian Mur valley floor (Maurin 1994).

All caves studied are located in the submontane vegetational zone that is dominated by deciduous forests (c. 4200–1100 m asl). As a result of the tectonic, geomorphological and hydrological development of the area most of the cave entrances are situated on steep slopes or cliffs. The nine caves described here are close together at altitudes of 416–525 m (Fig. 1). Four

exceptions are also considered: the Drachenhöhle near Mixnitz (949 m) is situated 12 km north; the Knochenhöhle near Kapellen (860 m) and the Grosse Ofenbergerhöhle (766 m) are located outside the Central Styrian karst area, 55 and 35 km NNW close to the Mürz valley; and finally the Luegloch near Köflach (550 m), 25 km to the SW on the western margin of the Central Styrian karst, provide further evidence of vertebrate faunas of the Würmian Late Glacial *s.l.* (Fig. 2).

The present climatic setting

In the Central European context, the Grazer Bergland is climatically protected against Atlantic marine influences from the southwest and north. It is more open to continental influences from the northeast and south. Thus, wind activity in lower and mid-altitudes is less significant than under the pronounced Atlantic regime of the western provinces. Temperature inversions, in which a layer of warm air traps cooler air near the land surface and prevents the normal gradient, are frequent. The annual mean precipitation is around 900–1000 mm. Lowest rainfall values are in January and the maximum in early summer indicates more continentality than western Austria or its central regions (Wakonigg 1978). Mean temperatures in Janu-

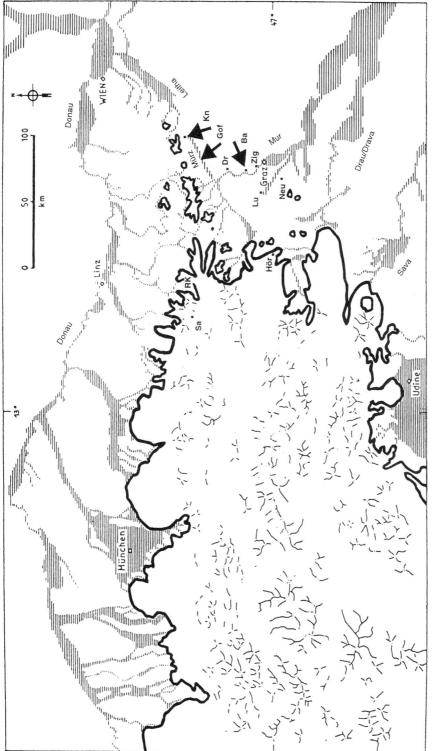

Fig. 2. Late Pleistocene sites in the southeastern periphery of the Alps (incomplete). Extension of ice streams, nunataks, major local glaciers in the east and river terraces (mainly 'Niederterrasse') during the last Glacial maximum (LGM, Würmian peak Glacial, Pleniglacial; after Van Husen (1997), modified). Arrows point to the palaeontological cave sites Knochenhöhle, Grosse Ofenbergerhöhle, Grosse Badlhöhle. Ba. Grosse Badlhöhle and caves near Peggau-Deutschfeistritz, Fröhnleiten and Semriach (Repolusthöhle, Steinbockhöhle, Grosse and Kleine Peggauerwandhöhle, Rittersaal, Tunnelhöhle, Tropfsteinhöhle am Kugelstein and Frauenhöhle); Dr, Drachenhöhle near Mixnitz; Gof, Grosse Ofenbergerhöhle; Hör, Hörfeld bog; Kn, Knochenhöhle; Lu, Luegloch near Köflach; Neu, Neurath gyttia; RK, Rameschknochenhöhle and Gamssulzenhöhle; Sa, Salzofenhöhle; Zig, Zigeunerhöhle near Gratkorn. All other caves of the Central Styrian karst cited in the text are situated within a radius of 5 km around the Grosse Badlhöhle (Ba). Horizontal shading, valley floors and flood plains.

Table 1. Middle to Late Pleistocene vertebrate taxa recorded from the Central Styrian karst area and the Mürzalpen, Austria

	M pleist	IG–IoMW	up MW	Plen glac	Kn	GOf	Ba	Late glac
Number of sites (assemblages)	1	8	8	6	1	1	1	5
Time span (absolute age) (ka BP)	250–130 approx.	130–40 approx.	40–24	24–13	14.1 OD	13.7 OD	12.4 Bö–D	14–10
Salmo trutta f. *fario* (river trout)				•			•	•
Thymallus thymallus (grayling)							•	•
Leuciscus leuciscus (dace)							•	•
Cyprinidae interminate					•			
Perca fluviatilis (river perch)							•	
Pisces indeterminate								•
Anura	?			•				•
Squamata				•				
Natrix sp. (adder)							•	•
Ophidia	?							
Anser erythropus (lesser white fronted goose)						•		•
Anas platyrhynchos (mallard)						•		•
Anas crecca A. qerquedulina (teal or garganey)						•	•	
Aegypius monachus (black vulture)			?	?				
# ***Gyps melitensis aegypioides* (fossil griffon vulture)**	•							
Aquila sp. (eagle)	•							•
Accipiter nisus (sparrowhawk)				?				?
Falco subbuteo (hobby)			•					
Falco peregrinus (peregrine falcon)				cf.				
Falco tinnunculus (kestrel)						•		•
Lagopus mutus (ptarmigan)					•	•	•	•
Lagopus lagopus (willow grouse)			•	•	•	•	•	•
Tetrastes bonasia (hazelhen)						•	•	
Lyrurus tetrix (black grouse)	•						cf.	•
Tetrao urogallus (capercaillie)			•					
Coturnix coturnix (partridge)			?	?			•	•
Gallinula chloropus (moorhen)			?	?	•	•		•
Crex crex (corncrake)					•			•
Scolopax rusticola (woodcock)						•		•
Gallinago media (great snipe)						•		•
Lymnocryptes minimus (jack snipe)						•		•
Tringa sp. (shank)						•		•
Nyctea scandiaca (snowy owl)					•	•		•
Asio flammeus (short-eared owl)						•	•	•
Strix uralensis (Ural owl)								•
Aegolius funereus (Tengmalm's owl)								•
Dendrocopus leucotos (white-backed woodpecker)								•
Eremophila alpestris (shore lark)						•		•
Hirundo rustica (swallow)						•		•
Delichon urbica (house martin)						•		•
Lanius sp. (shrike)								•
Cinclus cinclus (dipper)						•		•
Prunella collaris (Alpine accentor)						•	•	•
Acrocephalus arundinaceus (great reed warbler)						•		•
Phoenicurus ochruros (black redstart)						•		•
Oenanthe oenanthe (wheater)						•		•
Monticola saxatilis (rock thrush)								•
Parus caerulaeus (blue tit)								•
Emberiza sp. (bunting)						•		•
Montifringilla nivalis (snow finch)			•					•
Garrulus glandarius (jay)								•
Nucifraga caryocatactes (nutcracker)								•
Pica pica (magpie)								•
Pyrrhocorax graculus (Alpine chough)				•	•	•		•
Pyrrhocorax pyrrhocorax (chough)			•	•		•		•
Coloedus monedula (jackdaw)								•
Corvus corax (raven)			•					
Talpa europaea (incl. f. *magna*) (European mole)		•	•	•			•	•
# ***Talpa minor* (lesser mole)**	?							
Sorex araneus (common shrew)	?		•	•		•	•	•
Sorex alpinus (alpine shrew)						•		•
Sorex minutus (pigmy shrew)	?						•	•
Sorex minutissimus (least shrew)							•	•
# ***Sorex hundsheimensis* (undsheim shrew)**	?							
Neomys fodiens (watershrew)							•	•
Erinaceus euroaeus (incl. *Erinaceus* sp.) (hedgehog)			•					•
Myotis myotis (large mouse-eared bat)						•		•
Myotis bechsteini (Bechstein's bat)	?		?					•
Myotis nattereri (incl. aff. *nattereri*) (Natterer's bat)			?			•		•
Vespertilio sp. (incl. cf. *V. murinus*, parti-coloured bat)			•		•		•	•
Eptesicus serotinus (serotine)						•	•	•
Eptesicus nilssoni (northern bat)			?					

Table 1. (continued)

	M pleist	IG–IoMW	up MW	Plen glac	Kn	GOf	Ba	Late glac
Barbastella barbastellus (barbastelle)	?					•		•
# *Barbastella barbastellus schadleri* (Schadler's bat)			?					?
Plecotus auritus (incl. *Plecotus* sp.) (longeared bat)				•		•		•
# *Plecotus auritus abeili* (Abel's bat)			?					?
Marmota marmota (Alpine marmot)	•	•	•	•	•	•		•
Citellus citellus (European souslik) (incl. *Citellus* sp.)	cf.	•		•				•
Castor fiber (beaver)	•	•						
Glis glis (fat dormouse)	?					•		•
Microspalax leucodon (lesser mole rat)	?		?	•				
# *Cricetulus bursae* (Schaub's dwarf hamster)	?							
Cricetus cricetus (common hamster)	•						•	•
# *Cricetus cricetus major* (giant hamster)	•	•	•	•				
Apodemus sylvaticus (woodmouse) (incl. *Apodemus* sp.)	?						•	•
Apodemus flavicollis (yellownecked fieldmouse)	?							
Clethrionomys glareolus (bank vole)		•	•	•				
Arvicola terrestris (ground vole)	•	•	•	•		•	•	•
Microtus arvalis/M. agrestis (common vole or field vole)	•	•	•	•	•	•	•	
Microtus gregalis (gregarious vole, narrow-skulled vole)	•				cf.	cf.	cf.	
Microtus nivalis (snow vole)		•	•	•	•	•	•	•
Microtus oeconomus (northern vole, root vole)		cf.				cf.	cf.	cf.
Microtus (Terricola) sp. or *Pitymys* sp. (pine vole)				•				cf.
Microtus (Terricola) multiplex (Alpine pine vole)						cf.	cf.	cf.
Dicrostonyx torquatus f. gulielmi (Arctic lemming)					•			•
Sicista betulina (northern birch mouse)				•				•
Hystrix vinogradovi (small porcupine)	cf.	cf.						
Ochotona pusilla (steppe pika)	?		•	•		•	•	•
Lepus timidus (blue hare) (incl. *Lepus* sp.)	•	•	•	•	•	••	•	•
Lepus europaeus (brown hare)					•			•
# *Canis mosbachensis* (Mosbach wolf)	•							
Canis lupus (wolf)		•	•	•	•	•	•	•
Cuon alpinus (incl. *Cuon* aff. *alpinus*) (dhole, reddog)	•		?	?		?		
Vulpes vulpes (red fox)	•	•	•	•	•	•	•	•
Alopex lagopus (Arctic fox)				•	•	•	•	•
Mustela erminea (stoat)					•	•	•	•
Mustela nivalis (weasel)	•			•	•	•	•	•
Putorius sp. (polecat)	•							
Martes martes (pine marten) (incl. *Martes* sp.)	•		•	•				•
Meles meles (badger)	•							•
# *Gulo schlosseri* (Schlosser's wolverine)	cf.							
Gulo gulo (wolverine)			•	•				?
Ursus arctos (incl. # f. *priscus*) (brown bear)	•		•	•				•
# *Ursus deningeri* (Deninger's bear)	•							
# *Ursus spelaeus* (cave bear)		•	•	•				
# *Panthera spelaea* (cave lion)	•	•	•	•				
# *Crocuta spelaea* (cave hyena)	•	•	•					
Panthera pardus (leopard)		•						
Lynx lynx (northern lynx)	•							cf.
Felis silvestris (wild cat)	•							
Rangifer tarandus (reindeer)	•		•	•	•			•
Alces alces (moose)								cf
Cervus elaphus (red deer)	•	•	•	•				•
# *Megaloceros giganteus* (giant deer)	•	•	?	•				•
Capreolus carpreolus (roe deer)	•	•	?	?				
Sus scrofa (wild boar)	•	•						•
Capra ibex (ibex)	•		•	•		•		•
Rupicapra rupicapra (chamois)	•		?	cf.				•
# *Bison schoetensacki* (Schoetensack's bison)	cf.							
# *Bison priscus* (incl. *priscus* vel *bonasus*) (steppe bison)	•	•	?	•			•	
Bos primigenius or bison sp. (aurochs or bison)								•
# *Equus mosbachensis* (Mosbach horse)	cf.							
Equus sp. (horse)		•	•	•				•
# *Coelodonta antiquitatis* (woolly rhinoceros)			•	•				
# *Elephantide*	•	•						
# *Mammuthus primigenius* (woolly mammoth)				•				
Macaca sylvanus (European macaque)		•						

Compiled by the author after Mottl (1975), Bocheński & Tomek (1994), Mlíkovský (unpublished data), Reiner (1995), Rabeder & Temmel (1997) and the author's own data.

English names of extinct forms are generally after Kurtén (1968). Bias of the evidence by taphonomical differences caused by site formation processes, by age and by excavation dates should be noted (see text). Ba, Grosse Badlhöhle (lower entrance); Bö–D, Bölling–Older Dryas; Gof, Große Ofenbergerhöhle; IG, Riss–Würm–Interglacial; Kn, Knochenhöhle near Kapellen; Lateglac, Würmian Late Glacial in general; loMW, lower Middle Würmian; Mpleist, Middle Pleistocene; OD, Oldest Dryas; Plenglac, Würmian Upper Pleniglacial; upMW, upper Middle Würmian; #, extinct species or form; •, evidence; ?, uncertain stratigraphical position. Extinct species or forms, and species absent in the regional Holocene communities, are shown in bold type.

ary are *c*. –2 to –3°C and in July *c*. 16–18°C. Daily temperature ranges at low altitudes may exceed 20°C. About 55% of summer days and 35% of winter days have sunshine. In winter, the valley floors are generally cold and foggy. The small basins in the highland and the floors of the central Mur valley (e.g. the Basin of Peggau-Deutschfeistritz) correspond climatically to the Basin of Graz, more than 20 km to the south (Paschinger 1974).

Evidence of vertebrate taxa and palaeoclimatic significance

The actual list of vertebrate taxa from Middle to Late Pleistocene cave sediments in the Central Styrian karst area and the Mürzalpen comprises at least 47 species of birds and at least 78 species of mammals (Table 1). The quality of the information is strongly influenced by the current status of the investigations. Only one well-defined Middle Pleistocene faunal complex could be investigated (Repolusthöhle, excavated by Mottl in 1950–54; later finds are listed by Rabeder & Temmel (1997)). Eight assemblages are considered to be Riss–Würm-Interglacial (Eem) to lower Middle Würmian (*c*. 130–*c*. 40 ka BP) (Drachenhöhle, Repolusthöhle, Grosse Badlhöhle, Lurgrotte-Peggau, Kleine Peggauerwandhöhle, Rittersaal, Tunnelhöhle and Tropfsteinhöhle am Kugelstein), eight to be upper Middle Würmian (*c*. 40–24 ka BP) (Drachenhöhle, Frauenhöhle, Grosse Badlhöhle, Lurgrotte-Peggau, Grosse Peggauerwandhöhle, Tunnelhöhle and Tropfsteinhöhle am Kugelstein, Luegloch), six to be Pleniglacial (*c*. 24–14 ka BP) (Grosse Peggauerwandhöhle, Steinbockhöhle, Lurgrotte-Semriach, Tunnelhöhle and Tropfsteinhöhle am Kugelstein, Luegloch), and five to be Late Glacial *sensu lato* (*c*. 14–10 ka BP) (Knochenhöhle bei Kapellen, Grosse Ofenbergerhöhle, Drachenhöhle, Grosse Badlhöhle, Luegloch). The taxa are critically compiled after Mottl (1975), Bocheński & Tomek (1994), J. Mlíkovský, (unpublished data), Reiner (1995), Rabeder & Temmel (1997), and the author's own determinations.

Biological and archaeological biases

Taphocoenoses from cave sediments are regarded as sums of 'selected' samples of the local fauna (e.g. Andrews 1990). Some species need, or occasionally use, caves in their habitat (cave bears, wolves, foxes, martens, bats, owls, etc.). This increases the chance that their remains will become fossilized. Other species may not use

caves themselves (e.g. hares, deer), but their predators do. Some species may, or may not, be imported into the caves because their habitat is several kilometres away but within the range of avian predators. Absence of taxa in the site may reflect an original absence in the past environment. The record, therefore, represents random sampling over the region. Taphonomic patterns on the Late Glacial micro-mammalian remains from the Grosse Badlhöhle, such as the frequency and the intensity of molar corrosion, suggest Canidae, Mustelidae and owls, too, as predators (Reiner & Bichler 1996). The accumulation of bones in the Knochenhöhle was mainly produced by snowy owls, whereas even short-eared owls and kestrels played an important role in that of the Grosse Ofenbergerhöhle (Fladerer & Reiner 1996).

The site formation of cave sediments is highly complex (e.g. Kukla & Ložek 1958; Campy 1990). In this process mechanical, chemical and biological aspects are involved. The sediment profiles of the studied caves consist largely of gaps (Fuchs 1989; Fladerer 1994; Fig. 3) and, as a result, the palaeoclimatic record is fragmentary. All 22 radiocarbon dates from the studied karst regions were obtained after 1989. With two exceptions (charcoal samples from the early Upper Palaeolithic fireplace in the Drachenhöhle (Fladerer 1997*a*) and from a deep layer in the Rittersaal (Fig. 3)) bone material was used. The accuracy of the age determinations and the resolution rapidly decrease with the absolute age (see Table 1).

Palaeoecological implications

A few species or forms are first-rank climate indicators (e.g. Thenius 1976; Kowalski 1995; Rzebik-Kowalska 1995). The willow grouse *Lagopus lagopus*, the arctic lemming *Dicrostonyx torquatus*, the least shrew *Sorex minutissimus*, the wolverine *Gulo gulo*, the arctic fox *Alopex lagopus*, the reindeer *Rangifer tarandus* and the large form of the blue hare *Lepus timidus* are boreal species that now range from the northernmost temperate to the arctic climate in tundra, forest tundra or taiga habitats. They are absent in Holocene Central European natural communities. The phylogenetic affinities between the boreal *Lepus t. timidus*, the alpine *Lepus t. varronis* and the regional Pleistocene hare are not yet established. Cranial as well as postcranial features point to a closer relationship between the fossil and the Scandinavian form (e.g. Koby 1960). As a result of preliminary functional analyses some forelimb pecularities of the Alpine hare are interpreted as apomorphies

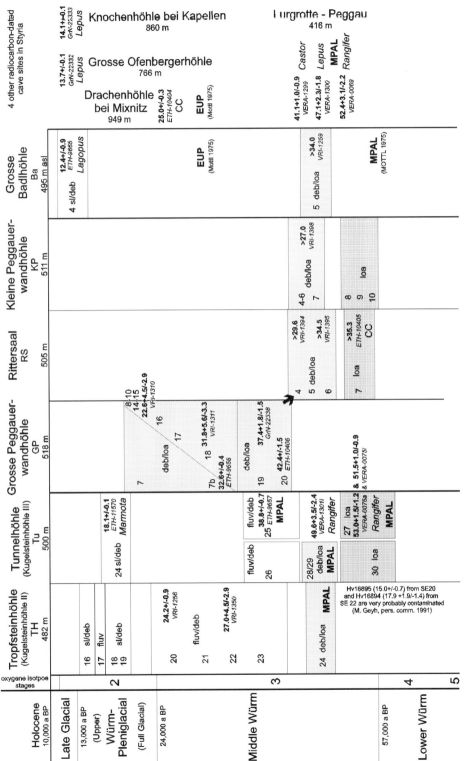

Fig. 3. Chronology of cave sediments in the Central Styrian karst area; excavations 1984-1999. Numbers indicate stratigraphic units as site entities. Radiocarbon dates in 1000 years, with sample number of laboratory. All data derive from cave bear remains (*Ursus spelaeus*), unless indicated otherwise (CC, genus name). The fillings indicate the generalized type of sediment: deb, debris, mainly autochthonous limestone; fluv, fluvial influenced limestone; sl, fluvial influenced sands and silts; loa, loam or hardened clay; sl, silty deposits, probably aeolian influenced. CC, charcoal; EUP, early Upper Palaeolithic; MPAL, Middle Palaeolithic. The small bold arrow indicates the connection of the two caves.

(Fladerer 1984, p.146). Not yet recorded from any Styrian site are the Norway lemming *Lemmus lemmus*, the steppe lemming *Lagurus lagurus*, the jerboa *Allactaga jaculus* and the *Saiga* antelope, as they have rarely been recorded from lowland sites in Lower Austria in the NE periphery of the Alps (Döppes & Rabeder 1997). The same is true for the musk ox *Ovibos*, but this is also reported from the Potočka Cave in northern Slovenia, 90 km to the SW of the Central Styrian caves, as an unstratified and somewhat doubtful find (Rakovec 1938).

Three bird species from Late Glacial taphocoenoses in Styria are worthy of note: the lesser white-fronted goose *Anser erythropus*, the snowy owl *Nyctea scandiaca* and the shore lark *Eremophila alpestris* breed in distinctly more northern biotopes (Peterson *et al.* 1985) but today they are very rare winter visitors in Styria (Höpflinger & Schliefsteiner 1981).

A second group of high habitat indicative species today inhabit Alpine heath meadows and rocky grounds at altitudes above the timber line: the ptarmigan *Lagopus mutus*, both chough *(Pyrrhocorax)* species, and the Alpine marmot *Marmota marmota*. The Styrian foothills lie outside their extant range. This only partly holds true for the snow vole *Microtus nivalis*, which is fairly common in Middle to Late Würmian strata. Today it cannot be fully excluded as a very rare local immigrant or prey species in the observed caves. The present-day distribution of the northern birch mouse *Sicista betulina* also has its core area at a higher altitude than the caves (Hable & Spitzenberger 1989).

Indicative species of a third group form a part of extant east Central European to Central Asian steppe communities. The animals are adapted to an arid climate with an annual precipitation of *c.* 200–500 mm and strong seasonality: the mole rat *Microspalax leucodon*, the steppe pika *Ochotona pusilla* and the narrow-skulled vole *Microtus gregalis* (e.g. Ognev 1966). It has to be noted that the very low percentage of the *gregalis*-morphotype in the *Microtus* samples does not prove the presence of the species *M. gregalis* (Reiner 1995). The northern vole, *Microtus oeconomus*, is a reed and marsh dweller within a core area between Northern and Eastern Europe and Central Asia. Isolated relict populations are still found in Central Europe (Tast 1982), the closest at Lake Neusiedl, which is 75 km E from the Mürz Alps. The European souslik, *Citellus citellus*, inhabits some open regions in northern and eastern Austria (Leitner 1988) but typically lives in eastern to south-eastern European grasslands (Ružić 1978).

All fossil taphocoenoses contain a certain percentage of species which have their core area within temperate woodlands: the capercaillie *Tetrao urogallus*, the woodcock *Scolopax rusticola*, the jay *Garrulus glandarius*, the common shrew *Sorex araneus*, the fat dormouse *Glis glis*, the bank vole *Clethrionomys glareolus*, the badger *Meles meles*, the pine marten *Martes martes*, *Lynx lynx,* and the wild boar *Sus scrofa*. A few of the abundant species of the extant fauna, such as the squirrel, are nearly absent in the local Late Pleistocene record. Faunal remains from humid phases or oscillations with a stronger Atlantic influence either are hardly recorded as a result of bone destruction in the more humic sediments in the proximal cave areas or form part of the palimpsest type of sediments (compare Kukla & Ložek 1958; Campy 1990).

Intra-site changes and diversity, and regional heterogeneity

Within the taxonomically studied remains from the Styrian foothill sites, *Gyps melitensis aegypiodes* (Jánossy 1989), *Canis mosbachensis* (Mottl 1975) and *Ursus deningeri* (Rabeder & Temmel 1997) are restricted to the Middle Pleistocene. Differences between Middle and Late Pleistocene taxa partly reflect speciation (*Ursus deningeri* into *U. spelaeus* (Fladerer in Fuchs 1999), *Canis mosbachensis* into *C. lupus* (Mottl, 1975)). Taphocoenosis spectra from Middle Würmian strata reveal differences from the preceeding time interval. The 'Mediterranean' temperate cercopithicids and porcupines are absent (see Von Koenigswald 1991). Boreal and montane species (polar fox, reindeer, ptarmigan, etc.) became more frequent. Nevertheless, species that today live in temperate deciduous forests (capercaillie, common shrew, bank vole and pine marten) are distinctly present (Table 1). At the infraspecific level within the Late Pleistocene (< 100 ka), the shift of skeletal morphological features becomes more obvious. In the cave bear, *U. spelaeus*, the complexity of the premolar and molar cusp pattern increases within a time interval of around 10 000–15 000 years. The evaluation of such phylogenetic shifts provides chronological information (Rabeder 1989; Rabeder in Döppes & Rabeder 1997).

Pleniglacial communities within the periphery of the Alps are dominated by boreal and alpine species. The cave bears' area diminished from its Middle Würmian alpine range to the foothills until its extinction in the upper Pleniglacial. Late Glacial *sensu stricto* (13–10 ka BP) communities

Table 2. Ratios of *Microtus*-morphotypes from three Late Glacial *s.l.* taphocoenoses from Styria, southeastern Austria

Site (altitude)	^{14}C (AMS) BP	arv/agr	greg	niv	oec	GLAC
Grosse Badlhöhle (495 m)	12.4±0.1 (ETH-9655)	60.9	1.0	37.1	1.0	interstadial/ Daun stadial phase
Grosse Ofenbergerhöhle (766 m)	13.7±0.1 (GrN-22332)	29.6	0.2	70.0	0.2	Gschnitz stadial phase
Knochenhöhle (860 m)	14.1±0.1 (GrN-22333)	35.0	–	65.0	–	Gschnitz stadial phase

arv–agr, *arvalis–agrestis*; greg, *gregalis*; niv, *nivalis*; oec, *oeconomus*; GLAC, suggested correlation with the glacier development in the Eastern Alps.

Table 3. Total number of bird and mammal taxa and portion of exclusive Pleistocene taxa within three Late Glacial *s.l.* taphocoenoses from the southeastern periphery of the Eastern Alps, and similarity between the three samples

Site	Taxa (*n*)	notHol (%)	Dice		
			Kn	Gof	Ba
Grosse Badlhöhle (Ba)	26	8 (31)	0.36	0.43	–
Grosse Ofenbergerhöhle (Gof)	47	13 (28)	0.44	–	0.43
Knochenhöhle (Kn)	21	12 (57)	–	0.44	0.36

n, number of taxa (mainly species ranked); notHol, taxa that are not present in the regional Holocene communities (species extinct in historical time, such as wolf, are counted as present in the Holocene period); Dice, *Dice* similarity coefficient.

lack cave bears. Presently, no finds of lion, hyena, giant deer and rhino have been identified. It is almost certain that these species had already become extinct in the foothills. Geographically adjacent analyses from nine sites in Slovenia record cave bears in two sites, giant deer in three sites and mammoth in one site up to *c.* 15 ka BP, but – as in Styria – no evidence of lion, hyena and rhino is reported from there after 20 ka BP (Pohar 1994, 1997).

The percentage of distinct tooth-morphotypes within the Microtinae of one taphocoenosis shows peculiarities that strongly correlate with the frequency of other climate indicators (Table 2). This is caused by the different ecological needs of the arvicolid species which reflect plant cover and humidity (e.g. Guthrie 1971; Nadachovski 1991). The low values of the *gregalis*- and the *oeconomus*-morphotype provides little evidence of *M. gregalis* or *M. oeconomus* (Table 3). The *gregalis*-morphotype can be observed in *M. arvalis* as well as in *M. agrestis*. The same holds true for the *oeconomus*-type in *M. nivalis* (see discussion in Reiner 1995). It is possible that the higher portion of the snow vole in the taphocoenoses from the Knochenhöhle and the Grosse Ofenbergerhöhle versus the Grosse

Badlhöhle is caused by the dominance of the snowy owl as predator in the first-mentioned cave. All three vole species (*M. arvalis, M. agrestis* and *M. nivalis*) prefer open environments but only the last is missing in wooded grounds. Other data support the picture of local and/or more widespread and/or more diversified woody habitats around the Grosse Badlhöhle locality (see below).

In the uppermost Pleniglacial–Late Glacial *sensu lato* (?Oldest Dryas) taphocoenosis from the Knochenhöhle near Kapellen (GrN-22333: 14 070 ± 100 BP), 57% of the taxa (*n* = 21) that are mainly species-ranked do not occur in the Holocene fauna (Table 2). In the Grosse Ofenbergerhöhle (GrN-22332: 13 690 ± 100 BP) (*n* = 47) the 'not-Holocene' part is *c.* 28%. It is approximately 30%, also, in the taphocoenosis from the Grosse Badlhöhle (*n* = 26), the AMS-date of which corresponds to the Bölling–Dryas II complex (ETH-9655: 12 430 ± 95 BP) (compare Table 3). The most striking feature of the assemblages is the permanence of the 'Pleistocene' species and the diversity within this group, rather than the appearance of 'Holocene' species (Table 2). Subarctic animals of today were sympatric with alpine animals as well as steppe

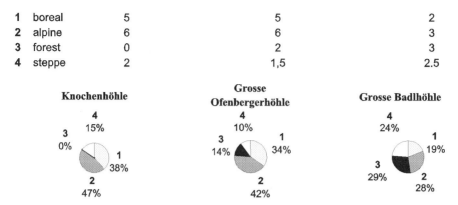

1	boreal	5	5	2
2	alpine	6	6	3
3	forest	0	2	3
4	steppe	2	1,5	2.5

Fig. 4. Percentage of well defined habitat-indicative bird and mammalian species in three Late Glacial taphocoenoses in the southeastern periphery of the Eastern Alps. From N (left) to S (right).
Note that the 0.5 value in the Grosse Badlhöhle and the Grosse Ofenbergerhöhle assemblages results from the gregalis-morphotype in the *Microtus* samples. The presence of the morphotype does not necessarily prove the presence of the species *M. gregalis. Lepus timidus* is calculated as an Alpine species.

residents in the eastern periphery of the Alps. Differences in the species spectra between both the northern caves, Knochenhöhle and the Ofenbergerhöhle, and the southern Grosse Badlhöhle, might be caused by the geographical distance of 30–50 km. In calculating the percentages, only well-defined habitat-indicative species were used (Fig. 4). In the fossil assemblage of the Knochenhöhle, which is situated within the extant sub-montane vegetational zone, no forest species could be observed. The percentage is generally very low in the other two taphocoenoses, supported by only the woodcock, the woodmouse and the bank vole. All three also live in boreal forests. The boreal portion (1) diminishes slightly from 38–34% towards the Grosse Badlhöhle (19%). If the mountain hare is counted as Arctic hare this would increase the values (see above). The Alpine portion (2), including *Lepus timidus*, diminishes from 47–42% in the north to 28% in the south. All counted Alpine species are absent in the local Holocene fauna (Table 1, Fig. 4). The faunal diversity in the periglacial foothills is greater in the Late Glacial *s. l.* than in the Holocene. It is concluded that the vegetational biota were more diversified, corresponding to the lower precipitation, the higher climatic seasonality and their impact on the relief of the landscape.

For evaluation of the similarity the *Dice* coefficient was used, as suggested by Archer & Maples (1987) and Maples & Archer (1988). It is calculated as $2A/(2A + B + C)$, where A is the taxa present in both faunas, B is the taxa present in fauna 1 but absent in fauna 2 and C is the taxa absent in fauna 1 but present in fauna 2 (Table 3). Although the samples differ in taxon richness the closer similarity (0.44) between the northern caves, Knochenhöhle and the Ofenbergerhöhle, which are 32 km apart, is evident. Within the comparison of the three sites, the ones most dissimilar (0.36) are the Knochenhöhle and the Grosse Badlhöhle, which are 48 km apart. This may mean a lower faunal similarity across the region and a much higher degree of heterogeneity within the region than is the case in the humid Holocene. Taking the difference of the radiocarbon dates as a proof (Table 2), the lower similiarity of the northern assemblages and the southern was caused by an increase in precipitation towards the Late Glacial interstadial (see below).

Contextual malacology

Pleniglacial molluscan taphocoenoses from silty, probably aeolian-influenced, sediments in the Tunnelhöhle near Peggau-Deutschfeistritz (around 18 ka BP; Fig. 3) contain *Vallonia tenuilabris, Succinella oblonga elongata, Chilostoma achates* and *Trichia hispida* (Fladerer & Frank 1997*b*). *Vallonia tenuilabris*, a north Central Asian species today, is a typical species of the mammoth steppe. The assemblage reflects the peak Glacial climate with similiar habitats for the Central Styrian karst area as well as in the lowland loess region of Lower Austria (150 km apart). Due to the rocky, and sometimes steep environments, a variety of petricole species is present. Vertebrate, molluscan and vegetational analyses produce Pleniglacial proxy data that suggest, for the periphery of the Eastern Alps, an annual precipitation of around 350–400 mm. Winter mean temperatures are set around –20 to –25°C although summer mean temperatures might have risen to $c. +15°C$.

In the uppermost Pleniglacial–Late Glacial vertebrate and pollen associations from the Knochenhöhle, near Kapellen (14 ka BP), the terrestrial snails indicate open (*Chilostoma achates*) as well as half-open to shady habitats (*Arianta arbustorum*) under a slightly less dry and cold climate (Fladerer & Frank 1997*a*).

The most striking feature of the Late Glacial fauna of the Grosse Badlhöhle (12.5 ka BP, ? Interstadial–Dryas II), which contains a high percentage of taxa (*Pupilla muscorum, Succinella oblonga, Trichia hispida*), reflecting open, but cool habitats, are aquatic molluscs including a *Congeria* species (Frank 1993). The precipitation rate had probably increased.

Palaeovegetation

Palynological analyses from the loamy Middle Würmian sediments in the cliff caves of Peggau (Grosse Peggauerwandhöhle, *c.* 518 m asl, radiocarbon dates *c.* 42–31 ka BP, Fig. 2: Ba) are in progress. The pollen was brought into the caves on the fur and in the dung of the cave bears (Draxler *et al.* 1986; Draxler 1995). It mainly reflects part of the forage plant spectrum of the animals and, hence, the samples are proved to be selective. The diverse herbaceous and perennial flora from the Grosse Peggauerwandhöhle mainly comprises a large proportion of Compositae and Dipsacaceae but only a few species of trees such as *Pinus, Betula, Alnus* and rarely *Tilia* (Draxler, pers. comm.). The communities strongly resemble those of extant subalpine heaths and meadows (1500–2000 m). The cave bear dens Ramesch-Knochenhöhle and Salzofenhöhle, near the Alpine crest at altitudes of 2000 m, *c.* 110 km apart (Fig. 2), yield 11 radiocarbon and uranium-series dates from cave bear bones of 64–31 ka BP. The context is interpreted by Rabeder as caused by glacial conditions (Draxler *et al.* 1986; Rabeder 1995). However these sites and the foothill sites (500–600 m), yield similiar plant spectra with abundant Compositae, together with a substantial amount of conifer material as well as a percentage of thermophilous trees (*Tilia*). This suggests a synchronous habitat area of the *Ursus spelaeus*

ranging from > 2000 m to < 500 m, and vegetational zonations and plant assemblage distributions that are quite different from the present one. Climate oscillations and plant cover shifts might have influenced the vertical range of the cave bears' area within *c.* 60–30 ka BP by several hundreds of years to around 1000 years, but this is beyond the limits of current dating accuracy. Interstadial conditions are evident but precise correlations with the Hengelo and the Denekamp–Les Cottés Interstadials of Northern and Western Europe (e.g. Weissmüller 1997) are not yet possible.

From the peak Pleniglacial (20–21 ka BP) a plant association was extracted from a gyttja horizon within terrace sediments 16 km SSW of Peggau-Deutschfeistritz in the West Styrian foothills, at a height of *c.* 380 m (Draxler & van Husen 1989; van Husen 1997) (Fig. 2). It consists of a diversified herb pollen flora with *Helianthemum* and dominant grasses and sedges. No thermophilous trees are recorded. This indicates tundra-like biotopes under a cold continental climate in the foreland region of the Alps at the peak of the last Glacial.

A preliminary study of the later, uppermost Pleniglacial (14 ka BP) pollen flora from the Knochenhöhle near Kapellen (860 m) shows that at this time there were close analogies to the extant Alpine vegetational zone at 2000–2400 m. The spectrum with Cichoriaceae, Asteraceae and Caryophyllaceae dominating, and with Ericaceae and *Helianthemum* sp., reflects a sparse and hardly closed herbaceous cover. Trees are represented by rare pollen from *Pinus* cf. *cembra* and *P. sylvestris* vel *mugo* (Draxler, pers. comm.). The spectrum partly matches the forage plants of the Arctic or mountain hare, whose remains dominate the vertebrate assemblage.

No fossil pollen could be extracted from the Late Glacial layer of the Grosse Badlhöhle (495 m) which records a radiocarbon date of 12.4 ka BP, measured on *Lagopus* bone (Table 2). Fragments of wood (*Pinus* sp. ?) from a coring in the Hörfeld bog (*c.* 920 m asl) 68 km west of the Grosse Badlhöhle (Fig. 2) yielded a date of 12 100 ± 400 BP (VRI-1408) (Posch 1995). The pollen analyses appear to indicate a taiga-like herbaceous and gramineous pine forest with low-density patches (Cichoriaceae, Chenopodiaceae). *Artemisia* is very rare. This, and a further eight dates from the Eastern Alps of 13.2–12.0 ka BP (van Husen 1997), suggest near interstadial conditions for that period.

Glaciology and palaeoenvironments

The last Pleniglacial Alpine ice-shield, which

stretched from Western Europe over the Eastern Alps, ended abruptly near 15° E, as shown by the end-moraines of the Enns, Mur and Drau glaciers (Fig. 2). Single mountain glaciers in the east intercede with the loess regions of the Middle Danube lowland in the north and the Styrian foothills and lowlands in the east and south. Apparently, the ice decayed rapidly after the 'Maximalstand' at c. 21 ka BP and the more prolonged 'Hochstand' until c. 17 ka BP. A distinct glacial re-advance and standstill ('Gschnitz-phase') probably corresponds to the Oldest Dryas of NW Europe (c. 16–14 ka BP) (van Husen 1997). Vertebrate open-air sites from Lower Austria yield equivalent 'continental' assemblages from these periods, but all micro-vertebrate taphocoenoses from cave sites within eastern Austria are dated to the Late Glacial s.l. (compare Döppes & Rabeder (1997) for the northern foothills). This raises the question as to whether the absence of a peak Glacial micro-vertebrate assemblage is the result of research deficiency.

In the periphery of the Alps, the deglaciation is accompanied by an *Artemisia*-pioneer phase. This is followed by a rapid increase of *Pinus*. There is no evidence for permafrost conditions on the floors of the main valleys within the Bölling at c. 13 ka BP (van Husen 1997). The following two glacier re-advances, 'Daun' and 'Egesen', again reflect drier and more continen-tal conditions. A precipitation c. 30% less than today and a lowering of the mean annual temperature by 2.5–4°C are suggested. The diversity of the faunal assemblages from the Grosse Badlhöhle reflect a heterogeneous envir-onment is probably located within these fluctua-tions (Table 2, Fig. 4).

Human evidence

Two thousand artefacts from the Middle Pleis-tocene sediments of the Repolusthöhle near Frohnleiten, north of Peggau, are typologically described as belonging to Middle Palaeolithic traditions (Mottl 1951; Fuchs & Ringer 1996). Most of the faunal remains, but few lithic artefacts, derive from the shaft in the distal part of the Repolusthöhle. They represent the oldest (c. 200–220 ka BP) proven lithic industry in the Eastern Alps (Mottl 1975; Fuchs et al. 1999). Faunal elements that have their core area within the forests of the temperate mid-latitudes include the beaver *Castor fiber,* a porcupine *Hystrix* sp., the badger *Meles meles,* the wild cat *Felis silvestris,* the roe deer *Capreolus capreolus* and the wild boar *Sus scrofa.* The broad mammalian spectrum of the Repolusthöhle

exhibits an interglacial character. This is under-lined by charcoal fragments of *Quercus, Fagus, Salix* and *Picea.* The bone pit of the distal part of the cave indicates a wide time and ecological range (*Cricetus cricetus, Marmota marmota, Rangifer tarandus* on the boreo-Alpine side). A somewhat more advanced technology, that closely resembles the Taubachien from Ganovce and Horka in Slovakia and Diósgyör-Tapolca in Hungary, is reported from upper layers in the Repolusthöhle, and as isolated finds from the Tunnelhöhle. The Taubachian is often corre-lated with the Late Riss to the Riss–Würm–Interglacial (Fuchs & Ringer 1996).

Isolated lithic artefacts from the Tunnelhöhle (Fuchs & Ringer 1996), the Tropfsteinhöhle am Kugelstein (Jéquier 1975) and from the Grosse Badlhöhle and the Lurgrotte-Peggau provide evidence of a local late Middle Palaeolithic (Neanderthalian) population. A culturally mod-ified reindeer upper foreleg bone from the Lurgrotte-Peggau (416 m asl), that has been dated to 52.4 + 3.1/–2.2 ka BP (VERA-0069), matches the first stratified archaezoological document of this cultural complex in Austria (Fladerer 1998). The caves are situated in a narrow gorge in the prominent N–S directed Mur valley (Fig. 2). Thus the geographical situation of the sites and their archaeological content give a first scanty insight into the hunting economy of regional Neandertals.

Upper Palaeolithic bone points derive from the Drachenhöhle and from the Grosse Badl-höhle (Mottl 1975; Fig. 3). A few unstratified, atypical, objects from a further three caves in the Middle Mur valley and from two sites in the westernmost part of the Central Styrian karst area (Fuchs 1989) provide only a blurred picture of Upper Palaeolithic hunter–gatherers' land use strategies. This reflects either former excavation methods or the very poor data density in modern excavations and prospection. Old excavations in the abris-like Zigeunerhöhle near Gratkorn (395 m), north of Graz (Fig. 2), have yielded bones and lithic artefacts of an advanced Late Palaeolithic culture. Its economy is osteologi-cally judged as having been based on fishing and red-deer hunting. This, and associated finds of fossil ptarmigan, indicates locally open conifer forest and the climate of a Late Glacial oscillation (?Younger Dryas to Preboreal).

Conclusions

Middle to latest Pleistocene mammalian and avian taxa from cave sites within a region of c. 4200 km^2 are summarized for the first time. All caves are situated in the extant sub-montane vegetational zone (400–950 m asl) close to the Middle and Late Pleistocene Alpine maximum glacial extension. The evidence of the 47 species of birds and at least 78 species of mammals, partly from three high-resolution assemblages, is strongly influenced by taphonomic differences and the actual state of research.

• Communities from the last Glacial period show, in general, a mixture of today's boreal–tundra species with those of East European to Central Asian steppe communities and, to a distinctly lesser percentage, with those of woodlands in the humid mid-latitudes. The well-known phenomenon of 'intermingled', 'disharmonious' associations of extant allopatric species during the Ice age of Europe was first noticed over130 years ago (Lartet 1867, cited by Lundelius 1989). Though palaeofaunal, intra-site ecological observations, and inter-regional conclusions have a long tradition in eastern Austria and the adjacent eastern and southern regions (Mottl 1975; Jánossy 1986; Döppes & Rabeder 1997; Pohar 1997), diversity comparisons based on AMS-dated high-resolution taphocoenoses are not available.

• Middle Würmian cave bear dens (c. 60–30 ka BP) from c. 2000 m asl (Draxler et al. 1986) to < 500 m (Fladerer 1997b), including contextual palynological data, suggest either a synchronous vertical habitat range of the specialized herbivorous carnivore over 1500 m or, more probably, climate oscillations that are below the limit of current dating accuracy. The data indicate vegetational zonations and plant assemblage distributions that are quite different from the actual, more oceanic-influenced, conditions. The sequence of vertical zones, as well as the asymmetry of vegetation according to the slope exposure, vary because of the climatic influence (Starkel 1994). Under cooler, but more humid climates, as well as under drier, cool, climates the sequence of vegetational zones differs from the actual zonation. Zonation reconstructions for the Middle Würm periods 50–40 ka and 35–30 ka in Central Europe (Starkel in Kozlowski 1994) generally agree with the case of the regional transect from Alpine sites to lowland valley floors. In that case the regional relief provides a great variety of habitats, comprising herbaceous meadows, heaths, tundra and forest–tundra vegetation in the upper altitudes, to taiga, grasslands and deciduous woodland at lower altitudes.

• Peak Pleniglacial assemblages (around 20 ka BP) from southeastern foothill sites point to a similar cold and strong seasonality generating climate conditions between the southeastern foothills (around 15°E/47°N) (Draxler & van Husen 1989; Fladerer & Frank 1997b) and the Middle Danube lowlands (around 16.5°E/ 48.5°N).

• Special attention has been paid to analyses of high-resolution uppermost Pleniglacial–Late Glacial taphocoenoses (c. 14–12 ka BP) in the Styrian foothills. They document a distinctly greater faunal diversity than in the temperate humid Holocene. They also reflect a more diversified vegetational biota, as suggested by malacology and palynology. The reconstruction of a close connection between increased influence of drier and more windy conditions (above all, cold and drier winters) and diverse plant assemblages reflecting variable growing seasons, has been described in a rather distant region by Guthrie (1982, 1990). The application of this model makes sense in understanding the Styrian foothills example. The noticable increase of woodland areas towards the south within the region corresponds with the reconstruction of the vertical zonations by Starkel (in Kozlowski 1994).

• Southeastern Austria was populated by humans for the first time not later than c. 200 ka BP (Fuchs et al. 1999). Rare artefact finds indicate a Late Middle Palaeolithic (Neanderthal) population (Fuchs & Ringer 1996; Fladerer 1998).

Caves and their sediments contain the most important regional proxy data for indirect measurements of Middle to Late Pleistocene environments and climates. The summarized results provide a basis for subsequent research and comparison studies in the adjacent regions.

The author especially thanks D. van Husen (Technische Universität Wien) and I. Draxler (Geologische Bundesanstalt, Wien) for their discussions during the writing of this paper. Special thanks go to G. Fuchs (Argis, Graz) for excellent archaeological companionship over many years. The Styrian Province Government provided financial support for the investigations in the Lurgrotte. Thanks are due to M. Hart and two anonymous referees for their comments and linguistic adjustments.

References

ABEL, O. & KYRLE, G. (eds) 1931. Die Drachenhöhle bei Mixnitz. *Speläologische Monographien*, **6–8**, (1–2).

ANDREWS, P. 1990. *Owls, caves and fossils.* University Press, Chicago.

ARCHER, A. W. & MAPLES, C. G. 1987. Monte Carlo simulation of selected binomial similarity coefficients (I): Effect of number af variables. *Palaios*, **2**, 609–617.

BOCHENSKI, Z. & TOMEK, T. 1994. Fossil and subfossil bird remains from five Austrian caves. *Acta zoologica cracoviensia*, **37**, 347–358.

CAMPY, M. 1990. L'engistrement du temps et du climat dans les remplissages karstiques: l'apport de la sédimentologie. *Karstologia, Mémoires*, **2**, 11–22.

DÖPPES, D. & RABEDER, G. (eds) 1997. Pliozäne und pleistozäne Faunen von Österreich. *Mitteilungen der Kommission für Quartärforschung der Österreichischen Akademie der Wissenschaften*, **10**.

DRAXLER, I. 1995. Palynologische Untersuchungen der jungpleistozänen Sedimente aus der Gamssulzenhöhle bei Spital a. Pyhrn (Oberösterreich). *In:* RABEDER, G. (ed.) Die Gamssulzenhöhle im Toten Gebirge. *Mitteilungen der Kommission für Quartärforschung der Österreichischen Akademie der Wissenschaften*, **9**, 37–49.

—— VAN HUSEN, D. 1989. Ein ^{14}C-datiertes Profil in der Niederterrasse bei Neurath (Stainz, Stmk.). *Zeitschrift für Gletscherkunde und Glazialgeologie*, **25**, 123–130.

——, HILLE, P., MAIS, K., RABEDER, G., STEFFAN, I. & WILD, E. 1986. Paläontologische Befunde, absolute Datierung und paläoklimatische Konsequenzen der Resultate aus der Ramesch-Knochenhöhle. *In:* HILLE, R. & RABEDER, G. (eds) Die Ramesch-Knochenhöhle im Toten Gebirge. *Mitteilungen der Kommission für Quartärforschung der Österreichischen Akademie der Wissenschaften*, **6**, 7–66.

FLADERER, F. A. 1984. Das Vordergliedmassenskelett von *Hypolagus beremendensis* und von *Lepus* sp. (Lagomorpha, Mammalia) aus dem Altpleistozän von Deutsch-Altenburg (Niederösterreich). *Beiträge zur Paläontologie von Österreich*, **11**, 71–148.

—— 1993. Neue Daten aus jung- und mittelpleistozänen Höhlensedimenten im Raum Peggau-Deutschfeistritz, Steiermark. *Fundberichte aus Österreich*, **31**, 369–374.

—— 1994. Aktuelle paläontologische und archäologische Untersuchungen in Höhlen des Mittelsteirischen Karstes, Österreich. *Český kras*, **20**, 21–32.

—— 1997a. Drachenhöhle bei Mixnitz. *In:* DÖPPES, D. & RABEDER, G. (eds) Pliozäne und pleistozäne Faunen von Österreich. *Mitteilungen der Kommission für Quartärforschung der Österreichischen Akademie der Wissenschaften*, **10**, 295–304.

—— 1997b. Grosse Peggauerwandhöhle. *In:* DÖPPES, D. & RABEDER, G. (eds) Pliozäne und pleistozäne Faunen von Österreich. *Mitteilungen der Kommission für Quartärforschung der Österreichischen Akademie der Wissenschaften*, **10**, 320–325.

—— 1998. Ein altsteinzeitliches Rentierjägerlager an der Murtalenge bei Peggau? *In:* FRITZ, I. (ed.) *Festschrift* Walter Gräf. *Mitteilungen des Referats für Geologie und Paläontologie, Landesmuseum Joanneum, Graz, Sonderheft*, **2**, 155–174.

—— & FRANK, C. 1997a. Knochenhöhle bei Kapellen. *In:* DÖPPES, D. & RABEDER, G. (eds) Pliozäne und pleistozäne Faunen von Österreich. *Mitteilungen der Kommission für Quartärforschung der Österreichischen Akademie der Wissenschaften*, **10**, 275–277.

—— & —— 1997b. Tunnelhöhle. *In:* DÖPPES, D. & RABEDER, G. (eds) Pliozäne und pleistozäne Faunen von Österreich. *Mitteilungen der Kommission für Quartärforschung der Österreichischen Aka-*

demie der Wissenschaften, **10**, 349–355.

—— & REINER, G. 1996. Hoch- und spätglaziale Wirbeltierfaunen aus vier Höhlen der Steiermark. *Mitteilungen der Abteilung für Geologie und Paläontologie am Landesmuseum Joanneum, Graz*, **54**, 43–60.

FLÜGEL, H. W. (ed.) 1975. *Die Geologie des Grazer Berglandes (2. Auflage)*. Mitteilungen der Abteilung für Geologie, Paläontologie und Bergbau am Landesmuseum Joanneum, Graz, Sonderheft, **1**, 1–288.

FRANK, C. 1993. Mollusca aus der Grossen Badlhöhle bei Peggau (Steiermark). *Die Höhle*, **44**, 6–22.

FUCHS, G. (ed.) 1989. *Höhlenfundplätze im Raum Peggau-Deutschfeistritz, Steiermark, Österreich. Tropfsteinhöhle, Kat.N. 2784/3. Grabungen 1986–87.* British Archaeological Reports, International Series, **510**.

—— & RINGER, A. 1996. Das paläolithische Fundmaterial aus der Tunnelhöhle (Kat. Nr. 2784/2) im Grazer Bergland, Steiermark. *Fundberichte aus Österreich*, **34**, 257–271.

—— FÜRNHOLZER, J. & FLADERER, F. A. 1999. Untersuchungen zur Fundschichtbildung in der Repolusthöhle, Steiermark. *Fundberichte aus Österreich*, **37**, 143–172.

GRAHAM, R. W. 1985. The response of mammalian communities to environmental changes during the late Quaternary. *In:* DIAMOND, J. & CASE, T. J. (eds) *Community Ecology.* New York, 300–313.

GUTHRIE, R. D. 1971. Factors regulating the evolution of microtine tooth complexity. *Zeitschrift für Säugetierkunde*, **36**, 37–54.

—— 1982. Mammals of the mammoth steppe as paleoenvironmental indicators. *In:* HOPKINS, D. M., MATTHEWS Jr., J. V., SCHWEGER, C. E. & YOUNG, S. B. (eds) *Paleoecology of Beringia.* New York, 307–326.

—— 1990. *Frozen Fauna of the Mammoth Steppe.* Universities Press, Chicago.

HABLE, E. & SPITZENBERGER, F. 1989. Die Birkenmaus, *Sicista betulina* PALLAS, 1779 (Mammalia, Rodentia) in Österreich. Mammalia austriaca 16. *Mitteilungen der Abteilung für Zoologie am Landesmuseum Joanneum, Graz*, **43**, 3–22.

HÖPFLINGER, F. & SCHLIEFSTEINER, H. 1981. *Naturführer Österreich: Flora und Fauna. Alle Wirbeltiere und die wichtigsten Pflanzengesellschaften der Ostalpenregion und des westpannonischen Raumes.* Styria, Graz.

JÁNOSSY, D. 1986. *Pleistocene vertebrate faunas of Hungary.* Elsevier, Amsterdam.

—— 1989. Geierfunde aus der Repolusthöhle bei Peggau (Steiermark, Österreich). *Fragmenta mineralogica et palaeontologica*, **14**, 117–119.

JÉQUIER, J.-P. 1975. Le Moustérien alpin. Révision critique. *Eburodonum*, **2**, 1–126.

KOBY, F. E. 1960. Contribution à la connaissance des lièvres fossiles, principalement de ceux de la dernière glaciation. *Verhandlungen der Naturforschenden Gesellschaft in Basel*, **71**, 149–173.

KOWALSKI, K. 1995. Lemmings (Mammalia, Rodentia) as indicators of temperature and humidity in the European Quaternary. *Acta zoologica cracoviensia*, **38**, 85–94.

KOZLOWSKI, J. K. 1994. Le rhythme climatique du Pléistocène supérieur et la présence humaine dans les montagnes. *Preistoria alpina–Museo Tridentino di Scienze Naturali*, **28**(2), 37–47.

KUKLA, J. & LOŽEK, V. 1958. On the problems of investigation of the cave deposits *ceskoslovensky kras*, **11**, 19–83 [in Czech with English summary].

KURTÉN, B. 1968. *Pleistocene Mammals of Europe*. Weidenfeld and Nicolson, London.

LEITNER, M. 1988. Ziesel (*Spermophilus citellus*). *In*: SPITZENBERGER, F. (ed.) *Artenschutz in Österreich*. Grüne Reihe des Bundesministeriums für Umwelt, Jugend und Familie, **8**, 177–179.

LUNDELIUS, E. L., Jr. 1989. The implications of disharmonious assemblages for Pleistocene extinctions. *Journal of Archaeological Science*, **16**, 407–417.

MAPLES, C. G. & ARCHER, A. W. 1988. Monte Carlo simulation of selected binomial similarity coefficients (II). *Palaios*, **3**, 95–103.

MAURIN, V. 1994. Geologie und Karstentwicklung des Raumes Deutschfeistritz-Peggau–Semriach. *In*: BENISCHKE, R., SCHAFFLER, H. & WEISSENSTEINER, V. (eds) *Festschrift Lurgrotte 1894–1994*. Landerverein für Höhlen-kunde, Graz, 103–137.

MOTTL, M. 1951. Die Repolusthöhle bei Peggau (Steiermark) und ihre eiszeitlichen Bewohner (mit einem Beitrag von V. MAURIN). *Archaeologia Austriaca*, **8**, 1–78.

—— 1953. Die Erforschung der Höhlen. *In*: MOTTL, M. & MURBAN, K. (eds) *Eiszeitforschungen des Joanneums in Höhlen der Steiermark*. Mitteilungen des Museums für Bergbau, Geologie und Technik am Landesmuseum Joanneum, Graz, **11**, 14–58.

—— 1975. Die pleistozänen Säugetierfaunen und Kulturen des Grazer Berglandes. *In*: FLÜGEL, H. (ed.) *Die Geologie des Grazer Berglandes (2. Auflage)*. Mitteilungen der Abteilung für Geologie, Paläontologie und Bergbau am Landesmuseum Joanneum, Graz, Sonderheft, **1**, 159–185.

NADACHOVSKI, A. 1991. Systematics, geographic variation, and evolution of snow voles (*Chionomys*) based on dental characters. *Acta theriologica*, **36**, 1–45.

OGNEV, S. I. 1966. *Mammals of the USSR and adjacent countries Vol. 4*, (*Rodents*). Israel Program for Scientific translations, Jerusalem.

PASCHINGER, H. 1974. *Steiermark: Steirisches Randgebirge, Grazer Bergland, steirisches Riedelland*. Sammlung geographischer Führer, **10**. Bontraeger, Berlin.

PETERSON, K., MONTFORT, G. & HOLLOM, P. A. D. 1985. *Die Vögel Europas* (German edition of *A Field Guide to the Birds of Britain and Europe*). Paul Parey, Hamburg.

POHAR, V. 1994. Great mammals descending from the culmination point of the last glacial in Slovenia. *Razprave IV. razreda SAZU*, **35**, 85–100 [in Slovenian with English summary].

—— 1997. Late Glacial mammal macrofauna in Slovenia. *Quartär*, **47–48**, 149–158.

POSCH, G. 1995. The development of the Hörfeld in Late Glacial and Holocene. *In*: BANER, S. J. (ed.) *Austrian Contributions to the International Geosphere–Biosphere Programme (Global Change)*, Austrian Academy of Sciences, **2**, 17–19.

RABEDER, G. 1989. Die Höhlenbären der Tropfsteinhöhle im Kugelstein. *In*: FUCHS, G. (ed.) *Höhlenfundplätze im Raum Peggau-Deutschfeistritz, Steiermark, Österreich. Tropfsteinhöhle, Kat.N. 2784/3. Grabungen 1986–87*. British Archaeological Reports, International Series, **510**, 171–178.

—— 1995. Evolutionsniveau und Chronologie der Höhlenbären aus der Gamssulzen-Höhle im Toten Gebirge (Oberösterreich). In: RABEDER, G. (ed.) Die Gamssulzen-Höhle im Toten Gebirge *Mitteilungen der Kommission für Quartärforschung der Österreichischen Akademie der Wissenschaften*, **9**, 69–81.

—— & TEMMEL, H. 1997. Repolusthöhle. *In*: DÖPPES, D. & RABEDER, G. (eds) Pliozäne und pleistozäne Faunen von Österreich. *Mitteilungen der Kommission für Quartärforschung der Österreichischen Akademie der Wissenschaften*, **10**, 328–334.

RAKOVEC, I. 1938. Ein Moschusochs aus der Höhle Potočka zijalka. *Prirodoslovne razprave*, **3**, 253–262.

REINER, G. 1995. Eine spätglaziale Mikrovertebratenfauna aus der Grossen Badlhöhle bei Peggau, Steiermark. *Mitteilungen der Abteilung für Geologie und Paläontologie am Landesmuseum Joanneum, Graz*, **52–53**, 135–192.

—— & BICHLER, H. 1996. Taphonomy in caves: two case studies. *In*: MÉLENDEZ HEVIA, G., BLASCO SANCHO, F. & PÉREZ URRESTI, I. (eds) *II Reunión de Tafonomia y fossilización*, Zaragoza, 347–352.

RUŽIC, A. 1978. *Citellus citellus* (Linnaeus, 1766). *In*: NIETHAMMER, J. & KRAPP, F. (eds) *Handbuch der Säugetiere Europas*, **1**, Nagetiere I, Aula-Verlag, Wiesbaden, 123–144.

RZEBIK-KOWALSKA, B. 1995. Climate and history of European shrews (family Soricidae). *Acta zoologica cracoviensia*, **38**, 95–107.

STARKEL, L. 1994. Regularities of mountain geoecosystems. *Preistoria Alpina–Museo Tridentino di Scienze Naturali*, **28**(1), 11–18.

TAST, J. 1982. *Microtus oeconomus* (Pallas, 1776). *In*: NIETHAMMER, J. & KRAPP, F. (eds) *Handbuch der Säugetiere Europas*, **2/I**, Nagetiere II, Aula-Verlag, Wiesbaden, 374–396.

THENIUS, E. 1976. Pleistozäne Säugetiere als Klima-Indikatoren. *Archaeologia Austriaca, Beihefte*, **13**, 91–112.

TRIMMEL, H. (ed.) 1978. *Die Karstverbreitungs- und Karstgefährdungskarten Österreichs im Massstab 1:50 000*. Wissenschaftliche Beihefte zur Zeitschrift Die Höhle, **27**.

VAN HUSEN, D. 1997. LGM and Late-Glacial fluctuations in the Eastern Alps. *Quaternary International*, **38–39**, 109–118.

VON KOENIGSWALD, W. 1991. Exoten in der Grosssäuger-Fauna des letzten Interglazials von Mitteleuropa. *Eiszeitalter und Gegenwart*, **41**, 70–84.

WAKONIGG, H. 1978. *Witterung und Klima in der Steiermark*. Institut für Geographie, Universität Graz, Graz.

WEISSMÜLLER, W. 1997. Eine Korrelation der Delta[18]O-Ereignisse des grönländischen Festlandeises mit den Interstadialen des atlantischen und des kontinentalen Europa im Zeitraum von 45 bis 14 ka. *Quartär*, **47–48**, 89–111.

Index

Note: **bold** type numbers denote illustrations and tables